Technology for a Quieter America

Committee on Technology for a Quieter America

NATIONAL ACADEMY OF ENGINEERING
OF THE NATIONAL ACADEMIES

THE NATIONAL ACADEMIES PRESS
Washington, D.C.
www.nap.edu

THE NATIONAL ACADEMIES PRESS **500 Fifth Street, N.W.** **Washington, DC 20001**

NOTICE: To arrive at the findings and recommendations of this report, the National Academy of Engineering has used a process that involves careful selection of a balanced and knowledgeable committee, assembly of relevant information, and peer review of the resultant report.

Support for this project was provided by a generous gift from NAE member William W. Lang, the National Academy of Engineering Fund, and the Federal Highway Administration. The opinions, findings, conclusions, and recommendations expressed in this publication are those of the author(s) and do not necessarily reflect the views of the supporting organizations.

Library of Congress Cataloging-in-Publication Data

Technology for a quieter America / National Academy of Engineering of the National Academies.
p. cm.
Includes bibliographical references.
ISBN 978-0-309-15632-5 (pbk.) — ISBN 978-0-309-15633-2 (pdf) 1. Noise control—Technological innovations—United States. 2. Noise pollution—United States. I. National Academy of Engineering.
TD893.T43 2011
363.740973—dc22
2010037657

Copies of this report are available from the National Academies Press, 500 Fifth Street, N.W., Lockbox 285, Washington, DC 20055; (888) 624-8373 or (202) 334-3313 (in the Washington metropolitan area); online at *http://www.nap.edu*.

Printed in the United States of America

THE NATIONAL ACADEMIES

Advisers to the Nation on Science, Engineering, and Medicine

The **National Academy of Sciences** is a private, nonprofit, self-perpetuating society of distinguished scholars engaged in scientific and engineering research, dedicated to the furtherance of science and technology and to their use for the general welfare. Upon the authority of the charter granted to it by the Congress in 1863, the Academy has a mandate that requires it to advise the federal government on scientific and technical matters. Dr. Ralph J. Cicerone is president of the National Academy of Sciences.

The **National Academy of Engineering** was established in 1964, under the charter of the National Academy of Sciences, as a parallel organization of outstanding engineers. It is autonomous in its administration and in the selection of its members, sharing with the National Academy of Sciences the responsibility for advising the federal government. The National Academy of Engineering also sponsors engineering programs aimed at meeting national needs, encourages education and research, and recognizes the superior achievements of engineers. Dr. Charles M. Vest is president of the National Academy of Engineering.

The **Institute of Medicine** was established in 1970 by the National Academy of Sciences to secure the services of eminent members of appropriate professions in the examination of policy matters pertaining to the health of the public. The Institute acts under the responsibility given to the National Academy of Sciences by its congressional charter to be an adviser to the federal government and, upon its own initiative, to identify issues of medical care, research, and education. Dr. Harvey V. Fineberg is president of the Institute of Medicine.

The **National Research Council** was organized by the National Academy of Sciences in 1916 to associate the broad community of science and technology with the Academy's purposes of furthering knowledge and advising the federal government. Functioning in accordance with general policies determined by the Academy, the Council has become the principal operating agency of both the National Academy of Sciences and the National Academy of Engineering in providing services to the government, the public, and the scientific and engineering communities. The Council is administered jointly by both Academies and the Institute of Medicine. Dr. Ralph J. Cicerone and Dr. Charles M. Vest are chair and vice chair, respectively, of the National Research Council.

www.national-academies.org

COMMITTEE ON TECHNOLOGY FOR A QUIETER AMERICA

GEORGE C. MALING, JR. (NAE) (*chair*), Institute of Noise Control Engineering of the USA, Inc. (retired), Harpswell, Maine
ROBERT J. BERNHARD, University of Notre Dame, Notre Dame, Indiana
ROBERT D. BRUCE, CSTI Acoustics, Houston, Texas
BETH A. COOPER, Glenn Research Center, NASA, Cleveland, Ohio
PATRICIA DAVIES, Purdue University, West Lafayette, Indiana
CARL E. HANSON, Harris Miller Miller and Hanson, Burlington, Massachusetts
ROBERT D. HELLWEG, JR. (consultant), Wellesley, Massachusetts
GERALD C. LAUCHLE, Pennsylvania State University (retired)
RICHARD H. LYON (NAE), RH Lyon Corp., Belmont, Massachusetts
IAN A. WAITZ, Massachusetts Institute of Technology, Cambridge

Project Staff

CAROL R. ARENBERG, Senior Editor, National Academy of Engineering
VIVIENNE CHIN, Administrative Assistant, Program Office, National Academy of Engineering
LANCE A. DAVIS, Executive Officer, National Academy of Engineering
PROCTOR P. REID, Director, Program Office, National Academy of Engineering
RICHARD TABER, Program Officer, Program Office, National Academy of Engineering (until February 2009)

Preface

Noise emissions are an issue in industry, in communities, in buildings, and during leisure activities. As such, the audience for a report on noise control is broad and includes the engineering community; the public; government at the federal, state, and local levels; private industry; labor unions; and nonprofit organizations. These stakeholders should find something of interest in this report.

In the past few decades advances have been made in noise control technology, instruments for noise measurement, and criteria for noise control. These advances need to be recognized in our approach to the control of noise and public policy designed to improve the noise climate in the United States. This, together with increasing worldwide interest in reducing noise, makes it necessary to examine American interests in the production of low-noise products with a view toward remaining competitive. Reducing product noise emissions and achieving noise reductions in our factories, office buildings, classrooms, homes, and the environment are challenging problems.

This study was undertaken by the National Academy of Engineering (NAE) to emphasize the importance of engineering to the quality of life in America, in particular the role of noise control technology making possible a quieter environment. This report was prepared by a study committee and five supporting panels of experts appointed by the NAE and reviewed by an independent panel appointed following NAE procedures. Implementation of the recommendations in the report will result in reduction of the noise levels to which Americans are exposed and will improve the ability of American industry to compete in world markets where increasing attention is being paid to the noise emissions of products.

Key areas where recommendations have been made include cost-benefit analysis of noise reduction, especially related to road traffic noise; improved metrics for noise control; lower limits for noise exposures in industry; "buy quiet" programs; wider use of international standards for noise emissions; airplane noise reduction technology; and noise control in structures such as schools, hospitals, and office buildings. Also recommended is improved cooperation between industry and government agencies involved with noise and, in particular, an expanded role for the Environmental Protection Agency, which can be undertaken under existing law.

George C. Maling, Jr.
Chair
Committee on Technology for a Quieter America

Acknowledgments

This report was reviewed in draft form by individuals chosen for their diverse perspectives and technical expertise, in accordance with procedures approved by the National Academy of Engineering (NAE). The purpose of this independent review is to provide candid and critical comments that will assist the committee and NAE in making its published report as sound as possible and to ensure that the report meets institutional standards for objectivity, evidence, and responsiveness to the study charge. The reviewers' comments and the draft manuscript remain confidential to protect the integrity of the deliberative process. We wish to thank the following individuals for their reviews of this report:

Lewis M. Branscomb, Emeritus, Harvard University, and Adjunct Professor, University of California, San Diego
Mahlon D. Burkhard, Consultant
William Cavanaugh, Cavanaugh Tocci Associates, Inc.
Malcolm J. Crocker, Sound & Vibration Research Laboratory
Tony F.W. Embleton, Retired, National Research Council of Canada
David K. Holger, Iowa State University
Alice Suter, Consultant in Noise and Hearing Conservation
István L. Vér, Consultant in Acoustics, Noise, and Vibration Control

Although the reviewers listed above provided many constructive comments and suggestions, they were not asked to endorse the conclusions or recommendations and did not see the final draft of the report before its public release. The review of this report was overseen by James L. Flanagan, Retired Vice President for Research, Rutgers, The State University of New Jersey. Appointed by NAE, he was responsible for making certain that an independent examination of this report was carried out in accordance with institutional procedures and that all review comments were carefully considered. Responsibility for the final content of this report rests entirely with the authoring committee and NAE.

In addition to the reviewers, the committee extends its sincerest gratitude to the members of the five expert panels that supported this study (Appendix K), and to the individuals who participated in the project's eight fact-finding workshops (Appendix L) for sharing their expertise, insights, and best ideas to the study. The committee also wishes to thank the consultants to the committee—Leo L. Beranek, Stephen H. Crandall, Kenneth M. Eldred, and William W. Lang—who provided invaluable advice throughout the project. The committee also thanks the project staff. NAE executive officer Lance Davis and NAE senior editor Carol Arenberg substantially improved the readability of the report. Study director Richard Taber managed the project through January 2009, and NAE program director Proctor Reid managed the project from February 2009 to completion. Vivienne Chin managed the committee's and panels' logistical and administrative needs.

Contents

Tables and Figures

TABLES

FIGURES

Executive Summary

Exposure to noise (i.e., unwanted or potentially hazardous sound) at home, at work, while traveling, and during leisure activities is a fact of life for all Americans. At times noise can be loud enough to damage hearing, and at lower levels it can disrupt normal living, affect sleep patterns, affect our ability to concentrate at work, interfere with outdoor recreational activities, and, in some cases, interfere with communications and even cause accidents. Clearly, exposure to excessive noise can affect our quality of life.

As the population of the United States and, indeed, the world increases and developing countries become more industrialized, problems of noise are likely to become more pervasive and lower the quality of life for everyone. Efforts to manage noise exposures, to design quieter buildings, products, equipment, and transportation vehicles, and to provide a regulatory environment that facilitates adequate, cost-effective, sustainable noise controls require our immediate attention. Specific recommendations that address these issues are included in this report.

This report looks at the most commonly identified sources of noise, how they are characterized, efforts that have been made to reduce noise emissions, and efforts to reduce the noise experienced by people in workplaces, schools, recreational environments, and residences. The report also reviews the standards and regulations that govern noise levels and the federal, state, and local agencies that regulate or should regulate noise for the benefit, safety, and wellness of society at large. This report also presents information on the cost-benefit trade-offs between efforts to mitigate noise and the improvements they achieve, information sources available to the public on the dimensions of noise problems and their mitigation, and the need to educate professionals who can deal with these issues.

Ubiquitous sources of noise include all modes of transportation—airplanes, trains, trucks, and automobiles; consumer products, such as lawnmowers, snow blowers, and leaf blowers; and manufacturing machinery in the workplace. Noise levels usually decrease as one moves away from a source, but people living close to the end of a runway or near a high-speed interstate highway cannot escape from highly annoying noise; lawn care equipment can annoy neighbors and at times can be hazardous to the user; and the requirements of operating noisy machinery can make it practically impossible for workers to retreat far enough to escape hazardous noise. Below are specific subjects addressed in this report.

IMPROVEMENT OF ENVIRONMENTAL NOISE METRICS

The committee looked in detail at the state of the technology with regard to noise metrics and concluded that modern advances in our ability to collect, store, and analyze noise data challenge us to reexamine current metrics that were developed in the 1970s or earlier with the objective of developing metrics better related to human response to noise.

HAZARDOUS NOISE AT WORK AND AT HOME

This report also provides information on noise, both occupational and nonoccupational, that can damage hearing. The committee recommends that current U.S. Department of Labor limits on occupational noise exposure be reviewed and changed. Engineering controls should be the primary means of controlling noise, and "buy quiet" programs will assist in the procurement of low-noise machinery and equipment.

TECHNOLOGIES FOR NOISE CONTROL

Technology alone will not solve all noise problems, but problems that *are* amenable to technical solutions can be solved by engineers with appropriate support from economists, psychologists, medical specialists, educators, and many departments in federal, state, and local governments. In this report the committee has made an assessment of transportation noise sources; noise from machinery, equipment, and consumer products that can affect U.S. competitiveness;

noise in the built environment; noise in the community; and hazardous noise. Some areas, such as aircraft noise reduction, have received a great deal of global attention, but other important sources of noise have received less attention, even though they affect many more people.

COST-BENEFIT ANALYSIS FOR NOISE MITIGATION

Cost-benefit analysis for different noise mitigation options is another area considered by the committee, both broadly and in the context of reducing noise generated by interactions between vehicle tires and road surfaces. At highway speeds this tire/road interaction noise dominates noise emissions from vehicles, and efforts are being made to design road surfaces that minimize this noise. The committee recommends that a formal analysis be performed to compare the costs and benefits of using pavement technology for noise reduction with the costs and benefits of installing noise barriers. This cost-benefit analysis would probably be a cooperative effort of the Federal Highway Administration, the Environmental Protection Agency (EPA), and several states. The efforts of the Federal Aviation Administration to develop a cost-benefit approach to analyze noise around airports could help in the development of a similar project to analyze options for reducing highway noise. European cost-benefit analyses, clearly much more extensive than similar American analyses, are also reviewed.

STANDARDS AND REGULATIONS FOR PRODUCT NOISE EMISSIONS

The European Union (EU) has been a leader in the development of noise regulations based on standards promulgated by the International Organization for Standardization (ISO) and the International Electrotechnical Commission (IEC). These regulations are more extensive than regulations in the United States, and consequently European manufacturers have gained an advantage over their U.S. counterparts in meeting demands for low-noise machinery and other products worldwide.

Regulatory and standards-setting activities regarding noise, especially in the EU, are examined, and their impact on the ability of U.S. manufacturers to compete in world markets is assessed. EU member states have placed significant emphasis on the need for noise emission standards and have exercised waxing influence within the ISO, and to some extent the IEC, on the development of international noise emission standards. Meanwhile, U.S. influence within ISO and IEC on noise-related issues has waned. Building on voluntary standards, noise emissions from consumer products are much more highly regulated in Europe than in the United States, and European requirements on noise levels in the workplace also are more stringent than in the United States. The role of the U.S. Department of Commerce, especially its National Institute of Standards and Technology, is reviewed, and several recommendations are made for strengthening U.S. manufacturers' participation in international standards-setting bodies related to noise control and for improving dissemination of information on noise emission requirements outside the United States.

Although noise requirements can sometimes be a burden, they can also encourage innovation. A manufacturer's desire for the design of a low-noise machine for sale in world markets is a positive force that could lead to the introduction of quiet products into American markets and be an incentive for manufacturers and purchasers to cooperate in "buy-quiet" programs. Indeed, at the time of purchase, consumers rank noise as one of the top five characteristics when comparing product performance. Yet noise levels for U.S. products are often buried in product literature and reported in different noise metrics, making it difficult for consumers to compare noise levels at the time of purchase. Thus, consumers are unable to make informed decisions about the noise emission of a product. This problem could be corrected if product noise levels were prominently displayed and manufacturers adopted a system of self-enforcement.

American manufacturers have the ingenuity to design quiet products. However, manufacturers and trade associations, as well as the voluntary-standards community, have been unable to agree on a uniform standard for measuring and labeling product noise.

THE ROLE OF GOVERNMENT

In some areas—notably aircraft noise, occupational noise, and highway noise that can be reduced by barriers—government regulation has played a major role. But this report shows that improvements can be made in other ways as well. For example, authority for cost-benefit analysis, interagency projects, and dissemination of public information was granted to the EPA by Congress. Because of a lack of funding, however, EPA has been unable to carry out these activities. The study committee recommends changes that will make it easier for the federal government to improve the lives of Americans.

EDUCATION OF NOISE CONTROL ENGINEERS

This report also examines the state of noise control engineering education and concludes that the nation needs to educate specialists in the field and provide basic knowledge of the principles of noise control engineering to individuals trained as specialists in other engineering disciplines.

PUBLIC INFORMATION

An informed public is an important element in efforts to create a quieter America, and the Internet is a low-cost avenue for dissemination of authoritative information on noise, noise control, and the effects of noise on people. The public would benefit from knowing that there are engineering solutions to many noise problems, and a uniform system of labeling the noise emissions of products would enable the public to make informed purchase decisions. EPA has the authority to do more than it is currently doing to create and disseminate public information, and engineering societies can contribute information on noise reduction that is accessible to the public. Citizens groups can also be a source of public information on noise. Specific recommendations to enhance public information efforts are given in this report.

NOISE AND HEALTH

The general relationship between noise and health is not covered in this report, although new information is becoming available (Babisch, 2008; DEFRA, 2009). However, it will take a multidisciplinary study committee to evaluate these results and determine their relevance to the health of the American people.

CONCLUSION

Reducing the noise levels to which Americans are exposed will require cooperation among engineers, industrial management, and government in many disciplines, and it will not be accomplished in a short time. Nevertheless, reduced noise levels will contribute to improved quality of life for many Americans, and the committee believes that the recommendations in this report, if implemented, will improve the current noise climate.

REFERENCES

Babisch, W. 2008. Road traffic noise and cardiovascular risk. Noise and Health 10(38):27–33. Available online at *http://www.noiseandhealth.org/article.asp?issn=1463-1741;year=2008;volume=10;issue=38;spage=27;epage=33;aulast=Babisch;type=0.*

DEFRA (Department for Environment Food and Rural Affairs). 2009. Estimating Dose-Response Relationships between Noise Exposure and Human Health in the UK. BEL Technical Report 2009-02. Available online at *http://www.defra.gov.uk/environment/noise/igcb/healthreport.htm.*

1

Introduction

In 2005 the National Academy of Engineering (NAE) held a workshop to review the present state of technology in noise control engineering. The workshop was organized by a steering committee charged with developing a prospectus for further studies of noise-related issues in the United States and to investigate how current technologies could be used to reduce exposures to noise. The issues framed by the steering committee were subsequently considered in a series of workshops held by NAE in 2007 and 2008.

These issues included a review of community noise and metrics for measuring community noise; new technologies for a quieter America; engineering controls and common descriptors for hazardous noise; the impact of noise on the competitiveness of U.S. products; a cost-benefit analysis of noise control technologies; the gap between industry demand for noise control specialists and the supply coming through the education pipeline; noise control activities at the federal and state levels; state and local community noise control programs; dissemination of information to the public on the benefits of low-noise products; and the adverse effects of excessive noise.

This report attempts to address these issues in the following ways:

- by summarizing the current state of the practice in noise control engineering;
- by recommending how existing knowledge can be applied to address current challenges;
- by presenting a research and education agenda that promotes the generation of new knowledge in fields that can provide the greatest benefit to society (ranging from employees to corporations and manufacturers to the public at large); and
- by recommending policies that agencies can develop and adopt to improve the American "soundscape" and to promote quieter products and living environments.

The following sections introduce the broad categories under which these issues are grouped in the body of the report.

A TAXONOMY OF NOISE

Americans are exposed to noise from many sources and in many environments. Almost anything or anyone can generate noise, but the major sources/categories of interest and/or concern include community noise in urban, suburban, and rural areas, as well as in workplaces, recreation areas, and classrooms; aircraft noise; noise from road traffic and other modes of surface transportation; hazardous noise; and consumer product noise.

As discussed in Appendix A, a common measure of noise is the sound pressure level in decibels. This level is almost always weighted according to the A-frequency weighting curve. The resulting value is expressed in dB(A).[1] Table 1-1 gives the reader an idea of sound pressure levels generated by various sources. Figure 1-1 shows the range of environmental sound pressure levels encountered outdoors. Metrics for assessment of noise are more complicated than this description indicates and are discussed in detail in Chapter 3.

Community Noise

Communities are made up of buildings and outdoor spaces of various types and uses, all of which are affected by exterior environmental sources over which an individual has little or no control. Some major sources of environmental noise in communities include aircraft, road and rail transportation systems, construction that for some large civil works projects may last for decades, outdoor stationary building air-conditioning units, electrical transformer substations and other equipment associated with individual buildings or utilities, and noise from nearby industrial plants.

Unlike occupational noise, which can cause hearing loss, community noise is usually an annoyance and a "quality-of-life" issue. In contrast to *emissions* of noise from the sources

[1]It is not the decibel that is A weighted but the level. However, the (A) is attached to the decibel for clarity and brevity and is widely used. Rather than say 50 dB(A), it is more correct to say the A-weighted sound pressure level is 50 dB.

TABLE 1-1 Sound Pressure Levels Generated by Various Noise Sources

Sound Pressure Level	dB(A)
Quiet library, soft whispers	30
Living room, refrigerator	40
Light traffic, normal conversation, quiet office	50
Air conditioner at 20 feet, sewing machine	60
Vacuum cleaner, hair dryer, noisy restaurant	70
Average city traffic, garbage disposals, alarm clock at 2 feet	80
Subway, motorcycle, truck traffic, lawn mower	90
Garbage truck, chain saw, pneumatic drill	100
Rock band concert in front of speakers, thunderclap	120
Gunshot blast, jet plane	140
Rocket launching pad	180

SOURCE: *http://www.nidcd.nih.gov/health/hearing/ruler.asp.*

noted below, community noise is an *immission* problem (i.e., what people hear).[2]

Today, the widely used criterion for assessing community noise levels in the United States is the day-night average sound level, or DNL (see Appendix A for definition). It is the sound pressure level averaged over 24 hours with the amplification of the measuring systems increased by 10 decibels during the nighttime hours. Since 1974, when the U.S. Environmental Protection Agency's (EPA) "Levels Document" and related documents were published, DNL and an exposure-effect relationship showing the percentage of respondents on social surveys who say they are "highly annoyed" by noise from various sources have generally been accepted as overall indicators of the impact of community noise. This exposure-effect relationship was first described in a classic analysis by Schultz (1978), who synthesized 12 major social surveys of reactions to transportation noise. The Schultz curve, which describes the results, essentially illustrates the percentage of the population predicted to be highly annoyed as a function of noise level (see Chapter 2 for further discussions of community and building noise criteria).

Noise in Quiet Areas

Areas in the United States that are relatively free of transportation noise and noise from most other sources are often used for recreation and are places where people value the absence of noise and the opportunity to hear "natural" sounds, such as the flapping of a bird's wings or wind rustling through trees. However, noise from aircraft, off-road vehicles, and other sources sometimes intrudes on these quiet environments. The DNL metric is generally inadequate to describe the "soundscape" in such areas.

[2]*Emission* and *immission* are defined in Appendix A. Briefly, *emission* is the sound directly emitted from a noise source essentially unaffected by the immediate environment around the source. *Immission* is the sound the receiver hears after it has traveled along a sound transmission path and has been affected by it.

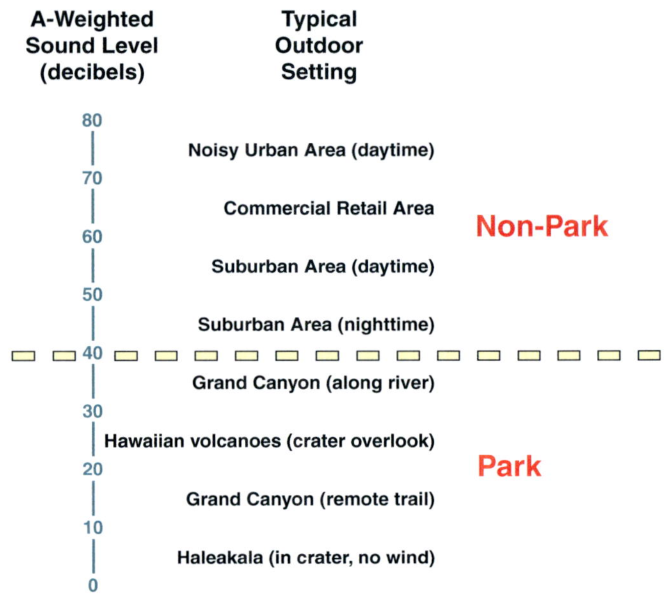

FIGURE 1-1 Comparison of A-weighted sound levels in common outdoor environments. Source: Miller (2003).

Aircraft Noise

Complaints about aviation noise have a long history. In an introduction to a review of current activities by the Federal Aviation Administration (FAA) related to aircraft noise, Burleson (2005) points out that 2003 was the 100th anniversary of flight and the 92nd anniversary of the first editorial complaining about aircraft noise.[3] The most serious problems arose in the late 1950s when commercial jet aircraft came into service.

In the past 50 years, considerable progress has been made in reducing noise emissions from aircraft—mainly through the introduction of high bypass ratio engines, which were driven by a desire to reduce noise emissions and increase fuel efficiency. A 2001 U.S. General Accountability Office (GAO) report stated: "We currently estimate that the airlines' costs directly attributable to complying with the transition to quieter aircraft noise standards ranged from $3.8 billion to $4.9 billion in 2000 dollars" (GAO, 2001). The transition, over a period of 35 years, led to a 95 percent reduction in the number of people impacted by aircraft noise in the United States (PARTNER, 2004).

Despite this progress, there are still noise issues around most of the nation's commercial airports. In a report to Congress in 2000, a survey of the nation's 50 busiest commercial airports indicated that noise was the number one concern for 33 airports and was of some degree of concern in areas around 49 of the 50 airports (GAO, 2000).

[3]Burleson, C. Aviation and the Environment: Navigating the Future. Presentation at an NAE-sponsored workshop, Technology for a Quieter America. Washington, DC., September 1, 2005.

Traffic Noise

No recent studies have been done in the United States on the extent of exposure to highway traffic noise, but in 1981 EPA estimated that 19.3 million people were exposed to "annoying" DNLs greater than 65 dB (Waitz et al., 2007).[4] Recent research has revealed that the interaction between road surfaces and tires is the main source of noise from vehicles traveling at highway speeds now that emitted engine and exhaust noise has been effectively reduced for most automotive vehicles. Research has also shown that this noise can be reduced by the proper design of highway road surfaces. Several states have initiated programs to determine the extent of noise reduction and the feasibility of building new road surfaces.

For the present, the primary solution has been to construct noise barriers in areas considered "noise impacted" by the Federal Highway Administration (FHWA). The federal government cooperates with state departments of transportation in the construction of these barriers, the costs of which depend on the materials used, their height, and the terrain. FHWA (2009) has reported that as of the end of 2004 more than 3,500 kilometers of barriers had been constructed in 45 states and the Commonwealth of Puerto Rico at a cost of more than $2.6 billion ($3.4 billion in 2004 dollars). Despite the high cost of roadside barriers, they are likely to remain an effective element in highway noise control. Reduction of tire/road interaction noise, however, generally provides more people with a smaller noise reduction—which is why cost-benefit analyses are needed. In some severe cases, both methods of noise reduction may be needed.

Nevertheless, traffic noise remains an issue both along the nation's highways and in urban areas. For example, in recent reports on noise by the New York City Council on the Environment, traffic noise in the city was rated high on the list of noise complaints (Bronzaft and Van Ryzin, 2006, 2007).

Consumer Product Noise

Americans are exposed to noise from consumer products both indoors and out. Although manufacturers have made considerable progress in reducing noise from dishwashers and other appliances, these and other product noises can be a source of annoyance. Outdoors, noise from lawnmowers, leaf blowers, and other lawn and yard care equipment is pervasive. Snowmobiles and other off-road vehicles also create noise problems, especially in wilderness areas.

Industrial and Other Potentially Hazardous Noise Sources

High levels of noise can cause noise-induced hearing loss (NIHL), and occupational exposure to hazardous noise is widespread (NIH, 2009a). In fact, NIHL is one of the most common occupation-related disorders in workplaces and in the U.S. military (IOM, 2005; Lang and Maling, 2007). The United States has no national surveillance program for reporting or monitoring the amount of compensation for costs related to treatment of occupational NIHL. As a result, no comprehensive data are available on the economic impact of noise exposure and hearing loss (NIOSH, 2001).

Children's toys, music conveyed through earphones or similar devices, loud music at concerts, the recreational firing of weapons, and similar sources also can be sources of hazardous noise. The National Institute on Deafness and Other Communication Disorders has estimated that more than 30 million people in the United States are exposed to hazardous noise on a regular basis (NIH, 2009b).

The U.S. Department of Labor, through the Occupational Safety and Health Administration (OSHA) and the Mine Safety and Health Administration (MSHA), has promulgated regulations to limit exposures to hazardous noise; the limits established by these agencies are similar but not identical.

Although in theory engineering controls are the preferred way to reduce noise levels, personal hearing protective devices (HPDs) are widely used. EPA's regulation for labeling the performance of HPDs is currently being updated.

TECHNOLOGIES

Noise control engineering and technology include a wide variety of measurement techniques and standards, engineering designs, and manufacturing techniques to control noise emissions, engineering controls and HPDs to mitigate exposures to hazardous noise in the workplace and elsewhere, and analysis techniques for determining the impact of noise over large areas.

Noise is measured in decibels (dB), designated as dB(A) when A-frequency weighting[5] is applied to the signal to make it more representative of the noise perceived by a listener. The basic quantities used in acoustics and noise control are described in considerable detail in Appendix A. Generally speaking, the level of noise (i.e., the sound pressure level in decibels) ranges from near 0 to 140 dB. For the most part, however, the public has little or no understanding of the decibel or A-frequency weighting and thus is unable to appreciate or participate in a discussion of quantitative levels of noise.

Efforts to control community noise frequently depend on controlling emissions of offensive noise from noise sources. For example, the National Aeronautics and Space Administration (NASA) Advanced Subsonic Transport Noise Reduction Program was a seven-year effort begun in 1994 to develop technology to reduce jet transport noise by 10 dB relative to 1992 levels. This program provided for reductions in engine source noise, improvements in nacelle acoustic treatments, reductions in noise generated by airframes, and modifications in the way aircraft operate in airport environs.

[4]The decibel is a unit of sound level (see Appendix A for definition).

[5]A-frequency weighting is defined in Appendix A.

The NASA Glenn Research Center also significantly reduced aircraft fan noise using active noise control methodologies. By the end of 2001, when the program ended, most of its objectives had been met.

Other technologies address the noise source/receiver as a system. The FAA's PARTNER (Partnership for Air Transportation Noise and Emissions Reduction) Program, founded in 2003, uses a systems approach to reducing noise from aircraft and its impact on airport environs. NASA and Transport Canada are cosponsors of PARTNER. The PARTNER Center of Excellence, located at the Massachusetts Institute of Technology (MIT), has undertaken the following projects: development of metrics to improve understanding of human response to aircraft noise; studies of land use around airports; analysis of the socioeconomic effects of noise and noise mitigation; cost-benefit analyses of technologies, operations, and policy alternatives for mitigating noise impacts; and development and testing of noise abatement flight procedures.

Technologies for controlling noise from road traffic are less well developed. In general, regulations to control noise emitted by vehicles have not been very effective in reducing community noise (Sandberg, 2001). Because many studies have shown that the major source of noise is the interaction between tires and road surfaces, several states have initiated programs to study how much noise reduction could be achieved by porous road surfaces (see Chapters 5 and 7).

A variety of methods can be used to control noise from rail-bound vehicles. If the United States embarks on an expansion of the rail system, planning for noise control, prediction tools, and the application of noise control technologies will become increasingly important (see Chapter 5).

COMPETITIVENESS

Noise control engineering can affect manufacturing competitiveness because, as the market for many industrial and consumer products becomes more globalized, U.S.-based firms must compete in both domestic and foreign markets. The latter are subject to noise standards and regulations that can impact competitiveness in two ways: (1) they can impose additional costs on U.S. manufacturers who want to enter foreign markets and (2) competitors' products that meet the more rigorous noise limits may enter the U.S. market with a competitive advantage over domestic producers. This advantage is evidenced in a growing trend by consumers who identify low noise as a desirable feature. In a 1999 survey, for example, 84 percent of consumers said that "ultra-quiet" operation was an important feature of a dishwasher (KBDN, 1999).

COST-BENEFIT ANALYSIS

As a practical matter, especially when large expenditures of public funds are involved, solutions to noise-related challenges must have a positive benefit for quality of life, and that benefit must be considered worth the cost of reducing the impact of noise. Therefore, the study committee has attempted to determine how economic analysis techniques could inform decisions about allocating scarce resources to achieving the greatest aggregate benefit for society.

In the case of environmental policy, using resources in a socially optimal way may mean limiting how one entity can use its resources in order to protect another entity from the consequences (e.g., a curfew on noisy flights from certain airports places limits on airlines, and ultimately on travelers, to protect people who live around the airports). Alternatively, it may mean deciding to invest public resources to mitigate the undesirable effects of others' activities (e.g., installing pavements that reduce traffic noise).

In both cases the study committee attempts to clarify the trade-offs involved. Chapter 7 provides an overview of cost-benefit analyses, a brief description of how they affect FAA decisions, and attempts to reduce tire/road noise on the nation's highways. The emphasis is on the need for cost-benefit analysis with respect to highway traffic noise.

THE ROLE OF GOVERNMENT

For many years the federal government has been involved with controlling noise, as have the European Union and the governments of most other industrialized nations. In the United States the control of occupational noise is the responsibility of the U.S. Department of Labor (DOL). Within DOL, OSHA and MSHA have regulatory authority with respect to noise. Under the Noise Control Act of 1972 and subsequent legislation, EPA was made responsible for addressing noise issues that included both regulatory authority and research. Activities were carried out through the Office of Noise Abatement and Control (ONAC). Funding for ONAC was discontinued in 1982, but many EPA responsibilities with respect to noise are still in the U.S. Code. The role of EPA is described in more detail in Chapter 8.

The U.S. Department of Housing and Urban Development, the Federal Housing Administration, the General Services Administration, and other federal government departments and agencies have promulgated policies and regulations for site selection for federally subsidized housing and for exterior building construction to meet minimum acoustical standards. The federal government also sets standards for noise in federal office buildings and leased spaces in commercially owned buildings used by federal agencies.

The U.S. Department of Transportation and its modal agencies (FAA, FHWA, Federal Railroad Administration, and Federal Transit Administration), have broad regulatory authority regarding noise issues. The U.S. Department of Defense and all of the armed services have noise programs and regulate noise. The U.S. Department of Health and Human Services, National Institutes of Health, National Science Foundation, and NASA all have noise programs related to the mission of each agency. In addition, the National Park

Service addresses noise issues in national parks and has the authority to both set noise limits and do research. However, noise-related activities by federal agencies are not well coordinated. (More details can be found in Chapter 8.) Reducing environmental noise will require that the development and support of noise control technologies be shared among government agencies and industry.

State and local governments can promulgate noise regulations as long as they do not conflict with federal government regulations. EPA still has some noise emission regulations "on the books" and has broad powers with respect to interstate commerce. Despite this, many states and municipalities have no noise regulations at all. Others have regulations, but they are poorly written or outdated. According to Hanson (2002), states and local municipalities would welcome better information and guidance, as well as financial and technical support, in enacting reasonable and effective environmental noise regulations.

EDUCATION AND THE WORKFORCE

Although acoustics—the science of sound—has a long history (Rossing, 2007), noise control engineering is a relatively new field. Noise problems emerged after World War II, with the building of the interstate highway system, the advent of jet airplanes, and the postwar building boom. MIT was a pioneer in noise control education (in the departments of aeronautics, architecture, electrical engineering, and physics), but even today not a single university in the United States has a department (or academic unit), nor is there a widely agreed-on curriculum, for noise control engineering.

Because noise control engineering is inherently a multidisciplinary field, noise control engineers must be knowledgeable in several subjects, including acoustics, aerodynamics, mechanical vibration, measurement, electronics, physiology, psychology, statistics, physics, and architecture. Today the demand for such individuals far exceeds the supply. Meeting this demand will require an emphasis on noise control engineering in the undergraduate curriculum as well as well-funded graduate programs.

The Public

Although people are quick to inform public officials when they are inconvenienced or oppressed by noise, they are poorly informed about how, or even if, noise can be mitigated in practical, cost-effective ways. It would be beneficial for people to have a better understanding of, for example, how noise is measured, so they could participate in informed debate on problems that affect them and recognize when sound pressure levels are likely to cause permanent hearing damage. Two studies in the 1990s included information related to public awareness of noise problems (ASHA, 1991; OECD, 1991); the Internet can also provide a great deal of information. Many organizations that provide public information are identified in this report, and the committee suggests how government might play a larger role in providing information to the public.

SUMMARY

All of these subjects are discussed in more detail in the following chapters of this report. In Chapter 11, findings are summarized, and a number of recommendations are given that the study committee believes will lead to the reduction of noise in the United States.

The decibel and other terms used in acoustics to describe noise are briefly described in Appendix A. A more complete description can be found in handbooks on acoustics and noise control, such as Rossing (2007), Vér and Beranek (2006), and Crocker (2007).

Sources of the many technical articles and Internet resources cited in this report include professional society journals and conference proceedings. Several are from a 2007 special issue of *The Bridge* (NAE, 2007). Others are from *Noise Control Engineering Journal* and the proceedings of national conferences (NOISE-CON) and international congresses (INTER-NOISE). Referenced papers from these sources are available on the Internet (*http://www.bookmasters.com/marktplc/00726.htm*) and through the Scitation platform hosted by the American Institute of Physics (*http://scitation.aip.org/*) and maintained by the Institute of Noise Control Engineering of the USA (*http://www.inceusa.org*). There are also references to papers published in *Noise/News International* (NNI), and these are available on the NNI website, *http://www.noisenewsinternational.net*. Reports from the International Institute of Noise Control Engineering are available on its website, *http://www.i-ince.org*.

REFERENCES

ASHA (American Speech Language-Hearing Association). 1991. Combatting Noise in the 90s: A National Strategy for the United States. Rockville, MD: American Speech Language-Hearing Association. Available online at *http://www.nonoise.org/epa/Roll16/roll16doc30.pdf*.

Bronzaft, A., and Van Ryzin, G. 2006. Neighborhood Noise and Its Consequences: A Survey in Collaboration with the Council on the Environment of New York City. Special Report #9. New York: Council on the Environment. Available online at *http://www.noiseoff.org/media/cenyc.noise.report.pdf*.

Bronzaft, A., and G. Van Ryzin. 2007. Neighborhood Noise and Its Consequences: Implications for Tracking Effectiveness of the NYC Revised Noise Code. Special Report #14. New York: Council on the Environment. Available online at *http://www.noiseoff.org/media/cenyc.noise.report.14.pdf*.

Crocker, M.J., ed. 2007. Handbook of Noise and Vibration Control. Hoboken, NJ: John Wiley & Sons.

EPA (U.S. Environmental Protection Agency). 1974. Information on Levels of Environmental Noise Requisite to Protect Public Health and Welfare with an Adequate Margin of Safety. Document 550/9-74-004. Available online at *http://www.nonoise.org/library/levels74/levels74.htm*.

FHWA (Federal Highway Administration). 2009. Summary of Noise Barriers Constructed by December 31, 2004. Available online at *http://www.fhwa.dot.gov/environment/noise/barrier/sintro.htm.*

GAO (Government Accountability Office). 2000. Aviation and the Environment: Results from a Survey of the Nation's 50 Busiest Commercial Service Airports. GAO/RCED-00-222. Available online at *http://www.gao.gov/archive/2000/rc00222.pdf.*

GAO. 2001. Aviation and the Environment: Transition to Quieter Aircraft Occurred as Planned, but Concerns About Noise Persist. GAO-01-1053. Available online at *http://gao.gov/new.items/d011053.pdf.*

Hanson, C.E. 2002. The Roles of State and Local Government Agencies in Noise Abatement and Control. Proceedings of INTER-NOISE 02, The 2002 International Congress and Exposition on Noise Control Engineering, Dearborn, MI, August 19–21. Available online at *http://www.bookmasters.com/marktplc/00726.htm.*

IOM (Institute of Medicine). 2005. Noise and Military Service: Implications for Hearing Loss and Tinnitus. L.E. Humes, L.M. Joellenbeck, and J.S. Durch, eds. Washington, DC: National Academies Press.

KBDN (Kitchen and Bath Design News). 1999. New Survey Pinpoints Dishwasher Usage Trends. Available online at *http://www.kitchenbathdesign.com/print/Kitchen-and-Bath-Design-News/New-Survey-Pinpoints-Dishwasher-Usage-Trends/2$1398.*

Lang, W.W., and G.C. Maling. 2007. Noise as a technological and policy challenge. The Bridge 37(3):4–10.

Miller, N.P. 2003. Transportation Noise and Recreational Lands. Proceedings of INTER-NOISE 2002, Dearborn, MI, August 19–21, 2002.

NAE (National Academy of Engineering). 2007. Special Issue on Noise Engineering. The Bridge 37(3).

NIH (National Institutes of Health). 2009a. Noise Induced Hearing Loss. Available online at *http://www.nidcd.nih.gov/health/hearing/noise.asp.*

NIH. 2009b. Statistics on Noise Induced Hearing Loss. Available online at *http://www.nlm.nih.gov/medlineplus/noise.html.*

NIOSH (National Institute for Occupational Safety and Health). 2001. Work-Related Hearing Loss. DHHS (NIOSH) Publication No. 2001-103. Cincinnati, OH: NIOSH.

OECD (Organization for Economic Cooperation and Development). 1991. Fighting Noise in the 90s. Paris: OECD.

PARTNER (Partnership for Air Transportation Noise and Emissions Reduction). 2004. Report to Congress: Aviation and the Environment, A National Vision Statement, Framework for Goals and Recommended Actions. Available online at *http://www.faa.gov/library/reports/media/congrept_aviation_envirn.pdf.*

Rossing, T. 2007. Springer Handbook of Acoustics. New York: Springer Science+Business Media LLC.

Sandberg, U. 2001. Noise Emissions of Road Vehicles—Effectiveness of Regulations. Final Report from the Working Party on Noise Emissions of Road Vehicles. International Institute of Noise Control Engineering. Available online at *http://www.i-ince.org/data/iince011.pdf.*

Schultz, T.J. 1978. Synthesis of social surveys on noise annoyance. Journal of the Acoustic Society of America 64(1):377–405.

Vér, I.L., and L.L. Beranek, eds. 2006. Basic Acoustical Quantities: Levels and Decibels. Chapter 1 in Noise and Vibration Control Engineering. Hoboken, NJ: John Wiley & Sons.

Waitz, I., J. Bernhard, and C.E. Hanson. 2007. Challenges and promises in mitigating transportation noise. The Bridge 37(3):25–32.

2

Community Noise

This chapter gives a few examples of sources of noise in the community (broadly defined). Noise from aircraft operations is an important source of community noise, and three examples are given: a noise program at a well-established airport, an example of "growing pains" around a relatively new airport, and a situation where a change in policy created problems at relatively low levels of noise. Other examples include noise along the nation's highways, noise from rail transportation, noise in urban areas (e.g., New York City), noise in national parks, noise from industrial facilities, and consumer product noise.

AIRCRAFT NOISE

The U.S. Environmental Protection Agency's "Levels Document" adopted 55 dB as the day-night average sound level (DNL) that would protect the public with an adequate margin of safety (EPA, 1974). However, according to the Federal Aviation Administration (FAA), a noise-impacted area is that inside the 65-dB DNL contour. The noise level in a community near an airport is generally defined by contours of equal DNL around the airport. Impacted areas are generally eligible for sound mitigation with insulation of homes and schools using FAA funding.

According to a report to Congress by the FAA, there has been a 95 percent reduction in the number of people affected by aircraft noise in the past 35 years (FAA, 2004). In 2000 approximately 500,000 people were exposed to a DNL of more than 65 dB and approximately 5 million people were exposed to a DNL of 55 dB.

Although the extent of exposure to aircraft noise has been greatly reduced, the 1974 EPA report stated that "aircraft noise is the single most significant local objection to airport expansion and construction." Some of the problems around the nation's airports are described below.

O'Hare International Airport

The O'Hare Noise Compatibility Commission (ONCC) works in the Chicago metropolitan area on problems regarding noise levels at O'Hare International Airport. ONCC considers and recommends operational programs to reduce aircraft noise impacts, evaluates reports from the Airport Noise Monitoring System, and works on noise abatement issues with air traffic controllers, FAA representatives, airline pilots, and community leaders. The ONCC issues a monthly report that contains information about runway usage, complaints, and noise levels at 37 different locations in the communities around the airport. The reports are linked to the ONCC home page (*http://www.oharenoise.org*).

ONCC has tracked phone call complaints about aircraft noise though its hotline since 1997, and its efforts have resulted in a reduction of calls from 25,773 in 1998 to 3,067 in 2004 (Mulder, 2005).[1] More recent data show that calls in 2005, 2006, and 2007 numbered 1,958, 1,362, and 1,248, respectively. More information about the activities of ONCC can be found on the commission's homepage and in Mulder (2005). ONCC also monitors the FAA's Residential Sound Insulation Program for communities around O'Hare. For a bibliography on noise issues around airports, see *http://airportnoiselaw.org/biblio.html*.

Denver International Airport

In 1997, 22 residents living near Denver International Airport (DIA) filed suit against the city of Denver for allowing what they claimed was excessive noise. The residents, who lived 2 to 6 miles from the airport, alleged that "high levels

[1]Mulder, A.J. Community Noise around Airports. Presentation at the NAE Workshop on Technology for a Quieter America, Washington, DC, September 13–15, 2005.

of noise, pollution, and vibrations on plaintiffs' property" devalued their homes (*Denver Post*, 1997).

In 2005 the developer of a $1.5 billion project called High Point near DIA planned to build the first of 3,000 homes about 2.5 miles from the end of a planned runway, making future noise complaints almost inevitable (*Denver Post*, 2005). Buyers were warned they would be moving into an area under aircraft flight paths, but demand for the homes was expected to be high. To protect buyers, homes were built with triple-pane windows, extra insulation, and central air conditioning.

A report in 2005 prepared by DIA's noise consultants showed that a fully constructed airport would put a noise limit line that would serve as a barrier to residential development right through the High Point project (*Denver Post*, 2005). The Denver City Council was concerned that occupants of the 3,000 homes in High Point might exert enough political pressure to keep the airport from being completed as envisioned.

More recently, the Denver International Airport Partnership has followed other airports (such as Washington Dulles International Airport) in providing guidelines for prospective homeowners about future development at the airport to head off potential problems (*Denver Post*, 2006). When DIA is fully built, it could expand from handling 43 million passengers in 2005 to more than 120 million passengers annually.

In 2006 about 320,000 people lived within 15 miles of the airport, and by 2030 the same zone is expected to have more than 500,000 people, according to the Denver Regional Council of Governments (*Denver Post*, 2006). Contracts for homes near the airport include a legal disclosure that buyers must sign, and some include easements for overhead traffic, precluding residents from suing because of aircraft noise.

East Coast Plan

The "East Coast Plan" developed by the FAA in 1987 to reduce flight delays changed flight paths across southern New Jersey and dispersed them over a larger area of the state. The plan affected small communities that had complained, not about high levels of noise but about the presence of a few aircraft where none had been previously. This is a good example of communities finding low levels of noise objectionable, when even lower levels existed before a change in policy.

SURFACE TRANSPORTATION NOISE

Corbisier (2003) states that "according to the most recent data available from 1987, noise from highway traffic affects more than 18 million people in the United States." This illustrates both the extent of highway traffic noise and the fact that it has been many years since a systematic study of such noise has been done. It is more difficult to assess the impact of railway noise (see Chapter 5). A reasonable estimate is that 10 million people are affected by railway noise, including 6.5 million by train horns at rail/highway crossings.

The current solution to specific vehicle noise emission problems on highways is to construct noise barriers—solid obstructions built between a highway and nearby homes. Effective barriers can reduce noise levels by 10 to 15 dB, cutting the loudness of traffic noise approximately in half (10 dB) for people who live in close proximity to the barrier. A barrier can be a mound of earth along the side of the road (an earth berm), a relatively high vertical wall, or an earth berm combined with a shorter vertical wall above. Earth berms can be attractively landscaped but require large land areas at the base on both sides of the barrier configuration. Walls are limited to about 8 meters in height and can be made of wood, stucco, concrete, masonry, metal, or other materials.

There are no Federal Highway Administration (FHWA) requirements for the type of material used in the construction of noise barriers. Materials are chosen by the state highway administration but must meet FHWA specifications in terms of rigidity and density. Whatever material is chosen must be rigid enough and of sufficient density to provide a transmission reduction of 10 dB compared with the noise diffracted over the top of the barrier.

FHWA uses the following additional criteria for determining the feasibility of noise barriers: a barrier must be high enough and long enough to block the view of the road; noise barriers do very little good for homes on a hillside overlooking a road or for buildings that rise above the barrier; a noise barrier can achieve a 5-dB noise reduction when it is tall enough to break the line of sight from the highway, and it can achieve an additional 1.5 dB of noise reduction for every meter of height above the line of sight (with a maximum theoretical total reduction of 20 dB). FHWA's rule of thumb is that a barrier should extend four times as far in each direction as the distance from the receiver to the barrier. Disruptions in noise walls for driveways or street intersections destroy their effectiveness. Moreover, in some areas, where homes are far apart, the cost of a barrier may be prohibitive.

The construction of noise barriers has always been a cooperative effort between state departments of transportation and FHWA, and states have a great deal of flexibility in designing and building noise barriers. Some states have built many noise barriers, and some have built none. Through the end of 2004, 45 state departments of transportation and the Commonwealth of Puerto Rico had constructed more than 2,205 linear miles of barriers at a cost of more than $2.6 billion ($3.4 billion in 2004 dollars). Five states and the District of Columbia have not constructed any noise barriers (FHWA, 2009).

Noise barriers tend to provide relief for a relatively small number of people in a given area, but the noise reductions are probably greater than those that could be achieved with modern pavement technology. However, the number of people who could potentially get relief from improvements in road surfaces is probably greater than the number of people who get relief from barriers.

FHWA currently does not recognize porous road surfaces as a solution to the highway noise problem; however, the agency does sponsor Quiet Pavement Pilot Programs to investigate their feasibility (Ferroni, 2007).[2] The costs and benefits of porous road surfaces are discussed in Chapter 7.

CONSTRUCTION NOISE

Noise from construction equipment has been a problem, especially in urban areas, for many years. Typical noise sources include jack hammers, compressors, pile drivers, excavators, electric generators, and various types of construction vehicles. Planning for noise control must start with planning for the project itself, and mitigation techniques include noise reduction at the source, construction of temporary noise barriers, and restriction of operating hours. The Federal Highway Administration has produced the *Construction Noise Handbook* (FHWA, 2006), which identifies many of the problems with construction noise. One recent example of control of construction noise is work done in connection with the Central Artery/Tunnel Project in Boston (Thalheimer, 2000, 2001). A second example is the recent New York City noise code described in the section below on urban noise. The code contains many limits on construction noise. The regulations have been described by Thalheimer and Shamoon (2007).

RAIL NOISE

Rail systems are a growing component of the transportation system in the United States because of their demonstrated efficiency in energy use for transporting people and goods. As oil becomes more expensive and interest in green economies increases, rail systems can be expected to expand. Commuters are opting for rail transit in urban areas, resulting in higher ridership each year. Amtrak's portion of intercity trips is growing in both the East Coast and the West Coast corridors. Freight railroads, which have been running at capacity, carry bulk cargo more efficiently than any other transportation mode. As rail transportation increases, an increase in noise exposure in and around transit and railroad facilities can be expected.

NOISE IN URBAN AREAS

Because there have been several recent surveys of noise in New York City, and because a new noise code went into effect in 2007, New York City is often used as an example of the problems associated with dealing with noise in an urban area. The first 10 noise sources in New York City that bother residents were found by Bronzaft and Van Ryzin (2007) to be:

- car alarms
- honking horns
- car stereos or boom cars
- rowdy passersby or people hanging out
- neighbors' activity or voices
- highway or street traffic
- sirens from police cars, fire trucks, etc.
- neighbors' music, TV, or radio
- motorcycles
- construction or repair work

It is difficult to describe many of these sources in terms of an environmental noise metric such as day-night average sound level. Consequently, the extent of noise impact is assessed in terms of the number of complaints received.

On August 17, 2005, Mayor Michael Bloomberg called a press conference to discuss the city's noise code. He said that between June 2004 and August 2005 the city's government services hotline received 410,000 noise complaints, making noise the number one complaint to the hotline. Online surveys conducted in collaboration with the Council on the Environment of New York City have been used to assess both the sources of urban noise and the number of complaints (Bronzaft and Van Ryzin, 2007). Surveys have focused on behavioral and emotional consequences of neighborhood noise, complaints about noise, specific sources of noise in communities, and general perceptions of neighborhood noise (Bronzaft and Van Ryzin, 2004, 2006). The surveys have shown that New Yorkers are bothered more frequently by noise and are more likely to lodge a complaint about it than respondents to similar surveys in other parts of the country (Bronzaft and Van Ryzin, 2004).

The top noise sources for New Yorkers and people nationwide that were most associated with behavioral and emotional consequences are rowdy passersby, neighbors' activities or voices, car stereos, car horns, motorcycles, and back-up beeps. NYC residents also report more frequent behavioral and emotional consequences from noise than respondents nationwide; they are more likely to close their windows, have trouble relaxing, lose sleep, and have trouble reading. Similarly, New Yorkers are more likely to feel annoyed, angry, helpless, upset, and tired because of community noise. "These findings should demonstrate to public officials that New Yorkers cannot find the peace and quiet in their homes that they deserve" (Bronzaft and Van Ryzin, 2006).

In response to NYC noise issues, Mayor Bloomberg asked the city's Department of Environmental Protection (DEP) to revise the noise code, and on December 29, 2005,

[2]Ferroni, M. FHWA Tire/Pavement Noise Policy and Programs. Presentation at the Workshop on Cost-Benefit Analysis and Transportation Noise, Washington, DC, February 22, 2007.

he signed a new version into law, which became effective on July 1, 2007. The city's DEP, which is responsible for noise regulation, has developed a brochure that provides a brief overview of the new code. (For the full text, see *http://www.nyc.gov/html/dep/pdf/law05113.pdf.*)

Important methods in the new code for controlling urban noise include specification of sound pressure level limits at certain distances for sources such as construction equipment, limits on operating hours, limits on noise from some sources that are "plainly audible" at a certain distance, and limits on some sources in terms of decibels above ambient noise. The metric day-night average sound level is not used in the code. Bronzaft and Van Ryzin note that the passage of a noise code will not have the desired impact unless it is supplemented by educational materials on the hazards of noise, noise protection, and protecting the rights of others to quiet (CENYC, 2009). They also recommended that the city council consider legislation that calls for the enforcement of apartment leases guaranteeing residents the right to quiet and include discussions of floor coverings, slamming of doors, and young children running around excessively.

NOISE IN QUIET ENVIRONMENTS

Recreational noise has been the subject of three special issues of *Noise Control Engineering Journal* (NCEJ, 1999). In an excellent article, Sutherland (1999) discusses how to measure, evaluate, and preserve naturally quiet areas. More recently, Miller (2003, 2008) has written about the effects of transportation noise in recreational areas and problems with the metrics used to measure noise levels in quiet areas.

Rossman (2005)[3] describes the Natural Sounds Program of the National Park Service (NPS) and lists the following issues that affect implementation of a noise policy in naturally quiet areas:

- There is no recognized "legal" standard for ambient noise in national parks.
- "Traditional" acoustic metrics are not adequate for measuring impacts on park "soundscape" resources.
- Many parks are already "noisy."
- Many sound sources originate outside park boundaries.
- Little information is available on the kinds of noise that disturb wildlife, and most sound data and models are weighted for human hearing.

Metrics used to measure noise levels in national parks have received much attention in the past few years, primarily because of inherent conflicts among park visitors—those engaged in activities that produce nonnatural sounds (e.g., users of personal watercraft and snowmobiles, air tours) and those who seek quiet and solitude (hikers, row boaters, campers). Nonnatural sounds also may conflict with the legislative mandates that established the park or with NPS management objectives of protecting natural sounds as a specific park resource. The NPS maintains a website devoted to natural sounds with a link to sources of human sounds (NPS, 2009).

Sources of intrusive noise include snowmobiles and other off-road vehicles, aircraft (including commercial aircraft, air tours, and private aircraft), and watercraft (including motor boats, personal watercraft, and other water vehicles). Aircraft overflights and their effects on national parks are discussed in a report to Congress (DOI, 1995).

In some parks, noise from road traffic is an issue, particularly when the only access to quiet areas is by means of a motor vehicle; not everyone is capable of hiking or skiing into naturally quiet areas. Thus, conflicts can arise even among groups of park users who have the same objectives—enjoyment of wilderness areas—but use different means of travel.

Although the day-night average sound level, DNL, is widely used to quantify community response to noise—usually in terms of the average percentage of the population likely to be highly annoyed—other factors must be considered in considering noise in naturally quiet areas. The overall goal might be described as protecting, maintaining, or restoring soundscapes appropriate to the park setting. Park soundscapes may be not only natural but also cultural (e.g., the drumbeat of a sacred tribal dance) or historical (e.g., cannon fire at a Civil War battlefield). Decisions about the appropriate soundscape for a given park area depend not only on visitor perception/satisfaction but also on judgments by park management.

One way to assess visitors' reactions to quiet spaces is to query them by means of a survey about their degree of satisfaction with the environment. A scale of "completely satisfied" to "completely dissatisfied" is more likely to yield valid results than questions about "annoyance." A promising alternative measure is the visitor-reported degree of "interference with the appreciation of natural quiet and the sounds of nature" (DOI, 1995; Miller, 1999).

Schomer has suggested an alternative; he argues that sound-quality techniques, which are widely used in the automotive industry and by product manufacturers (see Chapter 5), are more appropriate for measuring noise levels in parklike settings (Schomer, 2009; Schomer et al., 2008).

NOISE FROM INDUSTRIAL FACILITIES

Industry and industry regulators need better guidelines and standards to ensure that industrial plants operate as good

[3] Rossman, R. NPS Natural Sounds Program. Presentation at the NAE Workshop on Technology for a Quieter America, Washington, DC, September 13–15, 2005.

acoustical neighbors (Wood, 2005).[4] Uncertainty caused by the lack of well-defined, quantitative standards is a significant issue for companies planning and submitting plans for new industrial facilities. Significant delays caused by poorly defined standards have affected the construction of oil refineries, wind power farms, clean coal facilities, and nuclear power plants. Standards should set criteria and guidelines for low-frequency noise, tonal noise, and intermittent noise coming from, and within, new facilities; should define how equivalent noise levels should be used; and should establish protocols for documenting predevelopment, baseline, ambient noise levels in the surrounding environment.

New noise guidelines are needed for electricity-generating and transformer facilities and other industrial structures. Databases of industrial noise, similar to existing databases for cars, trucks, and airplanes, are needed to document industrial sound power and sound radiation, mufflers, building elements, and barriers, among other building characteristics. New noise modeling programs are also badly needed, as are improvements in noise control technologies (e.g., large axial fans) to reduce operating noise and more rapid technology transfer from government research programs to industry.

WIND TURBINE NOISE

Growing interest in renewable energy sources has led to the design and installation of large modern wind turbines for electric power generation. European nations such as Germany, Denmark, and Spain have developed wind power generation, and there is concern in Europe about the noise generated by these machines—as there is in the United States. Three conferences on wind turbine noise have been held in Europe, and a fourth is scheduled for 2011 (WTN2005, WTN2007, WTN2009, WTN2011).

There have been reports of adverse effects of noise in the United States, particularly downwind of turbines close to communities. Low-level audible noise is generated and sometimes modulates at the blade passage frequency of the turbines. Today's large modern wind turbines with blades rotating upwind of the tower are quieter than earlier wind turbines with blades that rotated through the turbulent wake downwind of the tower.

One concern is the adequacy of A-frequency weighting as a metric to assess the effect of noise on people. Another is the fact that these turbines are sometimes located in remote areas or quiet communities where the background noise level in the absence of ground-level wind can be very low. Thus, wind turbine noise is sometimes audible above the background at sound pressure levels that in most cases would be considered to have a minimal effect on people. A variety of state and local noise regulations have been developed to address concerns about noise from wind turbines (Bastasch, 2009; Barnes, 2007). Wind turbine research is ongoing at the National Renewable Energy Laboratory of the U.S. Department of Energy. The report *Wind Turbine Sound and Health Effects—An Expert Panel Review*, prepared by a panel of scientific and medical professionals from several countries, provides an assessment of plausible biological effects of exposure to wind turbine sound (Colby et al., 2009). There is also a short discussion of wind turbine noise in a National Research Council report (NRC, 2007).

NOISE IN BUILDINGS

There are many sources of noise in buildings—including noise from outside such as the transportation noise sources described above; noise from heating, ventilating, and air-conditioning systems; noise generated by equipment used by occupants, and noise generated by the occupants themselves. The section "Noise Control in Buildings" in Chapter 5 describes current problems and future challenges. The issues include noise in homes, noise in hospitals, and noise in business environments. There is widespread dissatisfaction with noise in buildings in which business is conducted. Postoccupancy evaluations by the Center for the Built Environment at the University of California at Berkeley (2007) indicate that occupants are generally dissatisfied with noise and sound privacy. It has also been shown that the move to design "green" buildings can further degrade the acoustical environment (Muehleisen, 2009; Razavi, 2009). Issues include windows that open, low-height screens, natural ventilation systems, and lack of "green" sound absorptive materials. Noise in hospitals and other buildings is covered in Chapter 5.

NOISE FROM CONSUMER PRODUCTS

Noise from consumer products can be a nuisance in the home, but it can be more than a nuisance when noisy products, such as lawnmowers and leaf blowers, are operated in suburban and urban environments. Reducing product noise is a significant factor in the market success of many consumer products. In addition, some foreign suppliers have made significant inroads in U.S. markets by making less noise a distinguishing feature of their products. Some U.S. manufacturers have also begun to take product sound seriously, but cost constraints frequently make it difficult for them to add engineering modifications to produce quieter products.

For some products, such as automobiles, sound is very important, and companies spend heavily to make their cars quiet and pleasant. Automobile companies have large staffs and good facilities for sound research and development, but most appliance/consumer products companies do not. One

[4]Wood, E. Community Noise from New Industrial Plants. Presentation at the NAE Workshop on Technology for a Quieter America, Washington, DC, September 13–15, 2005.

reason for this is that appliances, health care, and personal care products are subject to much more frequent changes than cars, and consumers regularly replace older products or choose to buy new ones because of a desired feature. The effect of frequent changes has been to compress development schedules and limit the transfer of improved noise suppression (e.g., a quieter way to support a small motor) to new models.

SUMMARY

Community and building noise is too broad a topic to be described in a single chapter. Nevertheless, a few examples can be given, organized according to the source of noise—and that is the approach used in this chapter.

Noise around the nation's airports has received a great deal of attention; still, many problems remain to be solved. Noise barriers have been the solution of choice at many locations along the nation's highways, but (as will be seen in later chapters) progress is being made to reduce the noise generated by the interaction between tires and the road surface.

Environmental, construction, and building noise is an increasingly widespread problem in densely populated urban areas, and the situation in New York City exemplifies that. As the rail network expands in the United States, rail noise will become more of an issue than it is today.

Finally, noise in quiet areas such as national parks is part of the community noise issue—broadly defined—and is introduced here and emphasized in the following sections of this report.

REFERENCES

Barnes. J.D. 2007. A Variety of Wind Turbine Noise Regulations in the United States—2007. Proceedings of the Second International Meeting on Wind Turbine Noise, Lyon, France, September 20–21.

Bastasch, M. 2009. Oregon's Wind Turbine Noise Regulations. Proceedings of the Third International Meeting on Wind Turbine Noise, Aalborg, Denmark, June 17–19.

Bronzaft, A., and G. Van Ryzin. 2004. Neighborhood Noise and Its Consequences: A Survey in Collaboration with the Council on the Environment of New York City. Special Report #4. New York: Council on the Environment. Available online at *http://www.cenyc.org/files/cenyc/noisesurvey04.pdf.*

Bronzaft, A., and G.Van Ryzin. 2006. Neighborhood Noise and Its Consequences: A Survey in Collaboration with the Council on the Environment of New York City. Special Report #9. New York: Council on the Environment. Available online at *http://www.cenyc.org/files/cenyc/noisesurvey06.pdf.*

Bronzaft, A., and G. Van Ryzin. 2007. Neighborhood Noise and Its Consequences: Implications for Tracking Effectiveness of the NYC Revised Noise Code. Special Report #14. New York: Council on the Environment. Available online at *http://www.noiseoff.org/media/cenyc.noise.report.14.pdf.*

Center for the Built Environment, University of California at Berkeley (2007). Acoustical Analysis in Office Environments Using POE (Post Occupancy Evaluation) Surveys. Available online at *http://www.cbe.berkeley.edu/research/acoustic_poe.htm.*

CENYC (Council on the Environment of New York City). 2009. Noise. Available online at *http://www.cenyc.org/greenlife/noise.*

Colby, W.D. 2009. Wind Turbine Sound and Health Effects—An Expert Panel Review. American Wind Energy Association and the Canadian Wind Energy Association. Available online at *http://www.awea.org/newsroom/releases/AWEA_CanWEA_SoundWhitePaper_12-11-09.pdf.*

Corbisier, C. 2003. Living with noise. Public Roads 67(1). Available online at *http://www.tfhrc.gov/pubrds/03jul/06.htm.*

Denver Post. 1997. Residents sue Denver Airport and Adams County over noise. June 7.

Denver Post. 2005. Homes jump DIA noise boundary. April 8.

Denver Post. 2006. Denver airport partnership prepares guides for new homebuyers. April 17.

DOI (U.S. Department of the Interior). 1995. Report on Effects of Aircraft Overflights on the National Park System. Washington, DC: National Park Service. Available online at *www.nonoise.org/library/npreport/intro.htm.*

EPA (U.S. Environmental Protection Agency). 1974. Information on Levels of Environmental Noise Requisite to Protect Public Health and Welfare with an Adequate Margin of Safety. EPA Report 550/9-74-004. Washington, DC: Office of Noise Abatement and Control. Available online at *www.nonoise.org/library/levels74/levels74.htm.*

FAA (Federal Aviation Administration). 2004. Aviation and the Environment. Available online at *http://www.faa.gov/library/reports/media/congrept_aviation_envirn.pdf.*

FHWA (Federal Highway Administration). 2006. Construction Noise Handbook. Available online at *http://www.fhwa.dot.gov/environment/noise/handbook/index.htm.*

FHWA (Federal Highway Administration). 2009. Summary of Noise Barriers Constructed by December 31, 2004. Available online at *http://www.fhwa.dot.gov/environment/noise/barrier/sintro.htm.*

Miller, N.P. 1999. Mitigating the Effects of Military Aircraft Overflights on Recreational Users of Parks. AFRL-HE-WP-TR-2000-0034.

Miller, N.P. 2003. Transportation noise and recreational lands. Noise/News International 11(1):9–21. Available online at *http://www.noisenewsinternational.net/docs/miller-2003.pdf.*

Miller, N.P. 2008. U.S. national parks and management of park soundscapes: A review. Applied Acoustics 69(2):77–92.

Muehleisen, R.T. 2009. Noise Problems and Opportunities in "Green Buildings." Presentation at the Second CAETS Forum on Worldwide Noise Sources, Ottawa, Canada, August 25, 2009. Available online at *http://www.noisenewsinternational.net/docs/caets-2009.*

NCEJ (Noise Control Engineering Journal). 1999. 47 (3–4), Special Issues.

NPS (National Park Service). 2009. Sources of Human-Caused Sounds in National Parks. Available online at *http://www.nature.nps.gov/naturalsounds/sources/.*

NRC (National Research Council). 2007. Environmental Impact of Wind Turbine Noise. Washington, DC: National Academies Press. Available online at *http://books.nap.edu/openbook.php?record_id=11935&page=160.*

Razavi, Z. 2009. Acoustical Improvements with Natural Air Ventilations in the Liu Institute at the University of British Columbia. Proceedings of INTER-NOISE 2009. The 2009 International Congress on Noise Control Engineering, Ottawa, Canada, August 23–26 Available online at *http://www.bookmasters.com/marktplc/00726.htm.*

Schomer, P.D. 2009. Visitor perception of park soundscapes: A research plan. Noise/News International 17(2):51–56. Available online at *http://www.noisenewsinternational.net.*

Schomer, P.D., G.F. Stanley, and W. Chang. 2008. Visitor Perception of Park Soundscapes: A Research Plan. Proceedings of INTER-NOISE 08, The 2008 International Congress and Exposition on Noise Control Engineering, Shanghai, China, October 26–29. Available online at *http://www.internoise2008.org/ProceedingsforSale.htm* or *http://scitation.aip.org/journals/doc/INCEDL-home/cp/.*

Sutherland, L.C. 1999. Natural quiet: An endangered environment. How to measure, evaluate, and preserve it. Noise Control Engineering Journal 47(3):82–86.

Thalheimer, E., and C. Shamoon. 2007. New York City's new and improved construction noise regulation. Proceedings of NOISE-CON 07. The 2007 National Conference on Noise Control Engineering, Reno, Nevada, October 22–24. Available online at *http://www.bookmasters.com/marktplc/00726.htm*.

Thalheimer, E. 2000. Construction noise control program and mitigation strategy at the Central Artery/Tunnel Project. Noise Control Engineering Journal 48(5):157–165.

Thalheimer, E. 2001. Proactive Construction Noise Control Policies Developed for the Central Artery/Tunnel Projects. Proceedings of NOISE-CON 01, The 2001 National Conference on Noise Control Engineering, Portland, Maine, October 29–31. Available online at *http://www.bookmasters.com/marktplc/00726.htm*.

WTN2005. First International Conference on Wind Turbine Noise, Berlin, Germany, October 17, 18. Available online at *http://www.confweb.org/wtn2005*.

WTN2007. Second International Conference on Wind Turbine Noise, Lyon, France, September 20–21. Available online at *http://www.confweb.org/wtn2007*.

WTN2009. Third International Conference on Wind Turbine Noise, Aalborg, Denmark, June 17–19. Available online at *http://www.confweb.org/wtn2009*.

WTN2011. Wind Turbine Noise 2011, Rome, Italy, April 12–14, 2011. Announcement online at *http://www.windturbinenoise2011.org*.

3

Metrics for Assessing Environmental Noise

Selecting a metric for assessment of environmental noise is no simple task, because it must reflect the impact on people. No single metric can describe all responses in all situations.[1] Context, expectations, and people's experiences and circumstances all affect their responses. Hence, levels of community response (e.g., annoyance) may vary from community to community, just as individual responses vary from person to person, even if noise levels do not change. However, one consistent finding has been that changes in noise exposure do affect individual and community responses and that increases in man-made noise usually have a negative impact. This is illustrated by the Schultz curves later in this chapter.

Thus, it is important to understand which characteristics of noise elicit a negative response and how exposure to noise with those characteristics affects people's lives. The metric chosen or developed for measuring community noise must reflect this human response and must be taken into account in making policy decisions.

Fifty years ago, when noise metrics were developed, the choices were based on simpler calculations and technologies and the acoustical quantities that could be predicted by sound propagation models used at the time. Although much more sophisticated measurements can be made today, many still consider these "older" metrics valid and continue to use them. However, with modern instruments (see Appendix E), much more accurate measurements and predictions can now be made of people's reactions to noise.

A meaningful metric, or set of metrics, translates sound pressure-time history measurements into a prediction of the effects of noise, such as annoyance, sleep disturbance, changes in health, interference with understanding speech, and ability to learn. Ideally, this translation should be based on context, expectations, and personal situations and preferences, in addition to noise information, and should account for a distribution of responses, including responses of vulnerable populations, such as children. Unfortunately, a holistic model of community response is still beyond present capabilities.

One fundamental issue that must be considered in the choice of an environmental noise metric(s) is the purpose for which the metric will be used:

- to implement public policy on noise immission from one or more sources
- to provide information about noise exposures in a form understandable to the public
- to assess a noise situation in terms of noise control engineering

The metrics to accomplish these purposes may differ, but all three relate directly to the impact of noise on the community. For example, a metric to inform decisions about noise control engineering strategies should result in reducing the noise impact, which would then be reflected in the policy metric(s) and the public information metric(s).

As new research results become available and accessible, they should influence the choice of metrics for the three purposes listed above. The results of such research may result in complex calculations that include many variables and may better quantify individual reactions to sound. Some modern procedures, such as calculation of loudness, are more complex than earlier methods, but available computational procedures make the results widely available.

Much of this chapter recounts the evolution of noise metrics and their applications to public policy. This history includes criteria originally used by the U.S. Environmental Protection Agency (EPA) to select a noise metric and the rationale Europeans have used for using a curve passed through highly variable data to determine what percentage of a population is "highly annoyed" by a given noise. In ad-

[1]This chapter considers only metrics related to the effects of environmental/community noise on people. The effects of noise on wild and domestic animals and on sensitive historical structures are not considered, even though these effects are often considered in environmental noise impact analyses.

dition, alternative metrics are described that may be easier for the public to understand than the day-night average sound level (DNL).[2]

LOUDNESS AND A-WEIGHTING

Arguably the modern history of noise metrics began in the 1930s with the search for a way to describe the loudness of sound. This led to the definition of weighting networks for sound-level meters and, because of limitations on the capabilities of calculating sound pressure levels at that time, a single frequency-weighted value—either A-weighted or C-weighted—came into common usage.

Loudness

In an early attempt to determine the loudness of sound (using discrete-frequency tones), Fletcher and Munson (1933) found that the loudness of a tone depends on both its amplitude and its frequency. Knowing this dependence, they were able to develop a set of equal-loudness curves. In modern terms the unit of loudness is the phon. For example, a 1,000-Hz tone with a sound pressure level of 40 dB has a loudness of 40 phon. At this loudness level the sound pressure level of tones between 1,000 and about 5,000 Hz is generally lower than 40 dB, and the sound pressure level of tones below 1,000 Hz and above about 5,000 Hz is higher than 40 dB.

The sound-level meter was standardized in the early 1930s when microphones and electronic circuits were being developed. Ideally, the standard sound-level meter would have a single-number description of the sound at a given point in space. The best description at the time came from the studies by Fletcher and Munson, who clearly showed that the shape of the equal-loudness curve was dependent on both the amplitude and the frequency of sound. Thus, using the linear electronic circuits of the time, a few curves had to be selected based on the amplitude of the sound. One of the curves selected, which is very close to the 40-phon curve, was designated as "A-weighting." Another, which was nearly independent of frequency, was designated as "C-weighting." A third curve, the "B-weighting" curve, which fell between the A and C curves, has long since fallen out of favor. A-weighting and C-weighting are still used today, although the shape of the curves has changed somewhat to provide a standardized mathematical description in terms of poles and zeros of a transmission network.

Work on improving the calculation of loudness based on measurement of the spectrum of sound continued. The best-known early work in the United States was by S. S. Stevens and in Germany by Eberhard Zwicker. Stevens's Mark VI and Zwicker's work on loudness were standardized by the International Organization for Standardization (ISO, 1975). Later work by Brian Glasberg and Brian Moore in the United Kingdom was the basis for the American National Standard on computation of loudness (ANSI, 2007).

Over the years, A-weighted levels were found to correspond reasonably well to human response, especially for noise spectra in typical offices. Single-number methods of rating noise in offices and other building spaces were also developed, including so-called noise rating curves (NR curves—a curve tangent method of obtaining a single number from an octave band spectrum) and ratings based on loudness and A-weighting.[3]

METRICS FOR MEASURING COMMUNITY REACTION TO NOISE

One early attempt to develop a metric for forecasting community response to noise was made by Stevens et al. (1955). Unlike the DNL, this metric included nonacoustical factors as well as noise levels and yielded a "composite noise rating." This rating was then plotted against a scale of community responses—vigorous community action, threats of community action, widespread complaints, sporadic complaints, and no observed reaction. A few case studies showed a reasonable correlation between the measurement and response but with considerable scatter. Community noise levels were determined by measuring the average octave band levels in the community averaged in space and time. A curve tangent method was used to reduce the octave band data to a single-number rating.

Day-Night Average Sound Level

After EPA established the Office of Noise Abatement and Control and after passage of the Noise Control Act of 1972, EPA was faced with the task of developing a metric for community noise with the following characteristics (EPA, 1974):

1. The measure should be applicable to the evaluation of pervasive long-term noise in various defined areas and under various conditions over long periods of time.
2. The measure should correlate well with known effects of the noise environment on the individual and the public.
3. The measure should be simple, practical, and accurate. In principle, it should be useful for planning as well as for enforcement or monitoring purposes.
4. The required measurement equipment, with standardized characteristics, should be commercially available.

[2]For information on how other countries measure noise in quiet areas, see Appendix B. More information on communication with the public can be found in Chapter 10.

[3]A-weighting is less useful for measuring human response to sound when the spectrum has a large low-frequency component, when high-amplitude peaks in the spectrum are in the 2- to 4-kHz range, and when the sound is tonal or impulsive.

5. The measure should be closely related to existing methods currently in use.
6. The single measure of noise at a given location should be predictable, within an acceptable tolerance, from knowledge of the physical events producing the noise.
7. The measure should lend itself to small, simple monitors that can be left unattended in public areas for long periods of time.

EPA also published its rationale for choosing A-weighting and for leaving open the possibility of using a different metric in the future (EPA, 1974; von Gierke, 1975):

> With respect to both simplicity *and* adequacy *for* characterizing human response, a frequency-weighted sound level should be used for the evaluation of environmental noise. Several frequency weightings have been proposed *for general use in* the assessment of *response* to noise, differing primarily in the way sounds at frequencies between 1000 and 4000 Hz are evaluated. The A-weighting, standardized in current sound level meter specifications, has been widely used for transportation and community noise description. For many noises, the A-weighted sound level has been found to correlate as well with human response as more complex measures, such as the calculated perceived noise level or the loudness level derived from spectral analysis. However, psychoacoustic research indicates that, at least for some noise signals, a different frequency weighting which increases the sensitivity to the 1000–4000 Hz region is more reliable. Various forms of this alternative weighting function have been proposed; they will be referred to here as the type "D-weightings." None of these alternative weightings [have] progressed in acceptance to the point where a standard has been approved for commercially available instrumentation.
>
> It is concluded that a frequency-weighted sound pressure level is the most reasonable choice for describing the magnitude of environmental noise. In order to use available standardized instrumentation for direct measurement, the A-frequency weighting is the only suitable choice at this time. The indication that a type D-weighting might ultimately be more suitable than the A-weighting for evaluating the integrated effects of noise on people suggests that at such time as a type D-weighting becomes standardized and available in commercial instrumentation, its value as the weighting for environmental noise should be considered to determine if a change from the A-weighting is warranted.

The decision to add 10 dB[4] in measuring nighttime levels and the selection of a two-period (day-night) metric rather than a three-period metric (day-evening-night) was based on community reaction studies at the time and tests that showed little difference between a two-period and a three-period metric. Thus, the DNL (A-frequency weighting for both daytime and nighttime levels and a 10-dB increase in measuring system gain at night) came into being for the evaluation of community noise.

In the United States, DNL and the percentage of persons highly annoyed (discussed in the next section) are widely used, especially by the Federal Aviation Administration (FAA). The Federal Highway Administration uses A-weighting and the average sound pressure level during the busiest traffic hour as a measure of community impact. The difference between C-weighted and A-weighted levels is used as an indication of the low-frequency content of the sound, and the sound exposure level (see Appendix A) is used to evaluate sounds of finite duration—for example, an aircraft flyover.

Day-evening-night sound level is widely used in Europe. In some countries, L_{day} and L_{night}, (average A-weighted sound pressure levels) are used in addition to or instead of a DNL-type metric. None of these metrics takes into account the time of night when the noise occurs, even though noise appears to cause greater sleep disturbance at the beginning and end of the night.

Several issues have arisen from the use of DNL and the percentage of persons highly annoyed: no one actually "hears" a DNL; there is a high variability from study to study around a nominal Schultz curve; and in many situations "highly annoyed" is not an appropriate measure of human response. Although the percent highly annoyed and DNL approach has been widely endorsed, variability around a nominal Schultz curve is troubling, and there are reports that this approach is not sufficient to predict community response (Fidell, 2002). Attitudinal and personal variables impact people's responses and are, to some extent, the reason for scatter (Fields, 1993; Flindell and Stallen, 1999; Miedema and Vos, 1999).

As shown in Figure 3-1, some researchers (Miedema and Oudshoorn, 2001) have found in their analyses of survey results that the nominal Schultz curve appears to depend on the noise source (e.g., aircraft, road traffic, rail traffic). In addition, DNL is a relatively insensitive measure of sleep disturbance and thus is not an appropriate metric for predicting awakenings in sleep disturbance studies. Finally, A-weighting is not the best weighting for measuring noises with unusual spectra (e.g., excessive high- or low-frequency noise or noise that has unusual peaks in its spectrum). For sounds with levels that evolve over time, the most appropriate

[4]A number of metrics have been developed to take into account daytime versus nighttime operations around airports. These include Noise Exposure Forecast, Community Noise Equivalent Level, and Noise and Number Index. The EPA rationale for selecting a 10-dB nighttime penalty (EPA, 1974) is as follows: "Methods for accounting for the differences in interference or annoyance between daytime/nighttime exposures have been employed in a number of different noise assessment methods around the world. The weightings applied to the nondaytime periods differ slightly among the different countries but most of them weight night activities on the order of 10 dB; the evening weighting if used is 5 dB. The choice of 10 dB for the nighttime weighting made in Section 2 was predicated on its extensive prior usage, together with an examination of the diurnal variation in environmental noise."

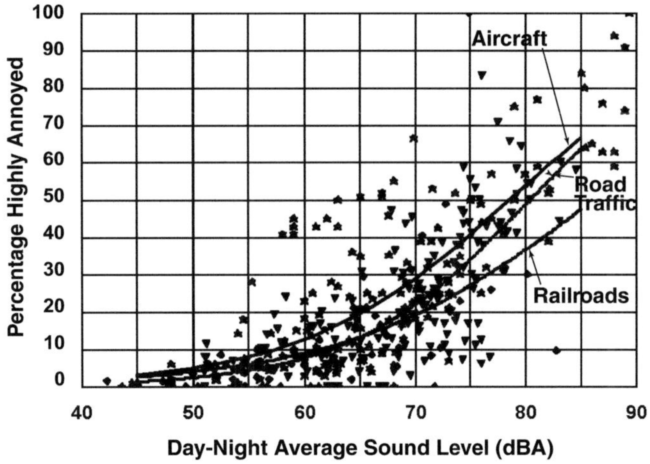

FIGURE 3-1 Variability in survey results. ▼ = road traffic. ★ = air traffic. ♦ = rail traffic. Curves are the results of fits to data associated with different modes of transportation. Source: Adapted from Schomer (2005) and Fidell and Silvati (2004).

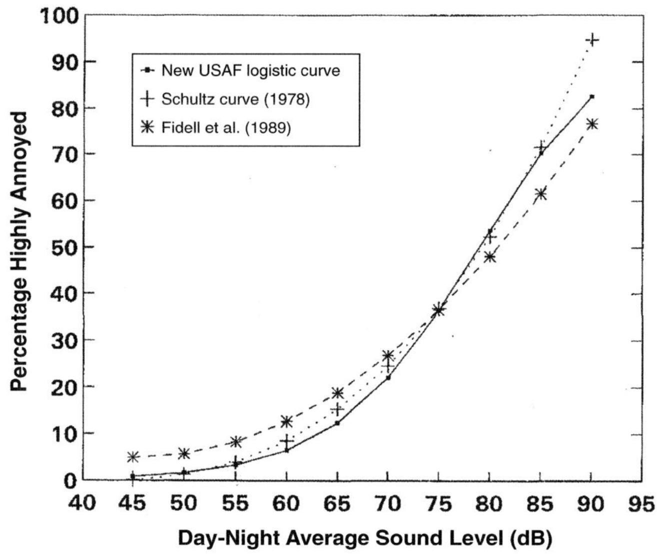

FIGURE 3-2 Three versions of a Schultz curve. ■ = the U.S. Air Force logistic curve. ✚ = the curve proposed by Schultz (1978). ✳ = a curve by Fidell et al. (1991). Source: Reprinted from Finegold et al. (1994).

weighting should change with the level; typically, however, only one weighting is used.

Percentage of Persons Highly Annoyed

The next major event in the selection of a noise metric was a study by Schultz (1978) of surveys of community reaction to noise. Schultz went back to original data to estimate the percentage of the population "highly annoyed" as a function of DNL. Even at that time, it was recognized that, for a variety of reasons, there was considerable scatter in the data. Nevertheless, Schultz proposed that a single curve (the Schultz curve) drawn through the data should be used as a measure of community response. Later studies led to modifications of the Schultz curve (Fidell et al., 1991; Finegold et al., 1994). In the latter study, three curves were compared (see Figure 3-2), and a U.S. Air Force logistic curve was defined

$$\%HA = 100/[1 + \exp(11.13 - 0.14L_{dn})] \quad (1)$$

The scatter in the highly annoyed response, compared to scatter in the average curve, was presented by Miedema and Vos (1998) and has been commented on by several subsequent researchers (e.g., Schomer, 2005). The first problem with scatter is that it causes great uncertainty in the prediction of community reaction. A second problem is that community reaction (percent highly annoyed) appears to depend on the source of the noise; for example, responses to aircraft noise, road traffic noise, and rail noise vary, even if the noises have the same DNL (see Figure 3-1). The question that must be answered is whether the variability in response is due to the nature of the noise source or reflects how the metric is calculated.

Consultants and other professionals are often asked to study community noise issues and recommend remedial action. Predictions of community response should not be based only on variations of the Schultz curve. It has been known for many years (Stevens et al., 1955) that nonacoustical factors influence community reaction to noise. Thus, at a minimum, temporal and spectral variations must also be taken into account.

Based on work by EPA, Schomer (2002) proposed modifications to DNL to account for tonality, impulsiveness, background noise, type of community, and other factors. Schomer also showed how this modified approach could be used to reduce variances in the survey data on which the Schultz curve is based.

The Federal Interagency Committee on Noise (FICON, 1992) endorsed the use of percent highly annoyed and DNL as metrics for assessing community noise around airports and recommended that the equation above be accepted as showing the definitive relationship between percent highly annoyed and DNL (see also Finegold and Finegold, 2002). Response curves for community annoyance have now been standardized nationally (ANSI, 2005) and internationally (ISO, 2003).

ALTERNATIVE METRICS

The science of measuring environmental noise has progressed rapidly in the past decade as computer technology has come on line to provide rapid data acquisition and analysis in small portable packages. The end result has been a revolution in the type and complexity of measurements and calculations that can be made in analyzing environmental noise. This section provides a more detailed description of presently used metrics and a variety of alternative metrics that are well within the capabilities of modern instrumentation.[5]

A Different Frequency Weighting

An alternative to A-weighting (i.e., D-weighting) could be considered. As noted earlier, this weighting was considered by the EPA in 1974 but rejected because there was no standard shape for the curve.

Perceived Noise Level

Community reaction to noise from jet planes led to important events in the development of noise metrics. The problem, which arose in 1956, is described in an autobiography by Beranek (2008). According to measurements made with a standard sound-level meter, the noise produced by a Boeing 707 jet airplane and that by a propeller airplane (Super Constellation) were equal. However, subjective testing showed that the 707 was considered much noisier; by subjective measures, the A-weighted sound pressure levels of the 707 would have to be significantly reduced to be considered as noisy as the Super Constellation. This early test of the usefulness of A-weighted levels in judging noisiness led to further evaluations of the relative noisiness of propeller-driven and jet airplanes and the development of the concept of "perceived noisiness" (Kryter, 1960; Kryter and Pearsons, 1962, 1963).[6]

Perceived noise level (PNL) was used in the development of specifications of noise *emissions* from airplanes for regulatory purposes in 1969 and is still used to certify airplanes today. When the perceived noise-level metric was adopted, it was possible to compute it only with a large amount of equipment. Today, it can be done with a handheld sound-level meter. D-weighting simplifies the PNL calculation, but neither PNL nor D-weighting solves the decibel issue, which relates to explaining noise to the public.

Loudness

Historically, the method of calculating PNL was similar to the method of calculating loudness. Today, several methods can be used to calculate loudness, all of them with a handheld sound-level meter. Loudness that exceeds some agreed-on value a given percentage of the time also can be calculated. On a linear scale (as opposed to a logarithmic scale), a doubling of the value of the calculated value corresponds to a doubling of the loudness. This may be easier to explain to the public than a metric that uses the phon (which uses a logarithmic scale) as a unit of loudness. For sounds in the midfrequency range, an increase in A-weighted level of 10 dB corresponds to a doubling of loudness.

Speech Interference

Standard methods of calculating speech interference are available, and the values may be translated into effects that are easier for the public to understand than DNL. For example, the difficulty of communicating over a given distance between speaker and listener may be quantified in terms of percentage of speech likely to be understood. Speech interference can be affected by the fact that hearing loss increases with age and usually starts at high frequencies. Thus, the ability to distinguish consonants that have high-frequency content such as "s" and "th" is diminished.

Nighttime Sleep Disturbance

In *Night Noise Guidelines for Europe*, published by the World Health Organization (WHO, 2007), sleep disturbance is related to the nighttime level designated as L_{night}, although researchers also use indoor L_{Amax} and indoor A-weighted sound exposure level (ASEL) when investigating the relationship between awakenings and noise. The temporal pattern of noise at night, however, is known to influence sleep disturbance. This problem is addressed to some extent in a new American National Standard (ANSI, 2008), in which terms such as the likelihood of awaking, are used; the new standard may be more understandable to the public than the day-night average level or the nighttime level used in Europe.

METRICS FOR COMMUNICATING WITH THE PUBLIC

An often-cited shortcoming of DNL is that the public does not understand what it means. Over the years, various people have advocated using supplemental metrics that describe noise in ways that are more understandable to the majority of people (FICAN, 2002). Metrics used to supplement DNL include time above (a certain level), number of events above a given value of the ASEL, number of loud events above a certain ASEL in a given period, and single-event descriptors such as L_{Amax} and ASEL. Most advocate using a group of metrics to give a fuller picture of the potential impact of the exposure and explaining that these measures supplement metrics such as DNL. The same argument can be made for using a group of metrics when addressing other measurements or predicting a variety of impacts (Eagan, 2007), such as the number of occurrences of speech interference; when noise levels inside buildings exceed recommended levels for a particular activity, such as learning in schools (ANSI, 2002); or the likelihood of being awakened based on predicted indoor single-event metrics (ANSI, 2008).

[5]For a description of the instruments, see Appendix E.

[6]For a general assessment of human reaction to aircraft noise, see Beranek et al., 1959.

The number of events has been recognized as an important factor in noise exposure, and it is included in metrics that are or have been used to predict annoyance; alternatives to DNL, such as the Noise Exposure Forecast (NEF) system used in Canada (Transport Canada, 2005) and elsewhere; and the Noise and Number Index (NNI) that was used prior to 1990 in the United Kingdom. The NEF metric is based on effective perceived noise level as well as the number of events; hence it takes into account some of the impact of tonal components and impulsiveness on annoyance. NNI is also based on a very basic loudness measure, perceived noise level in decibels and number of events. Analysis of data from a study at U.K. airports in 1982 and another study in 2005 showed that the relationship between annoyance and A-weighted equivalent level had changed. However, by combining a measure of average noise exposure with the number of events, it was possible to develop a metric that worked consistently for both studies (ANASE, 2007).

NOISE METRICS FOR RURAL/NATURALLY QUIET AREAS

Neither day-night average sound level nor percent highly annoyed is an appropriate metric for measuring noise in naturally quiet areas. Because of the logarithmic nature of the decibel, short-duration sounds of high amplitude compared with background noise can significantly increase the day-night level, even though the sound remains at the background level most of the time. As for percent highly annoyed, this is hardly the best measure of satisfaction for areas where quiet and solitude are valued. In addition, it can be difficult to measure very low sound pressure levels. A-weighted levels of 40 dB are at the upper end of the range, and lower levels can be at or even below the levels measurable with conventional sound-level meters.

Nevertheless, some quantification of noise impact is clearly needed in these areas as a basis for establishing public policy, which usually means regulatory action. The classic definition of noise is "unwanted sound," so the source of sound must be identified, either as part of the natural soundscape or not. Thus, simple metrics like sound pressure level are clearly not appropriate. For example, an airplane overflight may have a much lower sound pressure level and shorter duration than sound from a rushing stream, but the former is considered noise and the latter is considered sound. The method of assessment of the noise environment should also take into account the likely long-term impact on animals that use, for example, very low level sounds (perhaps inaudible or unnoticed by people) to locate prey or predators.[7]

[7]See Chapter 2 and Appendix B for more on noise metrics in quiet areas.

INTERNATIONAL ACTIVITIES RELATED TO NOISE METRICS

The International Commission for the Biological Effects of Noise holds meetings at five-year intervals. In 2008 the meeting was held in the United States, but most of the participants came from other countries, as did the presenters. Truls Gjestland of Norway presented a summary report on research in the past five years related to the effects of community noise, specifically annoyance. Although some research has been done in Japan, he said, not many significant projects had been undertaken. However, he noted that at least three different versions of the Schultz curve had been developed, all of them based generally on the same datasets (Gjestland, 2008). Around the same time Lawrence Finegold of the United States presented a review of major noise-related policy efforts around the world during the same time period (Finegold et al., 2008).

European Activities

In 1996 the European Union (EU) published *The Green Paper*, which established new noise programs that are used to address noise issues today (EC, 1996). European Directives have been issued concerning noise emissions from consumer products, and an EU Environmental Noise Directive (END) in 2002 led to the development of noise-mapping and, in a few cases, action plans that require noise metrics (EC, 2002a). Related activities include the HARMONOISE and IMAGINE projects (*http://www.imagine-project.org/*).

European Metrics (Indicators)

A-frequency weighting for determining sound levels that have been standardized in the United States and internationally is widely used in Europe. However, as discussed elsewhere in this chapter, frequency weighting alone is not enough to define a metric. A Working Group (WG1) that produced a report in 2000, *Position Paper on EU Noise Indicators*, in support of future European noise policies, identified five criteria for selecting an indicator: validity, practical applicability, transparency, enforceability, and consistency. Although this report was not an official EU document, the metrics recommended therein are now widely used (EC, 2000).

WG1 recommended that two indicators, both based on A-frequency weighting, be used for reporting data on noise exposure. These indicators were designated L_{EU} and $L_{EU}N$ but today they are widely known as the day-evening-night sound level, DENL, and the equivalent sound pressure level during the eight-hour nighttime period, L_{night}. The group explained, and questioned, the rationale for using 5 dB as level weighting for the evening period and 10 dB for nighttime. Nighttime was nominally designated as eight hours, from 11:00 p.m. to 7:00 a.m.; daytime, 12 hours; and evening, four hours (with some variation, depending on the country). For general purposes, the long-term average A-weighted sound pressure level, L_{Aeq}, was used. The WG1 report also recognized that the character of noise (impulsive, tonal, etc.) may affect human response. Thus, corrections to the metrics may be necessary, and A-frequency weighting may not be appropriate for measuring low-frequency noise.

The WG1 report was also the basis for metrics specified in the 2002 END that led to noise mapping. The directive also suggests supplemental metrics based on the WG1 report (EC, 2002a).

Dose-Effect Relationships (Exposure-Response Relationships)

Another Working Group (WG2) on Health and Socio-Economic Aspects of Noise also produced a report, again not official EU policy. In *Position Paper on Dose-Response Relationships Between Transportation Noise and Annoyance* (EC, 2002b), the group recommended that the percent highly annoyed (%HA) be used as a measure of community response to noise. Updated and modified Schultz curves, based on the work of Miedema and colleagues (e.g., Miedema and Oudshoorn, 2001) for the %HA as a function of day-evening-night sound level, are used to measure road traffic, rail traffic, and aircraft noise. WG2 also acknowledged the variability from study to study in the mean values in Schultz curves (e.g., Gjestland, 2008) but still supported the use of "norm" curves:

> Substantial deviations from the predicted percentage [of] annoyed persons must be expected for limited groups at individual sites because random factors, individual and local circumstances and study characteristics affect the noise annoyance. However, in many cases the prediction on the basis of a "norm" curve that is valid for the entire population is a more suitable basis for policy than the actual annoyance of a particular individual or group. For example, a "norm" curve is useful when exposure limits for dwellings and noise abatement measures are discussed. Equity and consistency require that limits and abatement measures do not depend on the particularities of the persons and their actual circumstances. For similar reasons, a "norm" curve also can be used to estimate the number of annoyed persons in the vicinity of an airport, road, or railway when different scenarios concerning, e.g., extension of these activities or emission reductions are to be compared. That the norm curve does not take local circumstances or reactions to a change in exposure itself into account, is considered to be an advantage for many purposes. Equity and consistency of policy would not be served if in each case the actual annoyance is taken as the (only) basis for these evaluations. The use of "norm curves" or "norm thresholds," which are valid for the entire population (or a particular sensitive subgroup), is common practice when exposures to other environmental pollutants, such as air pollutants or radiation, are evaluated. There they are used for the evaluation of an individual situation, irrespective of the population in that situation. It is recommended to take the same approach in the case of environmental noise and use the same curve irrespective of the population in the situation evaluated.

Nighttime Effects

In 2004, WG2 produced *Position Paper on Dose-Effect Relationships for Nighttime Noise*, again not an official EU document. In this paper the metric used was L_{night}, as defined above as the measure for sleep disturbance. Based on questionnaires, curves similar to Schultz curves were developed, the ordinate being the percent highly disturbed and the abscissa being the nighttime noise level. An effort was made to relate single events to the nighttime sound level (EC, 2004).

Annoyance and the Microstructure of Noise Exposure

Several studies have been published, mostly in connection with the EU-funded SILENCE project (*www.silence-ip.org*), on the importance of the microstructure of a noise exposure situation. The argument is that equivalent levels do not "tell the full story." Different traffic noise situations with the same equivalent level may be assessed differently with respect to annoyance. This is important information for decisions about how to reduce the negative impact of road noise through traffic management measures. Laboratory experiments have provided several examples:

- An even flow of traffic causes the same annoyance as when vehicles are clustered, but an even flow is more damaging to mental performance than clustered traffic.
- A large difference between equivalent level and L_{max} is more annoying than a small difference.
- Trams should receive a 3-dB "bonus" over buses.
- Different noises from a rail yard at equal equivalent levels may have a subjective difference of as much as 5 dB.

Recommendations for Future Research in Europe

Research for a Quieter Europe in 2020, a report produced under the auspices of the CALM Network (2007), provides a strategy for future noise research in the EU. The report includes an excellent review of EU activities related to noise and covers a wide variety of future needs, including noise emissions from various sources and the need for perception-based research into the effects of noise. There is one short section on metrics (indicators).

European Versus Japanese Results on Transportation Noise

A recent Japanese study by Yano et al. (2007) compared the effects of transportation noise in Japan with the EU results. The effects of road traffic noise are similar, but the effects of railway noise were quite different (see Figure 3-3). The authors suggest that the differences may be attributable to the proximity of Japanese homes to railroad tracks (where they are subject to vibration as well as noise). Differences in the construction of homes may also be a factor.

Japanese data for aircraft noise are based on one dataset of 410 responses around Kumamoto, a small airport, and may not be representative of noise around Japanese airports in general (Yano et al., 2007). There was also an active anti-noise group near Kumamoto airport. However, considering the scatter from study to study (e.g., Yano et al., 2007), the results of the Kumamoto study may be representative.

SUMMARY FINDINGS AND RECOMMENDATIONS

Established and New Environmental Noise Metrics

Use of the DNL metric has helped policy makers, road planners, airport managers, the public, and others understand potential noise impacts on communities and has helped guide noise mitigation efforts around airports, roadways, and rail systems. However, DNL has both strengths and weaknesses as a measure of noise.

The strengths of DNL are that it has become familiar over time, its calculation has been standardized, through experience it has become well understood, and it is now embedded in software used for planning. DNL has made it possible to communicate evaluations of noise to the public to provide people with a better understanding of how noise policy decisions are made and how changes in transportation systems, or choosing to live near an airport or a busy highway, might affect them. DNL has also been a mechanism by which people could be protected and systematically helped to address problems with environmental noise exposure fairly and equitably.

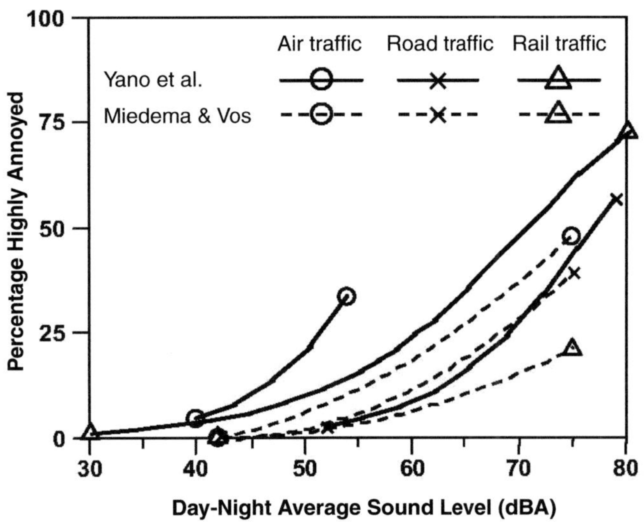

FIGURE 3-3 Comparison of the present dose-response curves with results from Miedema and Vos. Source: Adapted from Yano et al. (2007).

DNL also has some drawbacks. First, there is a great deal of variability from study to study in the percentage of the population believed to be "highly annoyed" as a function of DNL, which predicts only part of a community's response to noise. Efforts to develop metrics that can provide a more definitive assessment of community impact are still a topic for research and policy debate.

Many limitations of a DNL-type metric based on the average A-weighted sound pressure level used to assess environmental noise have been noted:

- DNL is insensitive to the impact of very loud, isolated events.
- Fewer loud events can have the same DNL as many quieter events; thus, the impacts of very different soundscapes are described as equal.
- DNL is insensitive to the time when an event occurs (e.g., noise early in the night causes different sleep disturbance than noise early in the morning).
- The only strong argument for using night and evening weightings in DNL is based on the fact that average nighttime ambient levels are lower than those during the day.
- Other metrics such as speech interference level and nighttime levels provide a better measure of annoyance with speech interference and conscious awakenings.
- DNL is an outdoor noise measure that may not reflect differences between outdoor sounds and the same sounds heard indoors.
- A-weighting does not reflect the results of research studies in psychoacoustics over the past 40 years.
- DNL does not take into account other sound characteristics (e.g., tonality and rate of loudness onset) that can influence annoyance and sleep disturbance levels.

Although DNL has limitations, it has served as the major environmental noise metric since the early 1970s. Despite the variability in community response, it is clear that the percentage of the population highly annoyed for a DNL of 65 dB is considerably greater than the corresponding percentage for a DNL of 55 dB. This supports the findings of EPA in the 1970s (EPA, 1974) that a DNL of 55 dB is the level necessary to protect the public health and welfare with an adequate margin of safety.

When new metrics are developed and values selected as a matter of public policy, the goal should be to protect a larger fraction of the population than is protected under the value now widely used—the DNL = 65 dB criterion. Many steps would have to be taken before a different metric (or set of metrics) could be recommended to policymakers. Changing to another metric would entail significant effort and cost (e.g., in conducting surveys and development of databases) and would be of limited value unless the new metric offers significant benefits over DNL, most importantly in providing a more transparent and definitive connection between noise level and annoyance or other effects on people's lives. Unfortunately, because of a lack of "real-world" data to test the performance of metrics, it is difficult to establish their advantages and disadvantages. The situation with respect to DNL has been recognized by the FAA, and two meetings have been held—one in August 2009 and one in December 2009—to discuss a "roadmap" to improve the situation regarding noise metrics.

A set of metrics, rather than a single metric, to describe different types of outcomes of environmental noise (e.g., number of interruptions of speech, learning impairment in schools, number of additional awakenings) would provide a multidimensional picture of noise impact and may be the best approach to informing the public. Supplementary metrics could make possible predictions of noise from transportation in sufficient detail to enable the development of noise maps.

When communicating with the public, it might be useful to translate metric values into words (e.g., categories such as no observed reaction, sporadic complaints, widespread complaints, threats of community action, vigorous community action) that can be more easily understood than DNL and other numerical metrics.

The ability to predict direct health effects of noise (e.g., hypertension, speech interference, cognitive impairment, sleep disturbance) and the relationship between these effects and annoyance requires further study in order to develop new metrics that account for health effects.

Recommendation 3-1: The federal government (e.g., agencies of the U.S. Department of Transportation with responsibilities related to noise and the U.S. Department of Housing and Urban Development) should adopt as a goal the 1974 recommendation of the Environmental Protection Agency (EPA, 1974) to limit the day-night average sound level (DNL) to 55 decibels (dB) to protect the public health and welfare. Currently, DNL (DENL in Europe), the accepted metric for characterizing the impact of community noise, shows that a large proportion of the population is highly annoyed at a DNL of 65 dB or higher.

Recommendation 3-2: Relevant agencies of the federal government (e.g., agencies of the U.S. Department of Transportation with responsibilities related to noise, the Environmental Protection Agency, and the U.S. Department of Housing and Urban Development) should fund the development of environmental noise metrics that are more transparent and more reflective of the impact of noise on an affected population than DNL. This will require improved tools for predicting community sound pressure time histories and the development of metrics that accurately reflect the sounds people hear. A more holistic model of annoyance is also needed that incorporates situational variables that can be used to generate predictions for overall response, as well as responses of vulnerable populations (e.g., elderly people, sick people, children, and noise-sensitive individuals). International cooperation in this effort will facilitate the development of national and international standards for calculating metrics and should include open-source code to facilitate broad implementation of the metrics. Certain measures should be taken to facilitate this development:

1. The international noise control engineering community should develop an open, collaborative data-sharing environment in which researchers can deposit and access data from community noise surveys (e.g., data from surveys of acoustic, environmental, community, and transportation systems to support comparisons of metrics and predictions by models).
2. Policy agencies should conduct extensive surveys around at least six U.S. airports to generate high-quality data to populate the database. These surveys should serve as models of good survey practices, including data recording and archiving to ensure that they are useful for future studies.

Noise Metrics for Quiet Environments

The impact of man-made noise in national parks and other quiet environments is another parameter that is not well modeled by the metrics used to assess the impact of noise around airports or roads. Detection of the sound and distinguishing between man-made and natural sounds are important because human reactions to man-made and natural sounds differ. If one goal of the national parks is to preserve places of natural beauty, then the natural soundscape of a park, which is an aspect of its beauty, should also be preserved.

In addition, predicting the impact of noise on wildlife in national parks may require a different kind of metric that reflects animals' hearing systems. Preserving wildlife is essential to preserving the ecostructure of a park. But wildlife preservation will require that animals' hearing also be preserved and protected, because many animals depend on their hearing to hunt and to detect potential predators. The U.S. Department of the Interior should fund the development of metrics to support noise management decisions in national parks and other quiet environments.

REFERENCES

ANASE (Attitudes to Noise from Aviation Sources in England). 2007. Attitudes to Noise from Aviation Sources in England (ANASE). Final Report. Prepared for United Kingdom Department for Transport, by P. Le Masurier, J. Bates, J. Taylor, I. Flindell, D. Humpheson, C. Pownall, and A. Woolley. Available online at *http://www.dft.gov.uk/pgr/aviation/environmentalissues/Anase*.

ANSI (American National Standards Institute). 2002. American National Standard: Acoustical Performance Criteria, Design Requirements, and Guidelines for Schools. ANSI/ASA S12.60-2002 (R2009). New York: ANSI.

ANSI. 2005. American National Standard: Quantities and Procedures for Description and Measurement of Environmental Sound—Part 4: Noise Assessment and Prediction of Long-Term Community Response. ANSI S12.9-2005/Part 4. New York: ANSI.

ANSI. 2007. American National Standard for the Computation of Loudness and Steady Sounds. ANSI S3.4-2007. New York: ANSI.

ANSI. 2008. Methods for the Estimation of Awakenings Associated with Outdoor Noise Events Heard in Homes. American National Acoustical Society of America Standard S12.19-2008, Part 6. Melville, NY: Acoustical Society of America.

Beranek, L.L. 2008. Riding the Waves: A Life in Sound, Science, and Industry. Cambridge, MA: MIT Press.

Beranek, L.L., K.D. Kryter, and L.M. Miller. 1959. Reaction of people to exterior aircraft noise. Noise Control 5:23–60.

CALM Network. 2007. Research for a Quieter Europe in 2020. An Updated Strategy Paper of the CALM II Network. Available online at *http://www.calm-network.com/SP_2020_final07.pdf*.

Eagan, M.E. 2007. Supplemental metrics to communication aircraft noise effects. Transportation Research Record: Journal of the Transportation Research Board 2011:175–183.

EC (European Commission). 1996. Green Paper on Future Noise Policy. COM(96) 540. Available online at *http://europa.eu/documents/comm/green_papers/com96_540/summary_en.htm*. See also *http://ec.europa.eu/environment/noise/greenpap.htm*.

EC. 2000. Position Paper on EU Noise Indicators. A Report prepared for the European Commission, Directorate-General Environment. Available online at *http://ec.europa.eu/environment/noise/pdf/noiseindicators.pdf*.

EC. 2002a. Directive 2002/49/EC of the European Parliament and of the Council of 25 June, 2002 Relative to the Assessment and Management of Environmental Noise. Official Journal of the European Communities, L189/12. Available online at *http://eur-lex.europa.eu/LexUriServ/LexUriServ.do?uri=OJ:L:2002:189:0012:0025:EN:PDF*.

EC. 2002b. EU's Future Noise Policy, WG2—Dose/Effect. Position Paper on Dose-Response Relationships Between Transportation Noise and Annoyance. Available online at *http://ec.europa.eu/environment/noise/pdf/noise_expert_network.pdf*.

EC. 2004. Position Paper on Dose-Noise Effect Relationships for Night Time Noise. European Commission Working Group on Health and Socio-Economic Effects. Available online at *http://ec.europa.eu/environment/noise/pdf/positionpaper.pdf*.

EPA (U.S. Environmental Protection Agency). 1974. Information on Levels of Environmental Noise Requisite to Protect Public Health and Welfare with an Adequate Margin of Safety. Document 550/9-74-004. Available online at *http://www.nonoise.org/library/levels74/levels74.htm*.

FICAN (Federal Interagency Committee on Aviation Noise). 2002. The Use of Supplemental Noise Metrics in Aircraft Noise Analyses. Available online at *http://www.wylelabs.com/content/global/documents/Metrics.pdf*.

FICON (Federal Interagency Committee on Noise). 1992. Federal Agency Review of Selected Airport Noise Analysis Issues. Available online at *http://www.wylelabs.com/services/arc/documentlibrary/federalandlocalguidanceonoise/ficon.html*.

Fidell, S. 2002. Reliable Prediction of Community Response to Noise: Why You Can't Get There from Here. Proceedings of INTER-NOISE 02, 2002 International Congress and Exposition on Noise Control Engineering, Dearborn, MI, August 19–12. Available online at *http://www.bookmasters.com/marktplc/00726.htm*.

Fidell, S., and L. Silvati. 2004. Parsimonious alternative to regression analysis for characterizing prevalence rates of aircraft noise annoyance. Noise Control Engineering Journal 52(2):56–68.

Fidell, S., D.S. Barber, and T.J. Schultz. 1991. Updating a dose-effect relationship for the prevalence of annoyance due to general transportation noise. Journal of the Acoustical Society of America 89:221–233.

Fidell, S., D.S. Barber, and T.J. Schultz, 1989. Updating a Dosage-Effect Relationship for the Prevalence of Annoyance Due to General Transportation Noise. (Technical Report HSD-TR-89-009), Human System Division (HSD/YAH), U.S. Air Force Systems Command, Brooks AFB, TX.

Fields, J.M. 1993. Effect of personal and situational variables on noise annoyance in residential areas. Journal of the Acoustical Society of America 93(5):2753–2763.

Finegold, L.S., and M.S. Finegold. 2002. Development of Exposure-Response Relationships Between Transportation Noise and Community Annoyance. Proceedings of the Japan Net-Symposium on "Annoyance, Stress, and Health Effects of Environmental Noise," November–December. Available online at *http://www.netsympo.com/2002/finegold/develop.pdf*.

Finegold, L.S., C.S. Harris, and H.E. von Gierke. 1994. Community annoyance and sleep disturbance: Updated criteria for assessing the impacts of general transportation noise on people. Noise Control Engineering Journal 42:25–30.

Finegold, L.S., C. Oliva, and J. Lambert. 2008. Progress on Development of Noise Policies from 2003–2008. Proceedings of the 9th Congress on the Biological Effects of Noise: Noise as a Public Health Hazard. Foxwoods, Connecticut. Available online at *http://www.bookmasters.com/marktplc/00726.htm*.

Fletcher, H., and W.A. Munson. 1933. Loudness, its definition, measurement, and calculation. Journal of the Acoustical Society of America 5:82–108.

Flindell, I.H., and P.J.M. Stallen. 1999. Non-acoustical factors in environmental noise. Noise and Health 3:11–16.

Gjestland, T. 2008. Research on Community Response to Noise—in the Last Five Years. Proceedings of the 9th Congress on the Biological Effects of Noise; Noise as a Public Health Hazard, Connecticut. Available online at *http://www.bookmasters.com/marktplc/00726.htm*.

ISO (International Organization for Standardization). 1975. Acoustics—Methods for Calculating Loudness Level. Geneva, Switzerland: ISO. Available online at *http://www.iso.org/iso/iso_catalogue/catalog_tc/catalogue_detail.htm?csnumber=4603*.

ISO. 2003. Acoustics—Description, Measurement, and Assessment of Environmental Noise—Part 1: Basic Quantities and Assessment Procedures. Geneva, Switzerland: ISO. Available online at *http://www.iso.org/iso/iso_catalogue/catalog_tc/catalogue_detail.htm?csnumber=28633*.

Kryter, K.D. 1960. The measurement and meaning of perceived noise level. Noise Control 6(5):12–27.

Kryter, K.D. 2009. Acoustical model and theory for predicting effects of environmental noise on people. Journal of the Acoustical Society of America 125(6):3707–3721.

Kryter, K.D., and K.S. Pearsons. 1962. Judgment tests of the sound from piston, turbojet, and turbofan aircraft. Sound 1:24–31.

Kryter, K.D., and K.S. Pearsons. 1963. Some effects of spectral content and duration on perceived noise level. Journal of the Acoustical Society of America 35:866–883.

Miedema, H.M.E., and C.G. Oudshoorn. 2001. Annoyance from transportation noise: Relationships with exposure metrics DNL or DENL and their confidence intervals. Environmental Health Perspectives 109:409–416.

Miedema, H.M.E., and H. Vos. 1998. Exposure-response relationships for transportation noise. Journal of the Acoustical Society of America 104(6):3432–3445.

Miedema, H.M.E., and H. Vos. 1999. Demographic and attitudinal factors that modify annoyance from transportation noise. Journal of the Acoustical Society of America 105(6):336–344.

Schomer, P.D. 2002. On normalizing DNL to provide better correlation with response. Sound and Vibration 36(12):14–25.

Schomer, P.D. 2005. Criteria for assessment of noise annoyance. Noise Control Engineering Journal 53(4):132–144.

Schultz, T.J. 1978. Synthesis of social surveys on noise annoyance. Journal of the Acoustical Society of America 64:377–405.

Stevens, K.N., W.A. Rosenblith, and R.H. Bolt. 1955. A community's reaction to noise: Can it be forecast? Noise Control 1:63–71.

Transport Canada. 2005. Aviation-Land Use in the Vicinity of Airports. Aircraft Noise, 8th ed. Available online at *http://www.tc.gc.ca/CivilAviation/publications/TP1247/Part4/Part4-2.htm*.

Von Gierke, H.E. 1975. Noise: how much is too much? Noise Control Engineering Journal 5(1):24–34.

WHO (World Health Organization). 2007. Night Noise Guidelines (NNGL) for Europe. Published by the World Health Organization in cooperation with the European Union. Available online at *http://ec.europa.eu/health/ph_projects/2003/action3/docs/2003_08_frep_en.pdf*.

Yano, T., T. Sato, and T. Morihara. 2007. Dose-Response Relationships for Road Traffic, Railway and Aircraft Noises in Kyushu and Hokkaido, Japan. Proceedings of INTER-NOISE 07, The 2007 International Congress and Exposition on Noise Control Engineering, Istanbul, Turkey, August 28–31. Available online at *http://scitation.aip.org/journals/doc/INCEDL-home/cp/*.

4

Control of Hazardous Noise

Hearing loss can result from exposures to high levels of occupational and nonoccupational noise. Occupational exposures are a by-product of working in close proximity to machinery and systems that are commonly used in industrialized societies. In most cases the equipment and the job of operating the equipment have been inadequately designed, so that the only way operators can perform their job is to be exposed to these noises. Nonoccupational exposures include listening to loud music, street conditions in urban areas, and other short-term and/or voluntary exposures. The main focus of this chapter is on exposures in occupational settings.

"Hearing loss due to occupational noise exposure is our most prevalent industrial malady," according to Robert Sataloff, a prominent physician, and it affects nearly every American household (Sataloff, 1993). The hearing loss related to exposure to excessive noise, which has been known since before the Industrial Revolution, was first documented by Bernardo Ramazzini in 1713 among millers and coppersmiths (Ramazzini, 1964). The dangers have been studied for well over a century, and many laws and regulations have been passed recognizing the hazards of noise. Today, concerns include how noise exposure can also impact nonauditory health, but these effects are beyond the scope of this report.[1]

Very few studies have been done recently on the number of people exposed to hazardous occupational noise. In 1981 the Occupational Safety and Health Administration (OSHA) estimated that 7.9 million workers were exposed to noise levels of, or exceeding, 80 dB(A). Also in 1981, the U.S. Environmental Protection Agency estimated that more than 9 million people were exposed to daily noise levels above 85 dB(A) (EPA, 1981). Table 4-1 shows the economic sectors included in the EPA study. These numbers have probably remained stable or even increased since these studies were conducted.

TABLE 4-1 Number of Workers Exposed to Noise of >85 dB(A)

Industrial Sector	Number of Workers
Agriculture	323,000
Mining	255,000
Construction	513,000
Manufacturing and utilities	5,124,000
Transportation	1,934,000
Military	976,000
Total	9,125,000

SOURCE: EPA, 1981.

The prevalence and long history of noise exposure on the job have often led to the realization by the working population that exposure to hazardous noise is an inevitable condition of employment. Since the onset of noise-induced hearing loss (NIHL) is typically slow and not painful, workers may be more accepting of hazardous noise than of other dangers in industrial settings.

Employers today are responsible for preventing NIHL by controlling hazardous noise exposures through the use of engineering controls, by monitoring the effects of exposure on employees through the administration of audiometric exams, and by providing hearing loss prevention programs that include hearing protection devices (HPDs). However, the effectiveness of HPDs and how well they comply with workplace rules are inconsistent, at best. As illustrated in the discussion below, the difference between reductions in noise levels in the laboratory and in the "real world" can be significant.

It is generally acknowledged that most large employers administer hearing loss prevention programs, although this is not the case with most small and many medium-sized companies. The effectiveness of these programs, when provided, is often inadequate. Engineering controls to suppress

[1]New information concerning the general relationship between noise and health is becoming available (Babisch, 2008; DEFRA, 2009). However, it will require a multidisciplinary study committee to evaluate these results and determine their relevance to the health of the American people.

noise sources are preferred but are less commonly used. Engineering solutions and expenditures that can actually reduce noise emissions are sometimes "sold to management" by highlighting their other advantages, such as gains in productivity or quality.

The cost of NIHL can be assessed in two ways. First, there are the impacts on quality of life, such as strained relationships, difficulty or inability to communicate, feelings of isolation, lost friendships, ridicule from peers, and a general inability to relate well to others. Accidents and absenteeism also should be included in the cost of NIHL. Second is the amount of money spent on compensation for NIHL. However, studies have shown that these costs are underrepresentative of the total cost of NIHL (Shampan and Ginnold, 1982; Suter, 1990).

Data for calculating the costs are difficult to come by. In a recent study by the Institute of Medicine (IOM, 2005), it was stated that disabilities of the auditory system, including tinnitus and hearing loss, were the third most common type of disability, accounting for nearly 10 percent of the total number of disabilities among veterans. For the roughly 158,000 veterans who began receiving disability compensation in 2003, auditory disabilities were the second most common type of disability. These veterans had approximately 75,300 disabilities of the auditory system, out of a total of some 485,000 disabilities. At the end of 2004, the monthly compensation payments to veterans with hearing loss as their major form of disability represented an annualized cost of some $660 million. The corresponding compensation payments to veterans with tinnitus as their major disability were close to $190 million on an annualized basis. A 1997 study by the World Health Organization estimated that the cost of NIHL in developed countries was in the range of 0.2 to 2.0 percent of a country's gross domestic product (GDP)—or roughly $28 billion to $280 billion for the United States (WHO, 2007). In a 2006 study in Australia, it was estimated that the real cost of hearing loss amounted to 11.75 billion AUD, or 1.4 percent of GDP (AE, 2006).

The remainder of this chapter focuses on criteria for acceptable risk of damage from hazardous noise in industry and government, hazardous noise from consumer products, research on impulsive noise, engineering controls in industry and the establishment of "buy-quiet" programs, HPDs, and the current status of HPD research.

CRITERIA FOR DETERMINING ACCEPTABLE RISK OF DAMAGE

Exchange Rate

Criteria for estimating the risk of damage from hazardous noise must be based on both the level of noise (almost always A-weighted sound pressure level) and the duration of noise exposure. In setting the level at which there is believed to be no hazard, the level at which action must be taken, and the level believed to be hazardous to hearing, it is common practice to define an *exchange rate* that takes into account exposure time.

Studies have shown that there is no exact value for the exchange rate (Stephenson, 2008).[2] An exchange rate of 3 dB, which corresponds to equal energy,[3] was first proposed by Eldred et al. in 1955. The 3-dB exchange rate, recommended by the National Institute for Occupational Safety and Health (1998a) is the most widely adopted and the most widely accepted rate by scientists (Stephenson, 2008; Suter, 1993), as well as by many government agencies in the United States, including the U.S. Department of Defense (DOD), the military services, and the National Aeronautics and Space Administration. It is also accepted by the American Conference of Government Industrial Hygienists. According to Beth A. Cooper, a member of the study committee, a 3-dB exchange rate is considered "best practice" among hearing conservation professionals (Cooper, 2009). This rate has also been standardized nationally (ANSI, 2006a) and internationally (ISO, 1990).

Not all U.S. government agencies, however, accept the 3-dB exchange rate. OSHA and the Mine Safety and Health Administration (MSHA), both part of the U.S. Department of Labor, use a 5-dB exchange rate (29 CFR 1910.95 and 30 CFR 62.0), as does the Federal Railroad Administration (49 CFR 229.121).

Using the 3-dB rate, an 85-dB level exposure for 8 hours would be considered as hazardous as an 88-dB level exposure for 4 hours. Using the 5-dB rate, the 85-dB exposure for 8 hours would be considered as hazardous as 90-dB exposure for 4 hours. As an example, it is estimated in American National Standard S3.44 (ANSI, 2006a) that an 8-hour daily exposure to 90 dB(A) would, after 20 years, result in a noise-induced threshold shift of 10 dB at 3,000 Hz for 50 percent of the population. Data for different levels, frequencies, and exposure times are given in Appendix F of the standard.

Considering the accuracy of sound-level meters and the difficulty of determining exposure over a period of eight hours, or even four hours, the difference is relatively small. However, it becomes significant for short exposure times, as shown in Table 4-2.

HAZARDOUS NOISE LEVELS IN GOVERNMENT AND INDUSTRY

For continuous hazardous noise, A-weighted sound pressure levels are used as the metric worldwide. Table 4-3 shows levels of 70 and 75 dB and higher that are known to pose some level of risk. The level in the first row (24-hour

[2]Stephenson, M.R. 2008. The Scientific Basis for the 85 dB Criterion and 3 dB Exchange Rate. Presentation at the NAE workshop on Engineering Responses to Hazardous Noise Exposures, Washington, DC, August 14–15.

[3]*Equal energy* means that when the level goes up by 3 dB (a doubling of energy), the exposure time must be reduced by a factor of 2.

TABLE 4-2 Hazardous Noise Exposures as a Function of Exposure Time for 3-dB and 5-dB Exchange Rates (based on exposure to 85 dB for 8 hours)

Exposure Time	Hazardous Levels (3-dB Exchange Rate)	Hazardous Levels (5-dB Exchange Rate)
8 hr	85 dB	85 dB
4 hr	88 dB	90 dB
2 hr	91 dB	95 dB
1 hr	94 dB	100 dB
30 min	97 dB	105 dB
15 min	100 dB	110 dB
7.5 min	103 dB	115 dB

TABLE 4-3 Action Points, References, and Type of Sound Level

A-Weighted Sound Pressure Level	Exposure Time	Explanation
70 dB	24 hr	Adequate to protect the most sensitive person at the most sensitive frequency (EPA, 1974). Equivalent sound level.
75 dB	8 hr	Adequate to protect the most sensitive person at the most sensitive frequency (EPA, 1974), assuming that the remaining 16 hrs are quiet. Equivalent sound level.
80 dB	8 hr	Required lower limit for beginning the integration to determine TWA (29CFR1910.95(d)(2)(i)). Instantaneous sound pressure level.
85 dB	8 hr	A widely used upper limit for exposure to hazardous noise (see Table 4-4; NASA, 2007; NIOSH, 1998b). Required "action level" in OSHA hearing conservation amendment (OSHA, 1981). Equivalent sound level.
87 dB	8 hr	European Union exposure limit value from 2003/10/EC (EC, 2003). Equivalent sound level.
90 dB	8 hr	OSHA (29 CFR 1910.95) and MSHA limits for exposure to hazardous noise (30 CFR 62.0). Time-weighted average level.
100 dB	8 hr	OSHA-specified level at which engineering controls should be used (OSHA, 2009). This is often known as the "100-dB Directive." It is the time-weighted average exposure level below which OSHA inspectors are encouraged not to issue citations for the absence of engineering or administrative controls, unless there is evidence that workers are losing their hearing.

NOTE: Except for the 80-dB level, the sound pressure levels in this table are related to how much risk is acceptable.

exposure) is 5 dB below the level in the second row (8-hour exposure; 10 $\log_{10}\{8/24\} = -5$ dB). This corresponds to a 3-dB exchange rate. Because sound pressure levels vary with time, the levels in the table are generally time-weighted averages (TWAs). OSHA and MSHA, however, have their own methods of calculating noise dose (29 CFR 1910.95, 30 CFR 62.0).

The International Institute of Noise Control Engineering (I-INCE) has studied regulations on exposure to noise worldwide (I-INCE, 1997). An updated version of the I-INCE data is shown in Table 4-4 (Suter, 2006). As the table shows, an 8-hour average A-weighted sound pressure level of 85 dB and a 3-dB exchange rate is used in many countries. The original references for the table are given in Suter (2006).

HAZARDOUS NOISE FROM CONSUMER PRODUCTS AND LEISURE ACTIVITIES

Noise from consumer products can sometimes be as hazardous as occupational noise in the workplace. In both cases the degree of hazard depends on noise level and exposure time. One major difference is that noise in the workplace tends to come from a number of sources, whereas noise from the products discussed in this section tends to come from a single source. When exposure to occupational noise is added to exposure to noise during a leisure activity, the degree of hazard increases.

Consumer Products

Noise from consumer products and other hazardous sounds can be generated by a variety of sources, including kitchen appliances, audio systems, power tools (both handheld and stationary), and all types of yard equipment. High levels of noise from these products and long exposure times can contribute to the risk of NIHL from occupational noise. It is reasonable to assume that the 3-dB exchange rate applies to exposures longer than 8 hours, from either or both. Noise levels from a variety of consumer products, including toys, have been published by Schwela (2006).

Recreational Noise

Sources of hazardous recreational noise include on-road and off-road vehicles, loud music at concerts, and small aircraft engines. Like exposure to noise from consumer products, the degree of risk depends on the sound pressure level and exposure time, as well as the amount of noise exposure in the workplace.

Noise from Personal Listening Devices

Personal listening devices are known to emit sound pressure levels that can be hazardous, if they are abused. It has been estimated that at least 5 percent of users of these devices

TABLE 4-4 Worldwide Regulations for Exposures to Hazardous Noise in the Workplace

Nation, Date (if available)	PEL (8-hr average) dBA	Exchange Rate dBA	dBA Level for Engineering Controls	dBA Level for Audiology Tests and Other HC Practices	Comments
Argentina, 2003	85	3	85	85	
Australia, 2000	85	3	85	85	Note[a]
Brazil, 1992	85	5	85		
Canada, 1991	87	3	87	84	[b]
Chile, 2000	85	3			
China, 1985	85	3	85		
Colombia, 1990	85	5			
European Union, 2003	87	3	85	85	[c]
				80	[d]
Finland, 1982	85	3	85		
France, 1990	85	3		85	
Germany, 1990	85	3	90	85	[e]
Hungary	85	3	90		
India, 1989	90				[f]
Israel, 1984	85	5			
Italy, 1990	85	3	90	85	
Mexico, 2001	85	3	90	80	
Netherlands, 1987	80	3	85		[g]
New Zealand, 1995	85	3	85	85	
Norway, 1982	85	3		80	
Spain, 1989	85	3	90	80	
Sweden, 1992	85	3	85	85	
United Kingdom, 1989	85	3	90	85	
United States, 1983	90	5	90	85	[d]
Uruguay, 1988	85	3	85	85	
Venezuela	85	3			

NOTE: PEL is the permitted exposure level in each country. dBA rather than dB(A) is used because it was used in the referenced table.

[a]Each Australian state and territory has its own legislation for noise, but all have now adopted the 8-hour PEL of 85 dBA and the 3-dBA exchange rate (ER).

[b]Despite the existence of a Canadian national standard, there is some variation among the standards in individual provinces: Ontario, Quebec, and New Brunswick use 90 dBA with a 5-dBA ER; Alberta, Nova Scotia, and Newfoundland use 85 dBA with a 5-dBA ER; and British Columbia uses 85 dBA with a 3-dB ER. Most require engineering controls to the level of the PEL. Manitoba requires certain hearing conservation practices above 80 dBA, hearing protectors and training on request above 85 dBA, and engineering controls above 90 dBA.

[c]The European Union (EU; Directive 2003/10/EC) puts forward three exposure values: an exposure limit value of 87 dBA; an "upper action" level of 85 dBA; and a "lower action" level of 80 dBA, all using the 3-dBA ER. The attenuation of hearing protectors may be taken into account when assessing the exposure limit value but not for requirements driven by the upper and lower action values. At no time shall an employee's noise exposure exceed the exposure limit value. When exposures exceed the upper action level, the employer must implement a program of noise reduction, taking into account technology and the availability of control measures.

[d]EU *continued*: Hearing protectors must be made available when exposures exceed the lower action value of 80 dBA. Hearing protectors must be used by workers whose exposures equal or exceed the upper action value of 85 dBA. Audiometric testing must be available to workers whose exposures exceed the upper action value, and when noise measurements indicate a risk to health, these measures must be available at the lower action value.

[e]The German standard (UVV Larm-1990) states that it is not possible to give a precise limit for the elimination of hearing hazard and the risk of other health impairments from noise. Therefore, the employer is obliged to reduce the noise level as far as possible, taking technical progress and the availability of control measures into account.

[f]India: This is a recommendation, not a regulation.

[g]The Netherlands' noise legislation requires engineering noise control at 85 dBA "unless this cannot be reasonably demanded." Hearing protection must be provided above 80 dBA, and workers are required to wear protection devices at levels above 90 dBA.

[h]These levels apply to the OSHA noise standard, which cover workers in maritime and general industries. The U.S. military has more stringent standards; DOD as a whole uses the 85-dBA PEL and the 3-dBA exchange rate. The Air Force and Army have similar requirements, and the Navy is about to adopt the 3-dB ER.

are exposed to A-weighted TWA levels of greater than 85 dB (Fligor, 2008).[4] Considering the very large number of such devices sold, and the frequent long exposure times, it has been estimated that 50,000 people develop NIHL from listening to such devices for more than 4 hours per day over a period of years (Fligor, 2009; SCENIHR, 2008). Recommendations for avoiding NIHL are listed below (Fligor and Cox, 2004):

[4]Fligor, B. 2008. Non-occupational hazardous noise. Recreational equipment, personal music devices, toys, buses, etc. Focus on children. Attribution of hearing loss/damage to variety of sources. Presentation at the NAE workshop on Engineering Responses to Hazardous Noise Exposures, Washington, DC, August 14–15.

- Limit listening level to 60 percent of the maximum volume.
- Limit listening time to 1 hour.
- Use a lower gain setting and shorter listening times when over-the-ear rather than in-ear earphones are used.

Exact limits are difficult to specify because the sensitivity of earphones varies with the manufacturer and because different earphones can be used with the same amplifier. Assessments of NIHL caused by personal listening devices have also been made in Switzerland (Hohmann et al., 1999) and the Netherlands (Passchier-Vermeer, 1999).

Noise from Toys

A survey of toy safety standards for noise levels (ASTM, 2008) and recommendations have been published by the U.S. Public Information Research Group (PIRG, 2005). Another survey that includes epidemiological studies as well as various national and international activities is available (Altkorn et al., 2005). The League for the Hard of Hearing provides examples of noise levels from toys (LHH, 2009). A report by the Institute of Sound and Vibration Research provides measurement data and recommendations about noise from toys (ISVR, 1997); for example, the recommendation for the C-frequency weighted instantaneous sound pressure level from cap-firing toys is that it not exceed 120 dB when measured 25 centimers from the ear and 125 dB when measured 2.5 centimers from the ear. Contrast this with an undated alert from the U.S. Product Safety Commission (CPSC, 2001a) that states, "CPSC reminds parents that caps may also pose a noise hazard. A current CPSC regulation limits the decibel level of caps to no more than 158 decibels. A warning label is mandatory on caps in the 138 to 158 decibel level as follows: WARNING—Do not fire closer than 1 foot to the ear. Do not use indoors." See CPSC (2001b) for the full requirement. The measurement method is specified in 16 CFR 1500.47.

IMPULSIVE NOISE

Physical Characteristics

In contrast to continuous noise, impulsive noise comes in many different forms and is much more difficult to describe. Impulsive noise may consist of a single burst, such as impact noise generated by a hammer hitting a nail, a sonic boom, or a single rifle shot. It may also consist of a series of bursts, either closely spaced or more or less isolated—such as a series of hammer blows. Thus, characterizing a particular impulsive noise has been a subject of interest to engineers for many years (IEEE, 1969). Characterizing impulsive noise and associated auditory hazards was the subject of a NIOSH workshop in 2005 and two presentations at a workshop sponsored by the National Academy of Engineering (NAE) in August 2008 (Dancer, 2008; Murphy, 2008).[5,6]

Physical measures of an impulsive noise depend on its character, and relating the degree of auditory hazard to a physical measure is a complex undertaking. For a single burst of noise, the instantaneous sound pressure is usually of most interest. However, measuring it requires a system with a wide frequency response, a wide dynamic range, and a small phase shift.

A conventional sound-level meter with a peak-reading circuit and C-weighting can provide a reasonably good measure of the peak value of the instantaneous sound pressure, and this peak value, expressed in decibels, has been widely used as a damage risk criterion. However, peak values cannot be measured using the fast or slow dynamic characteristics of a sound-level meter because of their long time constants. Also, A-frequency weighting does not satisfy the criterion of a wide bandwidth.

At one time, an impulse sound-level meter was standardized. The instrument had a time constant believed to approximate the loudness of a transient sound and a decay time constant long enough to obtain a peak reading. However, standards committees discouraged use of this meter, and it is no longer standardized.

Another quantity that can be measured with a sound-level meter is sound exposure, which is the integral of the squared instantaneous pressure over the duration of the burst. Expressed in decibels, this becomes the sound exposure level, which has been used to characterize, for example, aircraft flyovers. The relationship to auditory hazard is discussed in the next section.

One parameter that cannot be measured with a conventional sound-level meter is the *A-duration* (unrelated to A-frequency weighting). A-duration is the time from the beginning of the burst to the time that the instantaneous sound pressure is 20 dB below the peak value. A-duration and other characteristics of impulsive noise are described in an American National Standard (ANSI, 2006b).

For a series of bursts or for continuous noise with an impulsive component, two other parameters are of interest. The peak and root mean square sound pressure can be measured, and the ratio, expressed in decibels, is the *crest factor*. Impulsive noise is characterized by a crest factor higher than that of random noise, which is why a system with a wide dynamic range is necessary for making accurate measurements. High crest factors are related, in a statistical sense, to the ratio of the fourth moment and second moment about the mean

[5]Dancer, A.L. 2008. DRCs for High-Level Impulsive Noise and Validation Data. Presentation at the NAE workshop on Engineering Responses to Hazardous Noise Exposures, Washington, DC, August 14–15.

[6]Murphy, W.J. 2008. Impulsive Noise in Industry and in the Community: Considerations for Measuring Impulsive Noise. Presentation at the NAE workshop on Engineering Responses to Hazardous Noise Exposures, Washington, DC, August 14–15.

of the signal (*kurtosis*). Although not widely used, kurtosis is a measure of the impulsiveness of noise.

Auditory Hazard

The peak value of the instantaneous pressure, expressed in decibels and with C-frequency weighting, has been widely used as a measure of auditory hazard for impulsive noise. For example, OSHA has set a limit on the peak sound pressure level: "Exposure to impulse or impact noise should not exceed 140 dB peak sound pressure level" (OSHA, 1971c).

The European Union sets an upper limit of 200 Pa (1 Pa = 1 N/m^2) for instantaneous sound pressure using C-frequency weighting (EC, 2003). This limit is not expressed in decibels, presumably to avoid confusion with the lower limits for continuous noise. The 200-Pa peak sound pressure can be converted to a sound pressure level: 10 log $(p^2/p_{ref}^{\,2}) = 140$ dB, where the reference pressure, p_{ref}, is 20 micropascals.

World Health Organization guidelines state that peak sound pressure levels should not exceed 140 dB(A) for adults and 120 dB(A) for children (WHO, 1999). Note the use of A-frequency weighting in this case.

The use of peak sound pressure level as a measure of auditory hazard was questioned by Dancer at an NAE-sponsored workshop (Dancer, 2008). He showed data from French military studies comparing auditory hazard from howitzer and rifle rounds. Soldiers were exposed to 20 rounds at the same peak sound pressure level (159 dB) but with an A-duration of 9 and 0.25 milliseconds, respectively. Almost no temporary threshold shift (TTS) was observed for the howitzer rounds, but significant TTS was found for the rifle rounds. Since it is generally accepted that repeated exposure to noise that causes TTS leads to permanent threshold shift, these results lead one to question the peak level and emphasize the importance of the A-duration in determining auditory hazard. They also show that very high sampling rates are necessary when recording digital samples of a short burst of noise.

Because impulsive noise can be a series of bursts or continuous noise with an impulsive component, the question is whether such noise, generally recognized by its high crest factor, has the same auditory hazard as continuous noise when the two have the same A-weighted sound pressure level. This question has been raised for more than 30 years.

Brüel (1977) asked if damaging noise was being measured correctly. He noted that studies at the time showed that industrial noise, presumably with an impulsive component, appeared to be more damaging than music at a higher noise level. Similarly, he noted that pilots with no ear protection in certain airplanes do not suffer as much hearing loss as pilots who listen to radio communications with their attendant clicks and bursts. This suggests that the peaks in the time waveform are significant contributors to auditory hazard.

NIOSH (1998b) cites a number of studies that indicate that impulsive noise is more dangerous than continuous noise of the same level. However, NIOSH also cites studies that show that the equal energy rule adequately predicts hearing damage. Therefore, at these levels impulsive noise, even when superimposed on a background of continuous noise, can probably be treated similarly to continuous noise for the purposes of assessing auditory hazard.

Dancer (2008) presented results from a number of studies of hearing damage (i.e., Price 2007) and concluded that an 8-hour, A-weighted equivalent level of 85 dB (L_{Aeq8}) should be used as the damage risk criterion for both continuous and impulsive noise, whether military or occupational. If this conclusion is accepted, it would extend the equal energy concept for hazardous noise to even very short bursts of noise, which would greatly simplify the determination of auditory hazard. The sound exposure of an impulse would be determined, averaged over an 8-hour time interval, and compared with current damage risk criteria for continuous noise.

Another approach to assessing auditory hazard from impulsive noise is the Auditory Hazard Assessment Algorithm for Humans (AHAAH), developed by G.R. Price for the U.S. Army (U.S. Army Research Laboratory, 2010). This mathematical model of the human auditory system requires as input the incident sound pressure as a function of time. An assessment of this model is well beyond the scope of this report. We simply note that Dancer (2008) suggested it be used as a laboratory tool to clarify some specifics of weapons noise. Dancer also concluded that the AHAAH model provides better estimates of auditory hazard than L_{Aeq8} for the sound of air bags and high-level noise.

In the preceding discussion of physical characteristics of impulsive noise, kurtosis was identified as another measure of the impulsiveness of noise. Recent data from a series of animal experiments and at least one epidemiological study indicate that the kurtosis metric, with possible adjustments for frequency spectrum and bandwidth, in combination with equal energy would be an effective predictor of the traumatic effects of complex noise (Davis et al., 2009; Zhao et al., 2010).

The study committee concluded that damage risk criteria for impulsive noise need further study and that such studies and an agreement in the international standards community on optimal damage risk criteria for impulsive noise should serve as a basis for changing national, European Union, and international criteria for assessing auditory hazard from impulsive noise. Such studies will require the participation of both engineers and experts in the physiology of the ear. Military experience will be a very valuable input to the process.

ENGINEERING CONTROLS

HPDs and hearing protection programs are *not* the best way to protect the hearing of workers. The preferred way, often called "engineering controls," is to reduce workers' exposure by reducing the noise of the machinery or equipment that generates the noise. If it is not possible to reduce the noise from the source, then noise control along the path by which the noise propagates to workers, such as inserting a noise control element between the machinery and workers, can be used.

However, the most effective way, indeed, perhaps the only way, to eliminate NIHL from occupational noise exposure is well-designed engineering controls, which are permanent, are effective with or without worker/supervisor compliance, reduce absenteeism, make communication easier, reduce worker compensation costs, and reduce legal costs. For all of these reasons, engineering controls are the protection method of choice according to OSHA and MSHA.

Noise Mechanisms

The main differences between noise generated by industrial machinery and noise generated by other sources are size, complexity, and diversity. Compressors, for example, can be as small as a compact refrigerator or enormous, with a footprint of 20 × 30 feet. The main sources of noise are fluid flow, friction, magnetic forces, mechanical forced vibration, impact, and combustion.

Noise from fluid flow is generated by the movement of air (e.g., intake or exhaust air for engines, compressed air used to clean off a workbench), gas (e.g., process gas flowing through valves and piping), and liquids (e.g., fluid flow through pipes and valves). Mechanical vibration includes noise radiated from machinery casing compressing air or gas or pumping liquid. It includes noise radiated from surfaces mechanically attached to a noise source.

Friction sounds can be characterized as stick-slip sounds, like the screech of Styrofoam cups on a table top, chalk on a blackboard, or the squeal of tires when brakes are applied sharply; rubbing sounds, like sanding on a surface or pistons moving inside cylinders in an engine; and rolling friction, like ball bearings or tire noise. Sounds generated by magnetic forces can be emitted from electrically powered equipment, such as transformers, motors, and circuit breakers. Typically, these noises have strong tonal components.

Impact tools, like pneumatic chippers, generate impact noise when the equipment impacts the surface, but the response of the surface to the impact often dominates what is heard. Chipping on a rubber surface is obviously much quieter than chipping on a metal plate. Combustion noise, such as noise from a furnace or the sound from the ignition of fuel inside a gasoline or diesel engine, has strong low-frequency components.

Industrial machines have drivers (i.e., power sources), such as electrical, compressed air, or hydraulic motors; gasoline or diesel engines; or gas or steam turbines. The power sources drive blowers, fans, compressors, pumps, and countless other mechanisms.

Depending on the application, a gearbox might be placed between the power source and the driven equipment. An example of a complex noise source might be a compressor driven by an electric motor. These two items would probably be mounted on a metal frame called a skid. The noise sources would be those of the motor (including a cooling fan, motor casing that radiates magnetic and mechanical noise, and a skid that radiates structure-borne noise), and the compressor (including intake and discharge air, intake and/or discharge piping, engine casing of the compressor, and a supporting structure).

This brief description of noise sources illustrates that industrial equipment often has numerous sources of noise. Thus, noise control requires controlling the most powerful noise component first and then treating all of the other components in turn. For example, on a large fan with an open intake, the intake noise is dominant. Once the intake has been appropriately silenced, it is necessary to review other sources, such as the cooling fan for the motor, the casing-radiated noise of the fan and motor, and possible structure-borne noise from the skid.

Engineering Controls

Some engineering controls (Bruce, 2007) modify the noise source to reduce the amount of radiated noise. This can be accomplished in several ways:

- modifying the source so that it produces less noise
- changing the operating parameters so that less noise is generated
- adding mufflers or silencers to intakes and exhausts
- providing damping to reduce vibration
- isolating vibration to reduce excitation of other structures
- providing acoustical shielding from the source
- enclosing the source with lagging or a partial or total enclosure

A number of "obvious" engineering controls can usually be implemented in existing facilities to address 25 to 33 percent of noise problems in most workplaces (Driscoll, 2008).[7] Some are so obvious that they can easily be overlooked. Nevertheless, although these controls can be easily stated, their application requires careful selection. Some "obvious" controls are listed below:

- Maintain equipment properly (e.g., fix steam or air leaks). In operations that require high-pressure steam, steam leaks are often the dominant noise source.
- Change operating procedures (e.g., relocate the operator and controls to a quieter position).
- Replace equipment (e.g., buy a quieter version of the product).

[7]Driscoll, D.P. 2008. Noise Control Engineering: The Reader's Digest Version. Presented at the NAE workshop on Engineering Responses to Hazardous Noise Exposures, Washington, DC, August 14–15.

- Modify the room (e.g., install sound absorptive materials). If the noise source and worker are some distance apart, sound absorption in the intervening space can reduce noise in the reverberant field.
- Relocate equipment (e.g., put noisy equipment in areas that are often unoccupied).
- Use equipment at proper operating speed (e.g., the higher the speed, the louder the noise; run equipment at the lowest practical speed).

Noise from Fluid Flow

Noise sources with fluid flow include fans, compressors, engines, pumps, and valves. Controls for reducing intake and discharge noise include lining ducts, installing dissipative and reactive silencers, and installing special-purpose silencers.

The inlet or exhaust duct can be lined with a sound-absorbent material, such as fiberglass or mineral wool. Typical thicknesses range from 1 to 4 inches, depending on the strength of the low-frequency component. Dissipative silencers also use sound-absorbing materials to attenuate noise. A simple dissipative silencer might be a set of parallel baffles running lengthwise to direct airflow and reduce noise. The absorptive material might be mineral wool or fiberglass covered with glass fiber cloth to reduce erosion from airflow. In addition, a perforated or expanded metal facing could be added to the material to protect against contact damage. The longer the baffles and the closer they are together, the more effective they are as silencers. Reactive silencers operate on the principle of mismatching acoustic impedance. A change in acoustic impedance causes a portion of the sound energy to be reflected back to the source or back and forth within the silencer.

Special-purpose silencers are available to fit exhaust ports on pneumatic equipment, air wipes, and parts blowoffs. Recently, an innovative silencer called a duct resonator array (DRA) was developed based on the principles of a Helmholtz resonator (Liu, 2003). DRAs positioned at the diffusers in larger centrifugal compressors effectively reduce noise levels from these machines; DRAs can also be placed in discharge piping in a pipe spool. Basically, they reduce the A-weighted sound pressure level by at least 10 dB—which is similar to "halving" the loudness of the sound.

Lagging is a noise control treatment that consists of layers of treatment around piping to reduce radiated noise in refineries and noise from forced-draft and induced-draft fan ducts. The first layer wrapped around the pipe consists of glass fiber or mineral wool, typically 2 to 4 inches thick and 6 to 8 pounds per cubic foot. Next a mass-loaded vinyl layer weighing 1 to 2 pounds per square foot is wrapped around the glass fiber or mineral wool. The outer layer is a weatherproof covering. Depending on the details of the installation, lagging can reduce the A-weighted sound pressure level by 10 to 20 dB.

Radiated Noise from Machine Housings

Airborne noise can be radiated by any surface. For example, consider a piano. The keys strike hammers that strike the strings. The strings do not produce much sound by themselves, but they are attached to the much larger sound board that radiates the sound. In general, the larger the vibrating panel, the greater the sound radiated from the surface.

Another example is a parts bin into which metal parts are dropped. If the bin is made of perforated metal, the radiating area is smaller than if it is made of solid metal; thus, the level of radiated sound will be lower. Of course, materials with high internal damping radiate even less noise. If the bin were made of rubber (which has high internal damping), rather than metal (which has low internal damping), it would radiate even less sound.

Sometimes machinery is housed inside an enclosure provided by the original equipment manufacturer. In such situations it is desirable for the panels of the housing to be appropriately treated. Damping compound should be applied to the panels if there is any possibility that the resonance frequencies of the panels will be excited. If the machine inside the enclosure produces significant vibration into the enclosure housing and structure, the panels should be vibration isolated from the structure. In addition, it may be useful for the machinery enclosure to be mounted on vibration isolators to keep it from transmitting vibration to the floor.

Machinery Shields, Outdoor Barriers, and Enclosures

Shields. An acoustical shield may be inserted between the worker and a noisy section of a machine. Shields are often mounted directly on the machine and reduce noise by 8 to 10 dB under the following conditions:

- The worker is near the noisy operation.
- The smallest dimension of the shield is at least three times the wavelength of the dominant noise.
- The ceiling above the machine is covered with sound-absorptive material.

Shields can be manufactured from safety glass, one-quarter-inch clear plastic, metal, or wood. Durability, expense, need for visual observation of the operation, and need for access to the operation should all be considered in selecting a material. If possible, oil-resistant, cleanable, sound-absorptive materials should be incorporated into the machine side of the shield.

Outdoor Barriers. Any solid impervious wall that blocks the line of sight between a noise source and an observer will reduce the noise level at the observer. The reduction depends on the frequency of the noise, the distance of the source from the barrier wall, the distance of the receptor from the barrier wall, and the height of the wall. Low-frequency sound diffracts around the ends of the wall and over the top more readily than high-frequency sound. Thus, the wall has lower values of attenuation for low-frequency sound than for high-frequency sound. Typically, low-frequency sound is attenuated by less than 5 dB, whereas high-frequency sound can be attenuated by as much as 20 dB.

Partial Enclosures. A partial enclosure is a series of walls around a machine with the top left open. A partial enclosure can be effective inside a plant if located near a wall. However, some noise will still radiate out the top and contribute to the reverberant sound in the room. Reflections from the ceiling will increase the sound pressure level at distances farther from the enclosure. A sound-absorptive ceiling can reduce reflected sounds and thereby increase the effectiveness of the enclosure. For maximum effectiveness, the sound-absorptive ceiling should extend out to the location of the receivers. Sound-absorptive materials should also be applied to the inside of the enclosure walls.

For equipment handling flammable liquids/gases, appropriate fire-retardant systems and alarms will be required.

Total Enclosures. A total enclosure with a closed top provides better noise reduction than a partial enclosure. However, openings are usually necessary to provide (1) access by personnel, either for inspection, maintenance procedures, or operator usage, or (2) access to (or for) materials, such as raw materials, products, or scrap.

Sound leakage around doors, windows, and hatches make enclosures much less effective. Leaks can be handled with properly sealed doors, windows, and hatches. Closed-cell, elastomeric weather stripping with a pressure-sensitive adhesive can be effective seals. Special acoustical gaskets are available, as well as magnetic-strip gaskets similar to those used on refrigerator doors.

If workers must have visual access to machines, lighting may be required. If workers use the sound of the machinery to evaluate its performance, it may be necessary to retrain them or to place a rugged microphone inside the enclosure and send the signal to a small adjustable loudspeaker at the worker's position. Occasionally, it is possible to develop processors that incorporate the worker's knowledge to automatically adjust the machinery for optimal performance. Openings for raw materials, product, and scrap flow can be tunnels lined with sound-absorptive material. The noise reduction will depend on the length and cross section of the tunnel, as well as the thickness of the sound-absorptive material.

Ventilation is required for all total enclosures and some partial enclosures. Ventilation openings can be acoustically lined ducts, elbows, or mufflers, depending on the severity of the problem.

Enclosure panels and structures should not contact any part of the machinery. If the enclosure is mounted on the machinery, it should be vibration isolated.

Advantages of Designing for Noise Control

Good industrial hygiene (as well as common sense) involves removing hazards. In addition, workers may need personal protective equipment. For example, steel-toed shoes may protect workers from unexpected events, such as a large casting falling on a worker's foot. The same precautions should be taken to protect workers' hearing. Protecting hearing should not require constant intervention on the part of the worker, such as wearing earplugs or other HPDs. Workers' hearing can be protected by engineering controls designed into equipment or even added after the fact. With engineering controls, the noise level remains constant, whereas with HPDs, protection is dependent on the availability and proper selection of the HPD, proper training of the worker, proper action by the worker, and appropriate supervision.

Controlling noise in the workplace has many advantages, such as reducing absenteeism, improving communication among workers, reducing the number of accidents, improving efficiency, and increasing productivity. The two examples below show how designing engineering controls into a system can lead to process improvements. Both of these companies worked with their suppliers to develop engineering controls, which they then purchased.

An automobile company used a procedure published by the Association for Manufacturing Technology (formerly the National Machine Tool Builders Association) to measure noise levels in its facilities (AMT, 2006) and then proceeded to use engineering controls to control the noise. In one instance the company replaced some equipment in its metal assembly weld cells. According to Robert Anderson, an acoustical consultant, "Replacing pneumatic drives with servo drives has reduced noise from spot weld impacts, while extending weld tip life and saving energy" (Anderson, 2008).[8]

Michael Bobeczko, director of marketing, Sukut Construction, documented improvements in production that resulted from noise control measures in the manufacturing equipment industry (see Table 4-5). He described a typical manufacturing facility (Bobeczko, 1978).[9]

> A process line in a typical aluminum can plant produces approximately 800 cans per minute. This high-speed process starts at a cup press where sheet aluminum is blanked and drawn into cups. The cups are distributed to bodymakers where each machine redraws and irons the cup into a long seamless can. The can is then usually trimmed to a specific height and conveyed to a washer where it is cleaned, chemically treated and dried. The can exterior is decorated by dry offset methods and the can interior is sprayed with a protec-

---

[8]Anderson, R.R. 2008. Application of Sound Level Specifications for Industrial Equipment. Presentation at the NAE workshop on Engineering Responses to Hazardous Noise Exposures, Washington, DC, August 14–15.

[9]Bobeczko, M. 2008. Industrial Noise Control Solutions Have Improved Productivity. Presentation at the NAE workshop on Engineering Responses to Hazardous Noise Exposures, Washington, DC, August 14–15.

TABLE 4-5 Noise Reduction and Productivity in a Beverage Can Manufacturing Plant

Manufacturing Equipment	A-Weighted Sound Pressure Level at 1 meter (dB)			Operating Speed in Cans/Minute	
	Before	After	Noise Reduction	Before	After
Conveyors	110	77	33	600	2,400
Body maker	104	82	22	120	240
Trimmer	102	80	22	120	250
Necker/flanger	105	85	20	600	1,100
Scrap conveyor	105	80	25	600	2,400

SOURCE: Bobeczko, 1978.

tive coating. The last forming operation necks and flanges the open end of the can.

These same results might be possible through regulation (Porter and van der Linde, 2005):

> [R]egulation creates pressure that motivates innovation and progress. Our broader research on competitiveness highlights the important role of outside pressure in the innovation process, to overcome organization inertia, foster creative thinking and mitigate agency problems. Economists are used to the argument that pressure for innovation can come from strong competitors, demanding customers or rising prices of raw materials; we are arguing that properly crafted legislation can also provide such pressure.

In the early days of OSHA's regulation of noise exposure, many companies made considerable efforts to find ways to reduce noise levels, and trade associations conducted noise studies on behalf of their members. However, when OHSA compliance officers began citing companies for not having engineering controls in place, company attorneys turned the noise exposure questions into legal ones. Companies contended that economic feasibility was important and that engineering controls had to be both technically and economically feasible. The Occupational Safety and Health Review Commission (OSHRC), the federal agency that decides disputed citations and penalties issued by OSHA, decided that economic feasibility had to be considered. This decision caused OSHA to slow down on its citations and many companies to sue before OSHRC rather than pay penalties. For sample cases, see OSHRC, 2009.[10]

In 1983, OSHA put forward an enforcement directive (see OSHA, 2009, for the current version) setting a 100-dB action point for requiring engineering control of noise as long as workers' hearing was adequately protected by HPDs. This weak enforcement policy signaled the death knell for the engineering control of noise in all but the most progressive and innovative companies. As a result, original equipment manufacturers no longer had an incentive to manufacture quiet products because there was no more market for them in the United States.

"BUY QUIET" PROGRAMS

In general, retrofitting existing machinery for noise suppression, especially if it has already been installed in the workplace, can be very expensive. Even though many large manufacturers are acutely aware of the noise problem created by their equipment, many companies adopt hearing conservation programs in lieu of engineering controls.

There is some pressure, however, for companies to purchase quieter equipment. For example, EU regulations specify noise emission limits for many kinds of outdoor equipment. In addition, the Physical Agents Directive (EC, 2003) sets limits for workplace noise levels lower than the limits in the United States. In addition, one industry, the information technology industry, has requirements for noise emissions from their products defined by the Swedish government and spelled out in a Swedish standard (Statskontoret, 2004). Many purchasers of equipment—for example, in the automotive and oil refining industries—also have noise specifications for new equipment. Thus, U.S. companies that want to sell their products or equipment in the European market must meet these standards.

The federal government has the authority to purchase low-noise products (42 USC 65, Section 4914, Development of Low-Noise-Emission Products), and NIOSH and NASA both promote "buy quiet" programs and specifications for low-noise products.

NIOSH is involved in a number of efforts to promote noise declarations that can facilitate the implementation of buy-quiet programs.[11] One works within existing hand-arm vibration efforts by the U.S. Army, U.S. Navy, NASA, and the General Services Administration. The groups are collaborating to develop and disseminate a database of equipment sound and vibration exposure levels. This work includes vetting equipment through GSA acquisition channels for implementation throughout the Department of Defense. Additionally, the effort may persuade machinery and equipment manufacturers to make reduced noise a marketing feature and invest in developing, testing, and selling quieter products. NIOSH is also working to translate research on sound levels and engineering noise controls into practical information by making a revised NIOSH Noise Control Compendium

[10]Occupational Safety and Health Review Commission (OSHRC). 2009. For sample cases, see:
http://www.oshrc.gov/decisions/html_1978/7855_10561_12069_76-0025.html
http://www.oshrc.gov/decisions/html_1983/15647.html
http://www.oshrc.gov/decisions/html_1978/13773.html
http://www.oshrc.gov/decisions/html_1984/14131.html
http://cases.justia.com/us-court-of-appeals/F2/692/641/379249/
http://cases.justia.com/us-court-of-appeals/F2/594/566/480/

[11]Charles Hayden, personal communication, March 17, 2009.

(NIOSH, 1978) and a revised Compendium of Materials for Noise Control (NIOSH, 1975) available on the Internet. In addition, they are updating and expanding their existing powered hand tool database.

Stringent federal noise regulations in the mining industry have led to successful noise control collaborations between NIOSH and mining machinery manufacturers. Mining companies are committed to "buy quiet," and machinery manufacturers are amenable to any collaborative effort to assist them in reducing noise levels from their products. Similarly, noise regulation outside of mining would greatly assist in creating an environment of collaboration and openness to reduce levels, for example, in the construction industry (Hayden and Zechmann, 2009).

In 1996, NASA implemented a "buy quiet" program (Cooper and Nelson, 1996). The program was updated and reviewed in 1999 (Cooper et al.) and described in detail at the INTER-NOISE conference in 2009 (Cooper, 2009).

"Buy quiet" programs and a new "buy-quiet" criterion for industrial equipment have also been reviewed by a member of the NAE study committee, Robert D. Bruce (2009). He described how one global company had set a criterion of 80 dB(A) at 1 meter as the purchase requirement for new equipment. However, he said, this specification cannot stand alone; it must be accompanied by information about the acoustical environment, measurement locations, and machine operating conditions. Bruce also described how new equipment should be installed in existing facilities.

Buying and Selling Low-Noise Equipment

When considering the cost of purchasing quieter equipment, the potential buyer must take into consideration the costs of a long-term hearing conservation program for workers in environments with hazardous noise levels. Hearing conservation programs incur costs for annual audiometric monitoring, medical follow-up of hearing loss cases, monitoring of noise exposure, posting of warning signs and controlling access to high-noise areas, annual training for employees and supervisors, recurring purchases of personal hearing protectors, ongoing administration and record keeping, and inevitable workers' compensation claims for hearing loss.

Even if the buyer has made the decision to proceed, specifying a limit for the noise emissions of a product can be difficult. First, neither the seller nor the purchaser may be familiar with noise *emission* specifications. Second, the type of specification varies with the type of equipment, and the specification must be meaningful to the seller. For example, a specification such as "must meet OSHA requirements" is not adequate. If a manufacturer relies on vendors to supply subassemblies, the manufacturer/vendor relationship may be complicated. In addition, small vendors usually do not have the facilities to determine the noise emissions of components and subassemblies. Another complication is that the noise emitted by a machine may depend on the work being done, for example, when forming metal. The manufacturer may control the noise of motors, cooling equipment, etc., but the buyer must take some responsibility for the noise or the operating conditions of the machine must be very carefully specified when the noise emission levels are specified. Thus, manufacturer and buyer must work together closely to produce a satisfactory design.

Nevertheless, there are several good reasons for a buyer and seller to come to agreement on a noise specification:

- In areas with hazardous noise levels, the noise hazard can be reduced, saving the costs of a hearing conservation program.
- Speech communication in low-noise workplaces is much better than in high-noise workplaces. In addition, because no hearing protection is necessary, desired sounds, such as announcements via public address systems, are not attenuated.
- Low-noise workplaces promote safety (e.g., alarms are clearly audible).
- Low-noise workplaces make it easier for workers to concentrate and reduce fatigue.
- Low-noise workplaces are more productive and more comfortable.

Generally, low-noise equipment is easier to maintain than retrofitted equipment, and controls are easier to use. Although the energy radiated as noise is a very small fraction of the electrical or mechanical energy in any process, a low-noise machine is usually thought of as being more energy efficient and of good design. Guidelines for low-noise design can be found in international documents (ISO, 1995b, 1998).

Responsibilities of Buyers and Sellers

Both supplier and purchaser have responsibilities in implementing a "buy quiet" program. For individual pieces of machinery, such as compressors, motor generators, and similar equipment, the supplier must make available the noise specification for the equipment, usually in terms of the sound power level it emits. Several sets of standards have been established for determining sound power. For example, the International Organization for Standardization (ISO) has generic standards for noise emissions. In addition, many other international standards have been developed for noise emissions from specific kinds of equipment.[12] Other industry standards and other national standards may also be used. For example, ANSI Standard S12.15-1992 (R 2007) defines microphone positions and other information to determine noise emissions for a wide variety of equipment (ANSI, 2007a). In addition, ANSI S12.16-1992 (R 2007) provides information

[12]See Chapter 6 and Appendix C for more on ISO standards and other international standards.

on how to request noise emission data for machinery and equipment (ANSI, 2007b).

A variety of standards for noise emitted by household appliances have been published by the International Electrotechnical Commission (IEC, 2010). Information on preparing specifications and recommended noise emission levels for equipment that will be used in classrooms has also been published (Hellweg et al., 2006).

An alternative to a sound power level specification is an emission sound pressure level specification. Methods of making these measurements are described in international standards. The basic guidelines can be found in ISO 11200:1995 (ISO, 1995a), and the series is listed in the ISO Standards Catalog.

The relationship between sound power level (*emission*) and sound pressure level (*immission*) is discussed in most handbooks and is discussed in this report in Appendix A. Emission sound pressure level specifications cover the relationship between the two quantities for a diffuse field in a room and show, for a given amount of sound absorption, how sound decreases as one moves away from a noise source. Other examples of this relationship are techniques used in the information technology industry to calculate noise levels in a space with large numbers of machines (ECMA International, 1995).

In settings where a large number of pieces of equipment are on a manufacturing floor, the purchaser and supplier must work together to define noise emission specifications tailored to the specific pieces of equipment being purchased. The purchaser must translate the desired noise *immission* level in a given environment to a noise *emission* level that can be understood and validated by the manufacturer. The noise *emission* of equipment is generally specified in terms of sound power or emission sound pressure, whereas *immission*, the level measured in the workplace, depends on the number of machines in a given area, the size and shape of the space, the amount of sound absorption, and the scattering of sound from one or more machines. Thus, the connection between *emission* and *immission* must usually be resolved by a professional in noise control engineering who can define the relationship between the noise emission of equipment and the noise levels in the manufacturing environment. The purchaser can then determine whether or not a noise problem exists. A workplace with a sound pressure level of 75 dB(A) for an 8-hour exposure is generally considered safe but unpleasant. Levels of 80 dB(A) and higher raise concerns about damage to workers' hearing.

HEARING PROTECTION DEVICES

The First Hearing Protectors

HPDs, which protect the human ear from incurring NIHL, have been in existence at least since the early 1900s, although their use in U.S. workplaces was not regulated by law until 1971. In fact, in 1911 the famous band leader John Phillip Sousa complained to his friend and fellow skeet trapshooter J. A. R. Elliott that shooting traps "took a toll on his ears and was beginning to affect his livelihood [as a musician]." Elliott, an inventor, then developed and patented (in eight countries), the "Elliott Perfect Ear Protector," which became a commercial success. After using the Elliott Protector, Sousa wrote a letter to Elliott on January 20, 1913: "I consider your invention to lessen the shock of loud noises or overwhelming vibrations of sound of great comfort. The Elliott Perfect Ear Protector is a great success in affording protection from concussions to a sensitive ear. As a shock absorber it is invaluable" (Baldwin, 2004).

Unfortunately, although simple cotton plugs were known to be used in some workplaces before the turn of the 19th century (e.g., Barr, 1896), U.S. industrial workers did not routinely use effective HPDs until many years later. The lack of protection, coupled with high noise exposures, resulted in NIHLs and related problems, such as tinnitus, which were often viewed as accepted consequences of one's occupation. Terms such as "blacksmith's deafness" and "boilermaker's ear" were coined to describe these common afflictions (Fosboke, 1831; Holt, 1882).

Hearing Protection in the Military and Industry

In the first half of the twentieth century, HPDs were not commonly used in U.S. workplaces or for most leisure-time exposures. However, the U.S. military has recognized their importance at least since World War II for protecting against the effects of noise-emitting ordnance, as well as loud machinery, such as tanks and aircraft. In fact, one of the earliest regulations on hearing conservation was U.S. Air Force Regulation 160-3 of 1948 (Department of the Air Force, 1948), which called for the use of HPDs. In U.S. workplaces, however, although a few industrial hearing conservation programs appeared in the 1940s and 1950s (Berger, 2003b), hearing protection was not required by law until May 1971, with promulgation of the Occupational Noise Exposure Standard under the authority of the Occupational Safety and Health Act. This was the first requirement, based on exposure levels, for hearing protection in general manufacturing industry (OSHA, 1971a); a similar law was promulgated for construction workers (OSHA, 1971b), industry and construction being the settings where the U.S. workers were, and continue to be, most at risk for NIHL.

Later, in 1981, the legal advent of the OSHA Hearing Conservation Amendment for General Industry (OSHA, 1981) immediately caused the use of HPDs to proliferate in U.S. industrial workplaces. The amendment required that employers provide a choice of HPDs to any worker exposed to noise of 85-dB(A) time-weighted average (TWA), or 50 percent noise dose, for an 8-hour workday, with the measurement taken on the "slow" scale of the sound-level meter and using a 5-dB exchange rate between exposure dB(A) level and time of exposure.

Other industries, including airlines, truck and bus carriers, and railroads, have separate, generally less comprehensive, noise and hearing conservation regulations. Unfortunately, to date, no analog to the OSHA Hearing Conservation Amendment of 1981 has been adopted into law, although in 2002 OSHA issued an advance notice of proposed rulemaking for the construction industry (Federal Register, 2002) where noise levels can also be hazardous (Casali and Lancaster, 2008). Workers in agriculture and oil and gas drilling are not covered by noise regulations.

In the mining industry, hearing protection has been addressed, first under the Federal Coal Mine Health and Safety Act of 1969 and later under the Federal Mine Safety and Health Amendments Act of 1977. In 1999 the MSHA issued a more comprehensive noise regulation governing all forms of mining (MSHA, 1999). The regulation emphasized engineering control but did include the use of HPDs.

The major point about all of these historical milestones in federal regulatory development is that hearing protectors were not really addressed in U.S. occupational safety and health law until the late 1960s and early 1970s.

Hearing Protection Outside the Workplace

Although data on the use of HPDs outside the workplace are elusive, there are indications that they are becoming more popular as awareness of NIHL increases. HPDs can protect hearing from hazardous noises produced by some power tools, lawn care equipment, recreational vehicles, target shooting and hunting, spectator events, amplified music, outdoor construction equipment, and many other sources. In fact, some of these activities, such as using recreational firearms (Nondahl et al., 2000) and attending motorsport events, such as monster truck races (Casali, 1990), pose the risk of exposure to noise levels equal to, or even beyond, the levels experienced by workers in many industries. HPDs can reduce the energy to the ear from these noises and prevent hearing loss.

HPDs can also be beneficial in noisy environments that pose no real threat to hearing. For example, HPDs can reduce noise annoyance in passenger cabins of commercial aircraft and on subways and buses. They can also reduce noise in sleeping environments, such as traffic noise or snoring (although HPD use while sleeping is *not recommended* because it can interfere with the audibility of acoustic signals such as smoke alarms, telephones, and doorbells).

Hearing protection features are also incorporated into other products worn on the ears, such as headphones worn by music lovers and headsets worn by crew members for cockpit communications. Some hearing protection features can improve the fidelity of audio signals and the intelligibility of speech. Thus, an effective headphone/headset can not only improve the signal/speech-to-noise ratio but can also protect against the hazards of ambient noise.

When using HPDs, particularly earphones for listening to music or speech, wearers must understand that *external* signals, such as sirens or stall warnings in an aircraft cockpit, might be attenuated or even masked completely. The effects of headphones/headsets on the situational awareness of the wearer depend on the particular situation, the individual user, and the design of the HPD (e.g., the frequency spectrum of attenuation produced by passive and/or active noise cancellation features).

HEARING PROTECTION VERSUS (OR AS AUGMENTATION OF) ENGINEERING NOISE CONTROL

Systems Approach

In a straightforward systems approach to noise abatement, efforts to reduce or eliminate noise exposures are concentrated in three primary locations: the noise source, the sound propagation path, and the receiver (see Figure 4-1). From a systems perspective, HPDs are the last line of defense in the protective chain (Berger, 2003b; Casali, 2006; Gerges and Casali, 2007). This is because an HPD may not always be effective; it is an *active* countermeasure implemented *at the receiver, who must use it properly*, as opposed to a *passive* countermeasure, such as engineering controls that are implemented *at the noise source* or *in its propagation path*. Like a seatbelt, an HPD requires that a person use it at the right time and use it properly to provide protection (see Figure 4-2). Like a crashworthy design, the performance of an engineered noise control device does not depend on human behavior.

There are several other reasons for using a passive countermeasure:

- It is the employer's duty according to the Occupational Safety and Health Act of 1970 to provide a safe and

	Source ------------------>	**Path** ------------------>	**Receiver**
Countermeasures:	Engineering Controls	Engineering Controls	Hearing Protection
		Administrative Controls	

FIGURE 4-1 Systems approach to reducing noise exposures.

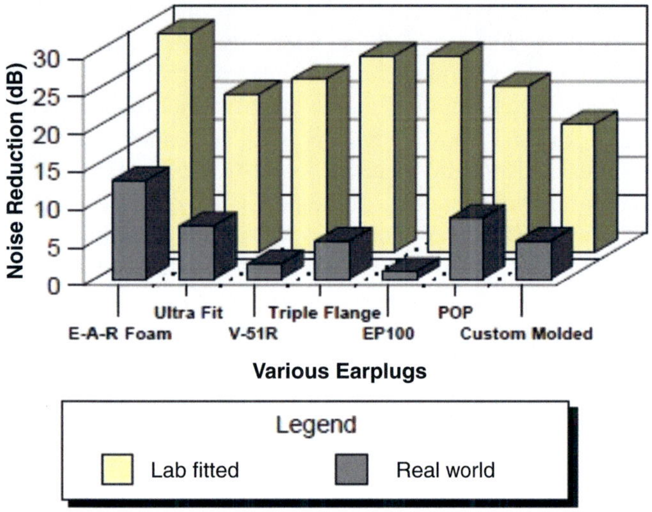

FIGURE 4-2 Comparative noise reduction ratings for various earplugs. Source: NIOSH, 1996.

healthy workplace, not the employee's duty to provide that protection.

- Hearing protectors often provide insufficient attenuation.
- Hearing protectors are often uncomfortable and therefore not accepted by workers.
- Hearing protectors are often counterproductive in that they reduce the worker's ability to hear speech communication and warning signals.
- Hearing protectors can have an adverse effect on the ability to localize sounds and therefore pose a safety hazard.

The focus of noise reduction efforts should certainly be on the development and purchase of quieter consumer products and industrial machinery and on reducing noise via engineering controls at the noise source or in the noise path. However, when engineering controls are not economically or technically feasible, or when an employer or manufacturer (unfortunately) does not consider them a high priority, hearing protection may supplant engineering and/or administrative controls (e.g., setting time limits to a worker's exposure).

For these reasons the use of HPDs has proliferated, especially in industrial and military settings. In situations where it is difficult, or even infeasible, to "engineer out" noise, such as that from weapons or aircraft, HPDs may be the only practical countermeasure. In addition, there are some cases in which engineering and/or administrative controls have been implemented but do not reduce the noise to an acceptable (or even legal) level. In these situations, HPDs may be used to "make up the difference" and protect against NIHL.

Limitations of Hearing Protection Devices

If exposures to noise at hazardous levels persist even after noise controls have been tried, HPDs may be the only way to protect hearing. Permanent NIHL is typically a progressive neural injury that damages or destroys the hair cells and neural pathways of the inner ear. However, NIHL can also be an immediate response to an acoustic trauma if the elastic limits of the tympanum, ossicular chain of the middle ear, or cochlear structures are overwhelmed by a powerful acoustical insult, such as an explosion. For the great majority of noises to which people are exposed, HPDs can mitigate NIHL, provided they are properly selected, fitted, and worn.

In industry, where 90 percent of noise exposures are at TWA levels less than or equal to 95 dB(A), compliance with the OSHA 1983 Hearing Conservation Amendment requires only a 10-dB reduction, which most HPDs can supply (Berger, 2003b). However, HPDs are not a panacea. They may be ineffective if the acoustical pathways, such as air leaks around the seal, HPD material transmission, HPD vibration, and flanking via bone conduction interfere with their noise reduction features. In addition, though rare, extremely high noise levels may overwhelm the attenuation capabilities of even the best HPDs. For example, for 8-hour daily TWA exposures that exceed about 105 dB(A), especially with dominant low-frequency content below about 500 Hz, *double* passive hearing protection (i.e., an earmuff worn over a well-fitted earplug) is advisable (Berger, 2003b). However, because of such factors as discomfort and concerns about safety, workers often complain about wearing even one earmuff or earplug, and convincing them of the necessity of wearing both during the course of a workday can be a challenge.

In even more severe noise environments, such as on aircraft carrier decks during flight operations, prevailing noise exposures can be extremely high, and short-term levels may range from 146 to 153 dB(A) at 50 feet from military jet aircraft on afterburner power (McKinley, 2001). These levels will overtax even the double passive HPDs. In those cases very specialized HPDs are required and are mainly of interest only in military applications. These devices are described briefly in the section on multicomponent systems for extreme noise later in the chapter.

Hearing Protection Devices in Nonoccupational Settings

Home and Recreational Use

In contrast to the use of HPDs in most occupational settings in the United States, the use of HPDs in recreational and home settings is generally up to the individual—and in most cases depends on awareness of what constitutes unsafe noise and a person's willingness to take risks. Despite programs to educate people about noise-related dangers and the importance of HPDs, public awareness about the hazardous effects of noise is low (ASHA, 2009; WHO, 1997).

In addition to conventional passive hearing protectors, which have been available for decades, HPDs styled and sized specifically for children, designed for spectator events (e.g., earmuffs that incorporate radios for sporting events),

HPDs with signal pass-through circuitry (e.g., electronic earmuffs for hunters), lightweight active noise cancellation HPDs for reducing low-frequency noise in aircraft cabins, uniform attenuation earplugs for musicians and concert attendees, and other innovative and attractive devices are now on the market.

Local Ordinances on Hearing Protection

In contrast to long-standing federal laws for occupational exposures, noise in community and recreational settings, if governed at all, is usually addressed in local ordinances, most of which relate to noise annoyance rather than to hearing risks. However, in venues having recreational exposures from amplified music, or gaming arcades, there may be warning signs stipulating that hearing protection is required upon entry. However, ordinances to protect hearing are in the minority, often passed in response to public complaints and/or civil litigation for premises' liability. Again, the committee questions why noise is not controlled to within safe limits by engineering means or by "turning down the volume," rather than by warning people to wear hearing protection and depending on them to have such protection at hand.

HEARING PROTECTION DEVICES: TECHNOLOGIES AND EFFECTS ON AUDIBILITY

Conventional (Passive) Hearing Protection Devices

The vast majority of HPDs are so-called conventional hearing protectors that attenuate noise by static passive means. Conventional, or passive, HPDs do not have dynamic mechanical elements, such as valves or reactive ports, or electronic circuitry, such as active noise cancellation or signal pass-through circuitry. The effectiveness of passive HPDs depends on a combination of acoustical factors, including the airborne sound transmission loss imposed by the construction materials, the reflection characteristics of the HPD for incident sound waves, the quality and integrity of the seal against the ear canal or outer ear or its surrounding tissue, the ability of the HPD to dampen vibrations of the ear canal wall, and the resonance frequency characteristics and acoustical impedances of the HPD. There are four general types of conventional HPDs—earplugs, semi-insert or ear canal caps, earmuffs, and helmets—each defined by the way the device interfaces with the ear or head.

Earplugs are vinyl, silicone, spun fiberglass, cotton/wax combinations, or closed-cell foam products inserted into the ear canal to form a noise-blocking seal. Proper fit to the user's ears and training in insertion procedures are critical to their effectiveness. A related, but different, category of HPD is the *semi-insert* or *ear canal cap*, which consists of an earplug-like pod positioned at the rim of the ear canal and/or in the concha bowl of the external ear (pinna). The device is held in place by a lightweight headband positioned under the chin, behind the head, or over the head. The headband of an ear canal cap can also be used to store the device around the neck when not in use.

Earmuffs are ear cups, usually made of a rigid plastic material with a noise-absorptive liner, that completely enclose the outer ear and seal around it with foam- or fluid-filled cushions. On some models the headband that connects the ear cups is adjustable so that it can be worn over the head, behind the neck, or under the chin, depending on the presence of other headgear, such as a welder's mask. *Helmets*, which enclose a large portion of the head, are usually designed to provide impact protection, but they can have integrated ear cups or a liner material that seals around the ears (Berger and Casali, 1997). Furthermore, for extreme noises that substantially transmit sound through bone conduction to the neural ear, helmets that cover the temporal and mandibular areas, as well as the cranium, can provide additional protection against bone-conducted noise (Gerges and Casali, 2007).

In general, earplugs provide better attenuation than earmuffs for noise below about 500 Hz and equivalent and better protection for sounds above 2,000 Hz. At intermediate frequencies, earmuffs generally provide better attenuation. Earmuffs are generally easier than earplugs or ear canal caps for the user to fit properly. However, in high temperatures and humidity, earmuffs can be uncomfortable; in cold temperatures they can be welcome insulators. Semi-inserts generally provide less attenuation and are more uncomfortable than earplugs or earmuffs. However, because they can be stored around the neck, they are convenient for workers who frequently move in and out of noisy areas. For a comprehensive review of conventional HPDs and their applications, see Gerges and Casali (2007) and Berger (2003b).

Although conventional HPDs provide adequate protection for most noise exposures, a potential disadvantage, due to the static, passive nature of the attenuation, is a deleterious effect on hearing quality and auditory performance. This effect varies with the user's hearing ability and the noise and signal conditions. For more specific information on the effects of HPDs on speech communication and signal audibility, the reader is referred to Casali (2006), Robinson and Casali (2003), and Suter (1992).

HEARING PROTECTION DEVICES: EFFECTS ON SIGNAL AND SPEECH AUDIBILITY

Overprotection Versus Underprotection

Safety professionals must select HPDs for the workplace that provide adequate attenuation for the noise threat but not so much attenuation that the worker cannot hear important signals and/or speech communications. Users may reject an HPD if it compromises hearing to the point that sounds seem unnatural, signals are undetectable, and/or speech is not understandable. Too much attenuation for a particular noise situation is commonly referred to as *overprotection*.

The selection of an overprotective or underprotective HPD can have serious legal ramifications. Here is a hypothetical statement by a workers' compensation plaintiff: "The hearing protector provided inadequate noise attenuation for defending my ears against the damaging effects of noise, so I lost my hearing over time." Also: "The hearing protector provided more attenuation than needed for the noise I was exposed to at work, and therefore was a causal factor in the accident when I could not hear the forklift backup alarm and was run over."

These are extreme examples, but in civil court such arguments can potentially lead to theories on which a legal foundation for recovery of damages may be based. Consider product liability, for example. The "failures" claimed in the statements above would typically fall under the category of defective design and/or availability of superior alternative design features, and/or breach of warranty. The threat of litigation is of great concern to both HPD manufacturers and employers. Thus, HPDs must be matched to workers and job requirements. The above paragraphs lend even more support to the principle that engineering noise controls should have priority over HPDs.

Effects on Audibility Leading to Technological Augmentations

Research on people with normal hearing suggests that conventional passive HPDs have little or no degrading effect on their understanding of external speech or signals in ambient noise levels above about 80 dB(A); they may even provide some improvements, with a crossover from disadvantage to advantage between 80 and 90 dB(A) (Berger and Casali, 1997; Casali and Gerges, 2006). However, in lower sound pressure levels, they often do increase misunderstanding and poorer detection. In these situations, HPDs are usually used not to protect hearing but to reduce annoyance. In the presence of intermittent noise, HPDs may be worn during quiet periods so that when a loud noise occurs, the wearer will be protected. However, during the quiet periods, hearing acuity may be reduced.

Technological enhancements are sometimes incorporated to create *level-dependent augmented HPDs* that provide minimal or moderate attenuation (or sometimes more amplification of external sounds) during quiet times and increased attenuation (or less amplification) as the noise level increases. However, commercially available versions of these devices have not been associated with a demonstrated improvement in signal detection over conventional HPDs in most situations (e.g., Casali and Lancaster, 2008; Casali and Robinson, 2003).

Noise- and age-induced hearing losses generally occur in the high-frequency range first, making it difficult to determine the effects of HPDs on speech perception for people with early impairment. Because their elevated thresholds for mid- to high-frequency speech sounds are elevated further by the protector, hearing-impaired individuals are usually disadvantaged by conventional HPDs. Comprehensive reviews (e.g., Suter, 1992) have concluded that people with sufficient hearing impairments usually experience additional reductions in communication abilities with conventional HPDs in noisy environments.

Moreover, because of the phenomenon known as the occlusion effect, people who wear HPDs lower their voices by about 2 to 4 dB, so that when both talker and listener wear protectors, the resulting decrease in speech recognition will tend to offset any benefits, even with normal-hearing listeners (Howell and Martin, 1975; Hoermann et al., 1984).

HPDs with electronic hearing-assistive circuits, sometimes called electronic sound transmission or sound restoration HPDs, can be offered to hearing-impaired individuals to determine if their hearing, especially in quiet-to-moderate noise levels below about 85 dB(A), can be improved and their hearing still somewhat protected. However, the benefits of electronic sound transmission HPDs for hearing-impaired users have not been empirically demonstrated in scientific studies.

Nonlinear Passive Attenuation

Conventional passive HPDs cannot selectively relay speech or nonverbal signals (or speech) energy rather than noise energy at a given frequency. Therefore, conventional HPDs do not improve the speech/noise ratio in a given frequency band, which is the most important factor for achieving reliable signal detection and speech intelligibility. As shown in Figure 4-3, conventional earplugs (labeled fiberglass, premolded, or foam) attenuate high-frequency sound substantially more than low-frequency sound; therefore, they attenuate high-frequency consonant sounds, which are important for word discrimination. They also attenuate frequencies that are dominant in many warning signals. This nonlinear attenuation profile, which generally increases with frequency for most conventional earplugs and nearly

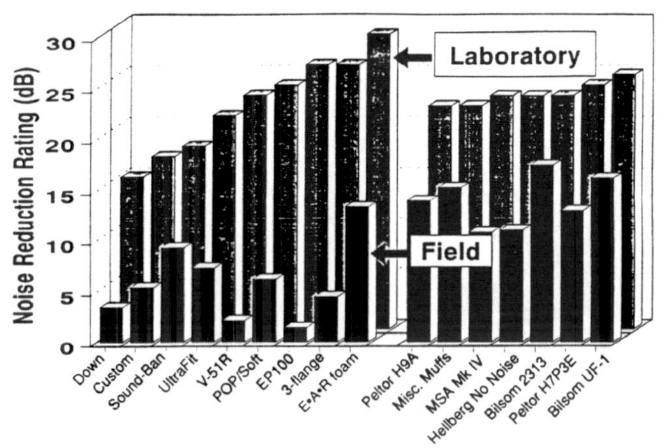

FIGURE 4-3 Comparative noise reduction ratings based on manufacturers' laboratory tests and real-world "field" performance of different types of hearing protection devices. Adapted with permission from Berger, 2003b, Fig. 10.18, p. 421.

all conventional earmuffs, allows more low-frequency than high-frequency noise through the protector, which causes an "upward spread of masking" if the penetrating noise levels are high enough (Robinson and Casali, 2003).

Certain augmented HPD technologies help overcome these weaknesses, particularly low-frequency attenuation. These include a variety of *active noise reduction* (ANR) devices that capture the offending noise and its electronic phase cancellation at the ear via superposition through feedback and/or feed-forward control loops. ANR devices boost low-frequency attenuation below about 1,000 Hz. ANR is especially effective in earmuffs, which are generally weakest in low-frequency attenuation and which also have enough space for the electronics of ANR circuitry to be packaged in/on the muff. ANR has also been used in earplug designs in the past decade (e.g., McKinley, 2001). The benefits of ANR-based HPDs can include reduction of upward spread of masking of low-frequency noise into the speech and warning signal bandwidths and reduction of noise annoyance in environments dominated by low-frequency noise, such as jet cockpits and passenger cabins (Casali and Robinson, 2003).

The tendency of conventional HPDs to exhibit a sloping nonlinear attenuation profile with changes in frequency creates an imbalance from the listener's perspective because the relative amplitudes of different frequencies are heard differently than they would be without the HPD. Thus, broadband acoustic signals are heard as spectrally different (more "bassy") from normal acoustic signals (Berger, 2003b).

People whose jobs depend on accurate sound interpretation (aural inspections by machinists, miners, engine troubleshooters), as well as people who perform or listen to music, may be adversely affected. Figure 4-4 shows attenuation curves for two *uniform (or flat) attenuation devices* (ER-15 and ER-20). These devices are more popular with musicians than conventional HPDs because they do not distort perceptions of the loudness of various pitches (Casali and Robinson, 2003).

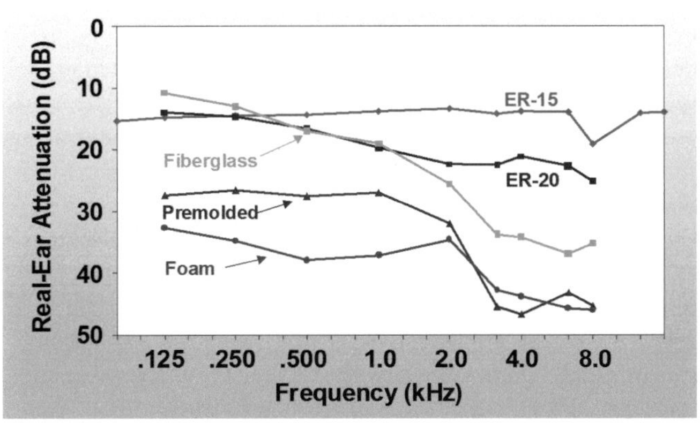

FIGURE 4-4 Spectral attenuation obtained with real-ear attenuation at threshold (REAT) procedures for three conventional passive earplugs (premolded, user-molded foam, and spun fiberglass) and two uniform-attenuation, custom-molded earplugs (ER-15, ER-20). Provided courtesy of E.H. Berger, AEARO—3M.

Some high-frequency binaural cues (especially above about 4,000 Hz) that depend on the external ears (pinnae) are altered by HPDs, compromising judgments of sound direction and distance. Earmuffs, which completely obscure the pinnae, may interfere with localization in the vertical plane and may also cause horizontal-plane errors in both contralateral (left-right) and ipsilateral (front-back) judgments (Suter, 1992). Earplugs may cause some ipsilateral judgment errors but generally cause fewer localization errors than earmuffs because they do not completely destroy cues from the pinnae.

Dichotic sound transmission HPDs compensate for lost pinnae-derived cues. These devices have an external microphone on each earmuff cup that transmits a specified passband of the noise incident on each microphone to a small loudspeaker under the earmuff cup. Binaural cues are thus maintained, as least partly, as long as the between-ear gain controls are properly balanced, the microphones are sufficiently directional, and the passband includes frequencies *outside* the range that cannot be typically localized (i.e., outside the range of about 1,000 to 3,000 Hz). However, a recent experiment with a dichotic sound transmission earmuff in an azimuthal localization task of determining the approach vector of a vehicular backup alarm demonstrated no advantage in localization over a conventional earmuff or earplug (Casali and Alali, 2009).

All of the augmentations described above—uniform attenuation, ANR, electronic sound transmission, and level-dependent or amplitude-sensitive attenuation—are effective in certain applications. For more information on these and other technologies that are commercially available, the reader is referred to a review by Casali and Robinson (2003). The major goals of these augmentations are (1) to encourage the use of hearing protection by producing HPDs that are more acceptable to the user population, amenable to the working environment, and adjustable to the noise exposure and (2) to improve hearing in a protected state, which may also make workers safer. Unfortunately, these noble goals are not always realized in practice. One reason is that these devices can be considerably more expensive than conventional passive HPDs, and most employers are reluctant to incur the cost.

EMERGING TECHNOLOGIES

To overcome some of the limitations of HPDs, several recent innovations have been developed or prototyped or in some cases made commercially available. Indeed, emerging technologies continue to be developed. The brief overview that follows includes examples of new technologies known to the committee. However, the list is not exhaustive, and the study committee does not advocate or promote these particular devices or technologies.

Adjustable Attenuation Devices

To "tailor" the attenuation of an HPD to a particular noise problem (i.e., in lieu of selecting *different* HPDs for different exposures), earplug designs have recently been developed that give the user some control over the amount of attenuation. These devices incorporate a leakage path that can be adjusted via a valve that obstructs a tunnel, or "vent," cut through the body of the plug (e.g., a Dutch earplug, the Ergotec Varifone) or by selecting a filter or dampers that can be inserted into the vent (e.g., Canadian devices, such as the Sonomax SonoCustom and the Custom Protect Ear dB Blocker).

Verifiable Attenuation Devices

To establish a quality fit of an HPD to a user, and in keeping with the OSHA (1983) Hearing Conservation Amendment requirement, per 29 CFR 1910.95 (i)(5), that "the employer shall insure proper initial fitting and supervise the correct use of all hearing protectors," several systems have been developed to verify the attenuation of a device worn by a given user. For example, the original SonoPass system, now called the AEARO/3M E•A•RFit, is sold as a system with a probe tube microphone test apparatus that verifies the attenuation level achieved via microphone-in-real-ear, noise reduction measurements on each user as he or she is fitted with the product. A similar system is the Sperian VeriPRO.

All of these verifiable-attenuation HPDs measure attenuation by placing a microphone or probe tube through a duct that runs lengthwise through an earplug and then measuring noise reduction in a sound field. "Verified attenuation" is thus established *for that ducted earplug for that particular fitting on the user*. Thereafter, for use in the field the ducted earplug is replaced with a solid (i.e., nonducted) earplug of the same type/model; alternatively, a noise-blocking insert is used to occlude the duct. The attenuation achieved by these systems is *not* recognized by OSHA as a means for determining the adequacy of the hearing protector for a given noise exposure, and it does not replace the EPA-required label of HPD attenuation data. However, with policy amendments, it may someday be recognized for either or both of these applications.

Devices with Enhanced Situational Awareness and Communication Capabilities

In the past few years several HPD-based devices have been developed that have multiple objectives, which include hearing protection from continuous noise, hearing protection from impulsive noise (particularly gunfire), measurement of protected noise exposure (i.e., at the ear under the HPD), improved hearing of ambient sounds and uttered speech, and improved communication capabilities. All of these products incorporate sound transmission circuitry to transduce ambient sounds via a microphone on the outside of the HPD; those sounds are then bandpass filtered to an amplified earphone inside the HPD.

Using elements of rapid-response automatic gain control with high pass-through gain capability, these devices can be used as assistive listening devices for military and other applications, as aids in threat detection, and as sound localizers. They can also improve hearing of low-level speech. When gunfire occurs, the amplification rapidly decays, causing the device to quickly revert to a passive hearing protector. These devices typically have more sophisticated and powerful pass-through filtering/gain circuits than industrial versions of sound transmission earmuffs.

Some of these systems incorporate elements that provide two-way communication capabilities, including versions with covert microphones located under the HPD in the ear canal, to pick up the wearer's voice by bone/tissue conduction. At least one device can transduce the noise level under the HPD, use it to determine cumulative noise exposure, and modulate the system pass-through gain based on these data. Examples of HPDs that provide enhanced situational awareness include the Communications and Enhancement Protection System from Communications & Ear Protection, Inc.; the QuietProby NACRE AS (Norwegian); and the Silynx QuietOPS. Because of their very recent development, some of these devices have not yet undergone experimentation to test their operational performance. Some, however, have already been deployed for use in combat settings. Further discussion of HPDs with situational enhancement, and experimentation on a subset of them, can be found in Casali et al. (2009).

Multicomponent Systems for Extreme Noise

Multicomponent HPD systems have recently been developed and tested for use in noise environments that greatly exceed the attenuation capabilities of even double passive protectors (i.e., earmuffs worn over earplugs). The most prominent of these environments is an aircraft carrier deck during flight operations, where flight deck personnel are subject to sound pressure levels as high as the mid-150-dB(A) range (McKinley, 2001). Some large-caliber weapons and explosive blasts can also produce exceedingly high exposures.

Specialized HPDs have been developed for use in these extreme conditions, including devices with multiple components for staged hearing protection for use on aircraft carriers. These HPDs provide both high passive attenuation through very deep insertion, custom-molded earplugs coupled with active noise cancellation in the in-canal sound field *under* the earplug, all covered with a tightly fitted earmuff with custom-fitted cushions (McKinley, 2001). Other devices are full-head-coverage helmets with circumaural active noise cancellation earcups inside, all worn over deeply fitted passive earplugs.

Composite Material Devices

A few HPDs have been produced from a combination of materials, typically in sandwich- or concentric-type construction, to take advantage of impedance mismatching (and the resultant attenuation benefit) that occurs with materials that differ in density, elasticity, and other physical parameters. This design practice has long been used in earmuffs, but composite structures in earplugs will require further investigation.

SUMMARY

Hearing protection is a very important component of a hearing conservation program and is currently essential to combating the problem of noise-induced hearing loss. However, there are disadvantages that accompany the use of HPDs, including but not limited to the high cost of certain augmented HPDs that are needed for specialized applications, situation- and user-specific interference with signal detection and speech communications, discrepancies between laboratory ratings and actual field performance of both conventional passive and electronic protectors, and a perception reported by some workers that HPDs may be "unsafe" (Morata et al., 2005). Furthermore, it must be noted that OSHA (1971a, 1971c, 1983) specifically assigns the responsibility for hearing protection to employers, not workers, and thus the performance of the hearing conservation program is largely dependent on employer commitment to exposure measurement, HPD selection, and training of workers to fit and use the devices properly.

Returning to the systems approach to noise abatement shown in Figure 4-1, it is important to reiterate that *hearing protection is a noise countermeasure that is only implementable at the very end of the noise propagation chain*, that is, at the receiver's ear. In the great majority of noise exposure situations the priority should be to reduce or eliminate the noise at its source or in its path through engineering controls and not to rely on hearing protection to curb the noise just before it enters the ears. Hearing protection, though effective when selected and applied properly, is not a panacea for combating the risks posed by noise, and its effectiveness will always be dependent on human behavior. It should thus not be viewed as a replacement for noise control engineering. However, in those cases where noise control engineering's afforded reduction is simply insufficient, or it is *truly* economically and/or technically infeasible (as perhaps with a personally shouldered, high-caliber weapon), hearing protection devices become the primary countermeasure.

FINDINGS AND RECOMMENDATIONS

Regulatory Changes in Damage Risk Criteria for Hearing

OSHA has promulgated a noise regulation for general industry, 29 CFR 1910.95. This regulation has been in effect since 1971 and calls for employers to use "feasible administrative or engineering controls" for all employees exposed above a daily time-weighted average level (TWA) of 90 dB(A). In 1981 and 1983 OSHA amended the regulation for hearing conservation requiring employers to supply hearing protection devices and other components of the hearing conservation program at and above a TWA of 85 dB(A). Since 1983, workers who have exhibited a "standard" threshold shift in hearing must be required to wear hearing protection devices above a TWA of 85 dB(A); in addition certain follow-up measures must be taken by the employer. The primacy of engineering and administrative controls over the use of hearing protection devices remains in effect, although OSHA has not enforced it in recent years. In practice, employers have often used technical and/or economic infeasibility as a justification for not implementing engineering controls.

In 1983 OSHA issued a policy directive advising its compliance officers not to issue citations to companies with "effective" hearing conservation programs until workers' TWAs exceed 100 dB(A). This policy still exists in the agency's Field Operations Manual (OSHA, 2009); however, it has never had the authority of regulation and could be revoked at any time.

The original OSHA noise regulation used an exchange rate of 5 dB per doubling or halving of exposure time, and this rule has not yet been changed, despite recommendations by EPA and NIOSH to change both the exchange rate to 3 dB and the permissible exposure limit from 90 to 85 dB(A). MSHA promulgated a revised noise regulation in 1999 that is similar but not identical to the OSHA regulation, although here the primacy of engineering noise control is clear. The agency discussed the issues of changing from the 5 dB to the more protective 3-dB exchange rate and from the 90-dB(A) permissible exposure limit to 85 dB(A) but failed to do so at the time, although the preamble to the rule stated that "[i]n both cases, the scientific evidence was strong." Ultimately, the change was not adopted because of significantly increased costs for small mine operators. The 85/3 limits are used by several other U.S. government agencies and are also written into most national and international standards.

The OSHA Field Operation Manual (OSHA, 2009) contains a statement widely known as the "100-dB Directive." With reference to issuing citations for noise violations, it states that "[h]earing protectors which offer the greatest attenuation may reliably be used to protect employees when their exposure levels border on 100 dB(A)." The effect of this statement has been to negate the well-recognized goal of using engineering controls as the primary means of controlling industrial noise.

Recommendation 4-1: To comply with the recommendation of the National Institute for Occupational Safety and Health, the policy of several other government agencies, and widespread national and international scientific opinion, the U.S. Department of Labor should adopt the

85-dB(A)/3-dB limit for exposure to hazardous noise. This would replace the current 90-dB(A)/5-dB requirement.

Measurement and Evaluation of the Hazard of Impulsive Noise

The peak sound pressure level is currently widely used to determine noise hazard, but recent studies have indicated that the duration of the impulse plays an important role. There is also evidence that impulsive noise and continuous noise can be included in a single measurement of equivalent sound level. The committee concluded that current damage risk criteria in the United States and internationally are inadequate and should be the subject of future research.

Recommendation 4-2: The National Institute for Occupational Safety and Health should be the lead agency and should be tasked by its parent agencies (U.S. Department of Health and Human Services/Centers for Disease Control and Prevention) to develop new damage risk criteria with assistance from the military services that have experience with high-amplitude impulsive noise.

Promoting Engineering Control to Reduce Hazardous Noise

Engineering noise controls provide significant long-term advantages over personal hearing protection. If workplace noise levels are limited by engineering controls, "buy quiet" programs, or other means to a level below the OSHA action level of 85 dB(A) TWA, the need for individual hearing protection devices (HPDs) is obviated from an OSHA standpoint. HPDs may still be desirable for reducing noise annoyance or ensuring that a noise hazard is fully mitigated.

In 42 USC 65, Section 4914, the federal government is required to encourage the procurement of low noise emission products:

> (1) Certified low-noise-emission products shall be acquired by purchase or lease by the Federal Government for use by the Federal Government in lieu of other products, if the administrator of General Services determines that such certified products have procurements costs which are no more than 1.25 percentum of the retail price of the least expensive type of product for which there are certified substitutes. (2) Data relied on by the administrator in determining that a product is a certified low-noise-emission product shall be incorporated in any contract for the procurement of such product.

The same principle applies in the consumer setting. Thus, the engineering of quieter products, such as power tools, toys, yard equipment, and recreational vehicles, would reduce the need for and reliance on HPDs. Even though an HPD may protect the wearer's hearing, it may create hazards if the user is unable to hear approaching vehicles or alarms.

The effectiveness of HPDs depends on human behavior, and the human factor is always a weak link in the "safety chain." HPDs must be comfortable, easily sized and fitted to the user, and straightforward to meet the hearing-critical needs of a particular job or situation. They must still provide situational awareness (e.g., communication enhancements where needed and attenuation performance labels that reflect the level of protection they provide in actual use). HPD technology has advanced greatly in the past 30 years, but HPD regulations have not kept pace. This discrepancy was recognized in the publication of EPA's proposed new rules (EPA, 2009), which should be adopted. However, despite the improvements of recent devices, they cannot be considered a substitute for engineering noise control because of the many factors cited above, and their efficacy simply has not been proven. Moreover, this represents an unacceptable shifting of the burden from employer to employee, which is contrary to both the letter and the intent of the Occupational Safety and Health Act of 1970.

Although an HPD is useful when the listener has no control over the noise level and when engineering controls cannot be applied, the committee concluded that engineering controls of noise in the workplace should be the primary method of protecting workers from hazardous noise exposure. Accordingly, the committee recommends the following actions by U.S. government agencies, engineering and trade societies, and other stakeholders to promote the development and use of engineering controls.

Recommendation 4-3: The U.S. Department of Labor should revoke the Occupational Safety and Health Administration (OSHA) "100-dB Directive" of 1983, which effectively raised the action point for engineering control of noise from 90 to 100 dB by allowing the substitution of hearing protectors for noise control up to 100 dB and thereby devastated the market for quiet machinery and equipment. At the same time, OSHA should reconfirm that engineering controls should be the primary means of controlling noise in the workplace.

Recommendation 4-4: The National Institute for Occupational Safety and Health and the U.S. Department of Labor should develop and distribute widely an electronic database of noise control problems, solutions, and materials—taking into account the many handbooks and articles devoted to industrial noise control.

Recommendation 4-5: Engineering societies and trade organizations should develop guidelines for defining the relationship between noise emission specifications in terms of sound power level and/or *emission* sound pressure level and noise *immission* levels in industrial situations. They should provide a primer for buyers and sellers of machinery and equipment that includes descriptions of how noise propagates in rooms; how to determine noise from a large number of machines; standards available to manufacturers and others for measuring noise emissions; and case histories of noise levels measured in *in situ* environments.

Recommendation 4-6: Government agencies should be instructed by a presidential directive or in congressional report language to show leadership in promoting "buy quiet" activities by developing and implementing programs for the purchase of low-noise products, as required by 42 USC 65, Section 4914. American industry should adopt "buy quiet" programs that require noise emission specifications on all new equipment and "declared values" in purchase specifications.

REFERENCES

AE (Access Economics Pty Ltd). 2006. Listen Hear! The Economic Impact and Cost of Hearing Loss in Australia. Available online at *http://www.audiology.asn.au/pdf/ListenHearFinal.pdf*.

Altkorn, R., S. Milkovich, and G. Rider. 2005. Measurement of Noise from Toys. Presented at NOISE-CON 05, The 2005 National Conference on Noise Control Engineering, Minneapolis, MN. Available online at *http://www.bookmasters.com/marktplc/00726.htm*.

AMT (Association for Manufacturing Technology). 2006. ANSI Technical Report for Machines: Sound Level Measurement Guidelines. McLean, VA: AMT.

ANSI (American National Standards Institute). 1974. Method for the Measurement of Real-Ear Protection of Hearing Protectors and Physical Attenuation of Earmuffs. ANSI S3.19-1974. New York: ANSI.

ANSI. 2002. Methods for Measuring the Real-Ear Attenuation of Hearing Protectors. ANSI S12.6-1997 (R2002). New York: ANSI.

ANSI. 2004. Microphone-in-Real-Ear and Acoustic Test Fixture Methods for the Measurement of Insertion Loss of Circumaural Hearing Protection Devices. ANSI S12.42-1995(R2004). New York: ANSI.

ANSI. 2006a. American National Standard Determination of Occupational Noise Exposure and Estimation of Noise-Induced Hearing Loss. S3.44. ANSI S3.44-1996(R 2006). New York: ANSI.

ANSI. 2006b. American National Standard Methods for the Measurement of Impulse Noise. ANSI S12.7-1986 (R2006). Melville, NY. Acoustical Society of America. Available online at *http://asastore.aip.org/shop.do?cID=10*.

ANSI. 2007a. American National Standards for Acoustics—Portable Electric Power Tools, Stationary and Fixed Electric Power Tools, and Gardening Appliances—Measurement of Sound Emitted. ANSI/ASA S12.15-1992 (R 2007). New York: ANSI.

ANSI. 2007b. American National Standard Guidelines for the Specification of Noise of New Machinery. ANSI/ASA S12.16-1992 (R 2007). New York: ANSI.

ASHA (American Speech-Language-Hearing Association). 2009. Noise and Hearing Loss—Noise Is Difficult to Define! Available online at *http://www.asha.org/public/hearing/disorders/noise.htm*).

ASTM (American Society for Testing and Materials). 2008. Standard ASTM F963-08E1, Standard Consumer Specifications for Toy Safety. West Conshohocken, PA: ASTM.

Babisch, W. 2008. Road traffic noise and cardiovascular risk. Noise and Health 10(38):27–33. Available online at *http://www.noiseandhealth.org/article.asp?issn=1463-1741;year=2008;volume=10;issue=38;spage=27;epage=33;aulast=Babisch;type=0*.

Baldwin, D. 2004. The road to yesterday: J. A. R. Elliott, premier trapshooter. Trap and Field (June-July):94–104. Available online at *http://www.traphof.org/roadtoyesterday/elliott-JAR.htm*.

Barr, T. 1896. Manual of Diseases of the Ear. Glasgow, Scotland: James Maclehose and Sons.

Berger, E.H. 2003a. Noise control and hearing conservation: Why do it? Pp. 2–17 in The Noise Manual, Revised 5th Ed., edited by E.H. Berger, L.H. Royster, J.D. Royster, D.P. Driscoll, and M. Layne. Fairfax, VA: American Industrial Hygiene Association.

Berger, E.H. 2003b. Hearing protection devices. Pp. 379–454 in The Noise Manual, revised 5th ed., edited by E.H. Berger, L.H. Royster, J.D. Royster, D.P. Driscoll, and M. Layne. Fairfax, VA: American Industrial Hygiene Association.

Berger, E.H., and J.G. Casali. 1997. Hearing protection devices. Pp. 967–981 in Encyclopedia of Acoustics, edited by M. Crocker. New York: John Wiley.

Berger, E.H., J.R. Franks, A. Behar, J.G. Casali, C. Dixon-Ernst, R.W. Kieper, C.J. Merry, B.T. Mozo, C.W. Nixon, D. Ohlin, J.D. Royster, and L.H. Royster. 1998. Development of a new standard laboratory protocol for estimating the field attenuation of hearing protection devices, Part III: The validity of using subject-fit data. Journal of the Acoustical Society of America 103(2):665–672.

Bobeczko, M.S. 1978. Noise Control Solutions for Can Manufacturing Plants. Pp. 413–418 in Proceedings of INTER-NOISE 78, The 1978 International Congress on Noise Control Engineering, San Francisco, CA, May. Available online at *http://www.noisenewsinternational.net/docs/bobeczko-1978.pdf*.

Bobeczko, M., and D. Landwith. 1982. The Evolution of Can Plant Noise Control Technology. Proceedings of INTER-NOISE 82, The 1982 International Congress on Noise Control Engineering, San Francisco, CA, May. Available online at *http://www.noisenewsinternational.net/docs/bobeczko-1982.pdf*.

Bruce, R.D. 2007. Engineering controls for reducing workplace noise. The Bridge 37(3): 33–39.

Bruce, R.D. 2009. A New Approach to Noise Control in the Workplace. Proceedings of INTER-NOISE 09, The 2009 International Congress and Exposition on Noise Control Engineering, Ottawa, Canada, August 23–26. Available online at *http://www.bookmasters.com/marktplc/00726.htm*.

Bruce, R.D., and E. Wood. 2002. National Noise Policy on Occupational Health in the U.S.A. National Industrial-Noise-Control Partnerships. Proceedings of INTER-NOISE 2002, The 2002 International Congress and Exposition on Noise Control Engineering, Dearborn, MI, August 19–21. Available online at *http://www.bookmasters.com/marktplc/00726.htm*.

Brüel, P.V. 1977. Do we measure damaging noise correctly? Noise Control Engineering Journal 8(2):52–60.

Casali, J.G. 1990. Listening closely to the noise of monster trucks. *Roanoke Times and World News*, March 5, p. A11.

Casali, J.G. 2006. Sound and noise. Pp. 612–642 in Handbook of Human Factors, 3rd ed., edited by G. Salvendy. New York: John Wiley.

Casali, J.G., and S. Gerges. 2006. Protection and enhancement of hearing in noise. Pp. 195–240 in Reviews of Human Factors and Ergonomics, Volume 2, edited by R.C. Williges. Santa Monica, CA: Human Factors and Ergonomics Society.

Casali, J.G., and G.S. Robinson. 2003. Augmented Hearing Protection Devices: Active Noise Reduction, Level-Dependent, Sound Transmission, Uniform Attenuation, and Adjustable Devices—Technology Overview and Performance Testing Issues. EPA Docket OAR-2003-0024. Washington, DC: U.S. Environmental Protection Agency.

Casali, J.G., W.A. Ahroon, and J.A. Lancaster. 2009. A field investigation of hearing protection and hearing enhancement in one device for soldiers whose ears and lives depend upon it. Noise and Health 11(42):60–69.

Casali, J.G., and K. Alali. 2009. Auditory alarm localization (or not): Effects of augmented passive and alarm spectral content. Spectrum 26(Suppl. 1), and CD of the 34th Annual National Hearing Conservation Association Conference, Atlanta, GA, February 12–14.

Casali, J.G., and J.A. Lancaster. 2008. Quantification and Solutions to Impediments to Speech Communication and Signal Detection in the Construction Industry. Technical Report 200803. Department of Industrial and Systems Engineering, Virginia Polytechnic Institute and State University, Blacksburg, April 15.

Cooper, B.A. 2009. Development and Implementation of Policy-Compliant Site-Specific Buy-Quiet Programs at NASA. Proceedings of INTER-NOISE 09, The 2009 International Congress on Noise Control Engineering, Ottawa, Canada, August 23–26. Available online at *http://www.bookmasters.com/marktplc/00726.htm*.

Cooper, B.A., and D.A. Nelson. 1996. A "Buy Quiet" Program for NASA Lewis Research Center: Specifying Low Equipment Noise Emission Levels. Pp. 465–470 in Proceedings of NOISE-CON 97, The 1997 National Conference on Noise Control Engineering, University Park, PA, June 15–17, 1997. Available online at *http://www.bookmasters.com/marktplc/00726.htm*.

Cooper, B.A., D.W. Hange, and J.J. Mikulic. 1999. Engineered Solutions to Reduce Occupational Noise Exposure at the NASA Glenn Research Center: A Five-Year Progress Summary (1994–1999). Pp. 1069–1074 in Proceedings of INTER-NOISE 99, The 1999 International Congress and Exposition on Noise Control Engineering. Fort Lauderdale, FL, December 6–8. Available online at *http://www.bookmasters.com/marktplc/00726.htm*.

CPSC (Consumer Product Safety Commission). 2001a. Consumer Product Safety Alert: "Don't Let Children Put Caps for Toy Guns in Their Pockets." Available online at *http://www.cpsc.gov/CPSCPUB/PUBS/5009.pdf*.

CPSC. 2001b. Text of the Regulation on Noise from Caps. 16 CFR 1500.121. Available online at *http://www.cpsc.gov/BUSINFO/label.pdf*.

Davis, R.J., W. Qiu, and R.P. Hamernik. 2009. The role of the kurtosis statistic in evaluating complex noise exposures for the protection of hearing. Ear & Hearing 30(5): 28–634.

DEFRA (Department for Environment Food and Rural Affairs). 2009. Estimating Dose-Response Relationships between Noise Exposure and Human Health in the UK. BEL Technical Report 2009-02. Available online at *http://www.defra.gov.uk/environment/noise/igcb/healthreport.htm*.

Department of the Air Force. 1948. Precautionary Measures Against Noise Hazards. AFR 160-3. Washington, DC: U.S. Air Force.

EC (European Commission). 2003. Directive 2003/10/EC of the European Parliament and of the Council of 6 February 2003, on the Minimum Health and Safety Requirements Regarding the Exposure of Workers to the Risks Arising from Physical Agents (Noise). Available online at *http://eur-lex.europa.eu/LexUriServ/LexUriServ.do?uri=OJ:L:2003:042:0038:0044:EN:PDF*.

ECMA International. 1995. Technical Report TR27: Method for the Prediction of Installation Noise Levels, 2nd Edition. Geneva, Switzerland: ECMA International. Available online at *http://www.ecma-international.org/publications/techreports/E-TR-027.htm*.

Eldred, K.M., W.J. Gannon, and H.E. von Gierke. 1955. Criteria for Short Time Exposure of Personnel to High Intensity Jet Aircraft Noise. U.S. Air Force, WADC Technical Note 55-355, Wright-Patterson AFB, OH.

EPA (U.S. Environmental Protection Agency). 1974. Information on Levels of Environmental Noise Requisite to Protect Public Health and Welfare with an Adequate Margin of Safety. Document Number 550/9-74-004. Available online at *http://www.nonoise.org/library/levels74/levels74.htm*.

EPA. 1979. Noise labeling requirements for hearing protectors. 40 CFR 211. Federal Register 44(190):56130–56147.

EPA. 1981. Noise in America: The Extent of the Noise Problem. Report No. 550/9-81-101. Washington, DC: U.S. Environmental Protection Agency.

EPA. 2009. Workshop on Hearing Protection Devices, Washington, DC, March 27–28. EPA Docket OAR-2003-0024. Available online at *www.regulations.gov*.

Federal Railroad Administration. Locomotive Cab Noise. 49 CFR 229.121. Available online at *http://frwebgate.access.gpo.gov/cgi-bin/get-cfr.cgi?TITLE=49&PART=229&SECTION=121&YEAR=2000&TYPE=PDF*.

Federal Register. 2002. Advanced Notice of Proposed Rule Making, Hearing Conservation Program for Construction Workers. Federal Register 67:50610–50818.

Fligor. B.J. 2009. Risk for noise-induced hearing loss from use of portable media players: A summary of evidence through 2008. Perspectives on Audiology 5(1):10–20.

Fligor, B.J., and L.C. Cox. 2004. Output levels of commercially available portable compact disc players and the potential risk to hearing. Ear and Hearing 25:513–527.

Fosboke, J. 1831. Practical observations on the pathology and treatment of deafness, No. II. Lancet VI:645–648.

Gelfand, S.A. 2001. Essentials of Audiology, 2nd edition. New York: Thieme.

Gerges, S., and J.G. Casali. 2007. Ear protectors. Pp. 364–376 in Handbook of Noise and Vibration Control, edited by M. Crocker. New York: John Wiley.

Hayden II, C.S., and E. Zechmann. 2009. NIOSH Power Tools Database. Available online at *http://wwwn.cdc.gov/niosh-sound-vibration/*.

Hellweg, R.D., Jr., E.K. Dunens, and T. Baird. 2006. Personal computer, printer, and portable equipment noise in classrooms. Noise/News International 14(3):96–100. Available online at *http://noisenewsinternational.net/archives_idx.htm*.

Hoermann, H., G. Lazarus-Mainka, M. Schubeius, and H. Lazarus. 1984. The effect of noise and the wearing of ear protectors on verbal communication. Noise Control Engineering Journal 23:69–77.

Hohmann, B.W., V. Mercier, and I. Felchin. 1999. Effects on hearing caused by personal cassette players, concerts, and discotheques and conclusions for hearing conservation in Switzerland. Noise Control Engineering Journal 47:163–165. Available online at *http://www.incedl.org/journals/doc/INCEDL-home/jrnls/*.

Holt, E.E. 1882. Boiler-maker's deafness and hearing in noise. Transactions of the American Otological Society 3:34–44.

Howell, K., and A.M. Martin. 1975. An investigation of the effects of hearing protectors on vocal communication in noise. Journal of Sound and Vibration 41:181–196.

IEC (International Electrotechnical Commission). 2010. IEC 60704 Series for Determination of Noise from Household Appliances. Available online at *http://webstore.iec.ch/webstore/webstore.nsf/mysearchajax?Openform?key=60704&sorting=&start=1*.

IEEE (Institute of Electrical and Electronics Engineers). 1969. Technical Committee Report on Recommended Practices for Burst Measurements in the Time Domain. IEEE 265. New York: IEEE.

I-INCE (International Institute of Noise Control Engineering). 1997. Final Report: Technical Assessment of Upper Limits on Noise in the Workplace. Noise/News International 5:203–216. Available online at *http://www.noisenewsinternational.net/archives_idx.htm*.

IOM (Institute of Medicine). 2005. Noise and Military Service: Implications for Hearing Loss and Tinnitus, edited by L.E. Humes, L.M. Joellenbeck, and J.S. Durch. Washington, DC: National Academies Press.

ISO (International Organization for Standardization). 1995a. ISO 11200, 1995, Acoustics—Noise Emitted by Machinery and Equipment—Guidelines for the Use of Basic Standards for the Determination of Emission Sound Pressure Levels at a Work Station and at Other Specified Positions. Geneva Switzerland: ISO. Available online at *http://www.iso.org/iso/iso_catalogue/catalogue_tc/catalogue_tc_browse.htm?commid=48474*.

ISO. 1995b. ISO TR 11688-1:1995, Acoustics—Recommended Practice for the Design of Low-Noise Machinery and Equipment—Part 1: Planning. Geneva, Switzerland: ISO.

ISO. 1998. ISO/TR 11688-2:1998, Acoustics—Recommended Practice for the Design of Low-Noise Machinery and Equipment—Part 2: Introduction to the Physics of Low-Noise Design. Geneva, Switzerland: ISO.

ISO. 1999. ISO 1999:1990, Acoustics—Determination of Occupational Noise Exposure and Estimation of Noise-Induced Hearing Impairment. Geneva, Switzerland: ISO.

ISVR (Institute of Sound and Vibration Research). 1997. Noise from Toys and Its Effect on Hearing. Report 5403 R02. Southampton, U.K.: ISVR. Available online at *http://www.isvr.co.uk/leisure/toys.pdf*.

LHH (League for the Hard of Hearing). 2009. Children and Toys: Noisy Toys. Available online at *http://www.lhh.org/noise/children/toys.html*.

Liu, Z. 2003. Acoustic Liner and a Fluid Pressurizing Device and Method Utilizing Same. U.S. Patent No. 6,550,574 B2, April 2.

McKinley, R. 2001. Future Aircraft Carrier Noise. Proceedings of the International Military Noise Conference, Baltimore, MD, April 24–26.

Morata, T.C., C.L. Themann, R.F. Randolph, B.L. Verbsky, D.C. Byrne, and E.R. Reeves. 2005. Working in noise with a hearing loss: Perceptions from workers, supervisors, and hearing conservation program managers. Ear and Hearing 26(6):529–545.

MSHA (Mine Safety and Health Administration). 1999. Health Standards for Occupational Noise Exposure; Final Rule. 30 CFR Part 62, 64. U.S. Department of Labor. Available online at *http://www.msha.gov/30cfr/62.0.htm*.

NASA (National Aeronautics and Space Administration). 2007. Environmental Health. Chapter 4.9 in Occupational Health Program Procedures. Available online at *http://nodis3.gsfc.nasa.gov/displayDir.cfm?Internal_ID=N_PR_1800_001B_&page_name=Chapter4*.

NIOSH (National Institute for Occupational Safety and Health). 1975. NIOSH Compendium of Noise Control Materials. Publication No. 75-165. June. Atlanta, GA: NIOSH.

NIOSH. 1978. Industrial Noise Control Manual. Publication No. 79-117. Atlanta, GA: NIOSH.

NIOSH. 1996. Preventing Occupational Hearing Loss—A Practical Guide, edited by J.R. Franks, M.R. Stephenson, and C.J. Mercy. Atlanta, Ga.: NIOSH. Available online at *http://www.cdc.gov/niosh/docs/96-110/pdfs/96-110.pdf*.

NIOSH. 1998a. Basis for the exposure standard. Chapter 3 in Criteria for a Recommended Standard: Occupational Noise Exposure. Publication 98-126. Available online at *http://www.cdc.gov/niosh/docs/98-126/chap3.html*.

NIOSH. 1998b. Criteria for a Recommended Standard: Occupational Noise Exposure. Publication 98-126. Available online at *http://www.cdc.gov/niosh/docs/98-126/*.

NIOSH. 2005. NIOSH/NHCA Best-Practices Workshop on IMPULSIVE NOISE. Noise Control Engineering Journal 53(2):53–60.

Nondahl, D.M., K.J. Cruickshanks, T.L. Wiley, R. Klein, B.E.K. Klein, and T.S. Tweed. 2000. Recreational firearms use and hearing loss. Archives of Family Medicine 9(4):352–357.

OSHA (Occupational Safety and Health Administration). 1971a. Hearing Protection (Construction Industry). 29 CFR 1926.101. Federal Register.

OSHA. 1971b. Occupational Noise Exposure (Construction Industry). 29 CFR 1926.52. Federal Register.

OSHA. 1971c. Occupational Noise Exposure (General Industry). 29 CFR 1910.95. Federal Register.

OSHA 1981. U.S. Department of Labor: Occupational Noise Exposure; Hearing Conservation Amendment; Final Rule. Federal Register (46)4078–4179.

OSHA. 2002. Occupational Noise Exposure. 29 CFR 1910.95. Revised. Available online at *http://edocket.access.gpo.gov/cfr_2002/julqtr/29cfr1910.95.htm*.

OSHA. 2009. Field Operations Manual, Directive #CPL02-00-148 Effective March 3, 2009. Pp. 4-40 and 4-41. Available online at *http://www.osha.gov/OshDoc/Directive_pdf/cpl_02_00_148.pdf*.

Park, M.Y., and J.G. Casali. 1991. A controlled investigation of in-field attenuation performance of selected insert, earmuff, and canal cap hearing protectors. Human Factors 33(6):693–714.

Passchier-Vermeer, W. 1999. Pop music through headphones and hearing loss. Noise Control Engineering Journal 47:182–186. Available online at *http://asastore.aip.org/shop.do?cID=10*.

PIRG (U.S. Public Interest Research Group Education Fund). 2005. Trouble in Toyland: 20th Annual Toy Safety Survey. (See "Dangerously Loud Toys" and "Excessively Loud Toys." Available online at *http://toysafety.net/2005/troubleintoyland2005.pdf*.

Porter, M.E., and C. van der Linde. 2005. Conception of the environment-competitiveness relationship. Pp. 92–114 in Economics and the Environment, Selected Readings, fifth edition, edited by R.N. Stavins. New York: W.W. Norton.

Price, G.R. 2007. Validation of the auditory hazard assessment algorithm for the human with impulse noise data. Journal of the Acoustical Society of America 122(5):2786–2802.

Ramazzini, B. 1964. Diseases of workers (De morbis artificum, 1713). Trans. Wilmer Cave Wright. New York: Hafner Publishing. Orig. pub. 1713.

Robinson, G.S., and J.G. Casali. 2003. Speech Communications and Signal Detection in Noise. Pp. 567–600 in The Noise Manual, Revised 5th Ed., edited by E.H. Berger, L.H. Royster, J.D. Royster, D.P. Driscoll, and M. Layne. Fairfax, VA: American Industrial Hygiene Association.

Sataloff, R.T. 1993. Occupational Hearing Loss, 2nd edition. New York: Marcel Dekker.

SCENIHR (Scientific Committee on Emerging and Newly Identified Health Risks). 2008. Potential Health Risks of Exposure to Noise from Personal Music Players and Mobile Phones Including a Music Playing Function. European Commission, Brussels, Belgium.

Schwela, D. 2006. Noise of consumer products: Consequences for environmental health. Noise/News International 14:102–111. Available online at *http://www.noisenewsinternational.net/archives_idx.htm*.

Shampan, J., and R. Ginnold. 1982. The status of workers' compensation programs for occupational hearing impairment. In Forensic Audiology, edited by M.B. Kramer and J.M. Ambruster. Baltimore, MD: University Park Press.

Statskontoret. 2004. Statskontoret Technical Standard 26:6: Acoustical Noise Emission of Information Technology Equipment. Swedish Agency for Public Management. Available online at *http://www.statskontoret.se/upload/2619/TN26-6.pdf*.

Suter, A.H. 1990. Popular misconceptions about occupational noise exposure. Proceedings NOISE-CON 90, The 1990 National Conference on Noise Control Engineering, Austin, TX. Available online at *http://www.noisenewsinternational.net/docs/suter-1990.pdf*.

Suter, A.H. 1992. Communication and job performance in noise: A review. ASHA Monograph No. 28. American Speech-Language-Hearing Association, Rockville, MD.

Suter, A.H. 1993. The relationship of the exchange rate to noise-induced hearing loss. Noise/News International 1:131–151. *http://www.noisenewsinternational.net/docs/suter-1993.pdf*.

Suter, A.H. 1994. Current Standards for Occupational Exposure to Noise. Presentation at the 5th International Symposium on Effects of Noise on Hearing, Gothenburg, Sweden, May.

Suter, A.H. 2006. Position paper for the International Safety Equipment Association (ISEA). Available from ISEA, 1901 N. Moore St., Arlington, VA 22209.

USARL (U.S. Army Research Laboratory). 2010. Auditory Hazard Assessment Algorithm for Humans (AHAAH). Available online at *http://www.arl.army.mil/www/default.cfm?Action=31&Page=343*.

WHO (World Health Organization). 1997. Prevention of Noise-Induced Hearing. Report of an informal consultation held at the World Health Organization, Geneva, Switzerland, October 28–30. No. 3 in the series "Strategies for Prevention of Deafness and Hearing Impairment." Available online at *http://www.who.int/pbd/deafness/en/noise.pdf*.

WHO. 1999. Guideline Values. Chapter 4 in Guidelines for Community Noise, edited by B. Bergland, T. Lindvall, and D. Schwela. Available online at *http://www.who.int/docstore/peh/noise/guidelines2.html*; *http://www.who.int/docstore/peh/noise/Comnoise-4.pdf*.

WHO. 2007. Prevention of Noise Induced Hearing Loss, Report of an Informal Consultation. Geneva, Switzerland: WHO. Available online at *http://www.who.int/entity/pbd/deafness/en/noise.pdf*.

Zhao, Y-M., W. Qiu, Z. Lin, S-S. Chen, X-R. Cheng, R.I. Davis, and R.P. Hamernik. 2010. Application of the kurtosis statistic to the evaluation of the risk of hearing loss in workers exposed to high-level complex noise. Ear and Hearing 31(4):527–532.

5

Technology

Although noise can be generated by turbulence in high-speed flows, most noise is generated by mechanical motion caused by forces acting on structures. The motion can be very complex—for example, in the case of a panel on a machine. Consequently, the coupling between the moving structure and air, required to generate noise, depends on details of the motion as well as frequency. Generally, low-frequency vibration is less efficient in generating sound than high-frequency vibration. Sound reaches the ear by propagation from the source to the receiver and can be complicated by reflections from nearby surfaces as well as atmospheric conditions outdoors. Some motion is unavoidable; for example, fan blades must move and vehicle tires must rotate. In many cases the function of the machine is unrelated to the noise generated. An example is the mechanical suspensions that attach airplane engines to an airplane but also allow the transmission of vibrations to the fuselage. This transmission into the fuselage and subsequent radiation of sound can be (and is) minimized by good design—which can also save money by reducing wear and fatigue.

This chapter is concerned with new technologies in materials and systems to reduce noise, the modeling and analytical tools used to design products for reduced noise, and experimental methods of gathering and interpreting data to test and determine how much noise is generated by different product designs. It will be immediately obvious that there are enormous disparities among programs, facilities, and resources for addressing noises of different types. For example, although engineering tools may be available for reducing aircraft noise and highway noise, the former has been deemed a national priority, while the latter has received less attention. Resources allocated for noise reduction are not always commensurate with noise exposures and impacts.

Many tools for designing and developing quieter products have become available in the past few decades, driven largely by increases in computational power and reductions in computational costs. Even so, access to new tools is as uneven as the allocation of resources; corporate budgets for capital equipment are generally tight and there is competition between departments for available funds. Furthermore, organizations that are doing only routine testing of products according to national and international standards find expensive new tools hard to justify. Thus, even though noise mechanisms in aircraft, automobiles, rapid transit and trains, consumer products, and industrial machinery are fundamentally similar, the availability and application of tools for addressing them are not. The question is whether ways can be found to give industry and academia access to these tools for the benefit of manufacturers, workers, and the public.

AEROSPACE AND AEROACOUSTICS

SOURCES OF AIRCRAFT NOISE

Noise from aircraft includes both noise from airplanes and noise from helicopters. At commercial airports, airplanes are the major noise source and will be emphasized here because of the widespread annoyance issues that have affected the quality of life for many persons. Noise from helicopters is also an important issue and affects people living near heliports and in densely populated areas where helicopter flights are not uncommon. The Federal Aviation Administration was asked to prepare a report to Congress on nonmilitary helicopter noise (FAA, 2004). One important issue relates to noise metrics; the impulsive character of the noise requires that metrics in addition to the widely used day-night average sound level (DNL) be used to assess its effects on people.

The noise heard when an airplane flies overhead comes from many sources, but the main contributors are engine noise and airframe noise. Engine noise comes from the fan/propeller, compressor, turbine, combustor, and jet exhaust. Airframe noise is produced mostly by airflows around lifting and control surfaces, such as flaps and slats, and around landing gears.

The relative contribution of these sources depends on the engine and airframe designs and the operating conditions. For example, during takeoff, when the engines are at full thrust, jet noise is the largest contributor to the noise signature of an aircraft. At approach, when the engine is throttled back, noise comes more from the airframe. Other sources, such as the fan, are significant contributors during both takeoff and landing.

Typical noise sources for a fixed-wing aircraft are shown in Figure 5-1. The noise received by an observer depends on the sources and propagation effects. The noise sources for a propeller-driven aircraft are shown in Figure 5-2.

RESEARCH IN AEROACOUSTICS

The aeroacoustics community has made significant progress over the years in understanding and reducing aircraft noise. Figure 5-3 shows comparative contributions from different noise sources for 1960s and 1990s engines. The figure, which originally appeared in *Rolls-Royce* (2005a), shows that the development of the turbofan engine and reduction in noise from individual engine components resulted in smaller, more evenly matched noise contributions from engine sources (SBAC, 2009).

Over a period of 30 years, these improvements, coupled with advances in aircraft aerodynamics and weight technologies, have reduced aircraft noise by about 20 dB, which corresponds to a reduction in noise annoyance of about 75 percent (EU, 2007). The new Airbus A380, the largest commercial aircraft ever produced (average of 525 passengers), has takeoff and approach noise levels comparable to those of heavy road traffic, a lower noise level than in an underground train. The noise footprint of the A380 is about half that of older, large commercial aircraft (Rolls-Royce, 2005b).

Despite these impressive results, airport community noise continues to be a significant environmental problem, and research and development (R&D) continue in the United States and Europe to meet increasingly stringent noise requirements set by regulatory bodies, such as the Federal Aviation Administration (FAA), the International Civil Aviation Organization (ICAO), and individual airports (*Rolls-Royce*, 2005b). Over the years, the FAA and ICAO have required comparable reductions.

INFRASTRUCTURE AND PROGRAMS THAT SUPPORT RESEARCH AND APPLICATIONS

A Note on Test Facilities

Both the United States and Europe have first-class aeroacoustics test facilities. Anechoic flight simulation facilities, the most useful for testing both jet noise and airframe noise, are available on both sides of the Atlantic on a rental basis. In the United States, high-quality anechoic chambers for model-scale testing are available at the National Aeronautics and Space Administration (NASA) Langley and Glenn Research Centers, as well as at Boeing, General Electric, United Technologies, and some U.S. universities, such as Georgia Institute of Technology, which inherited Lockheed Georgia's aeroacoustics facilities. Rolls Royce in England has used the NASA Glenn jet noise acoustic chambers, and Boeing researchers have used facilities in England. NASA

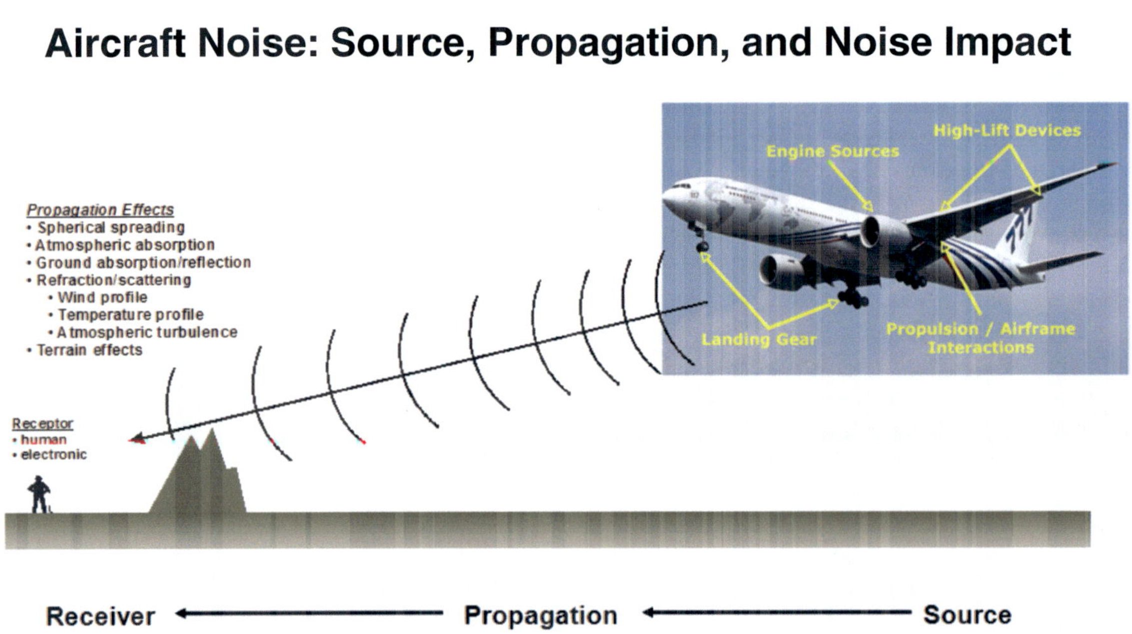

FIGURE 5-1 Breakdown of typical noise sources for fixed-wing aircraft. Source: Posey (2008).

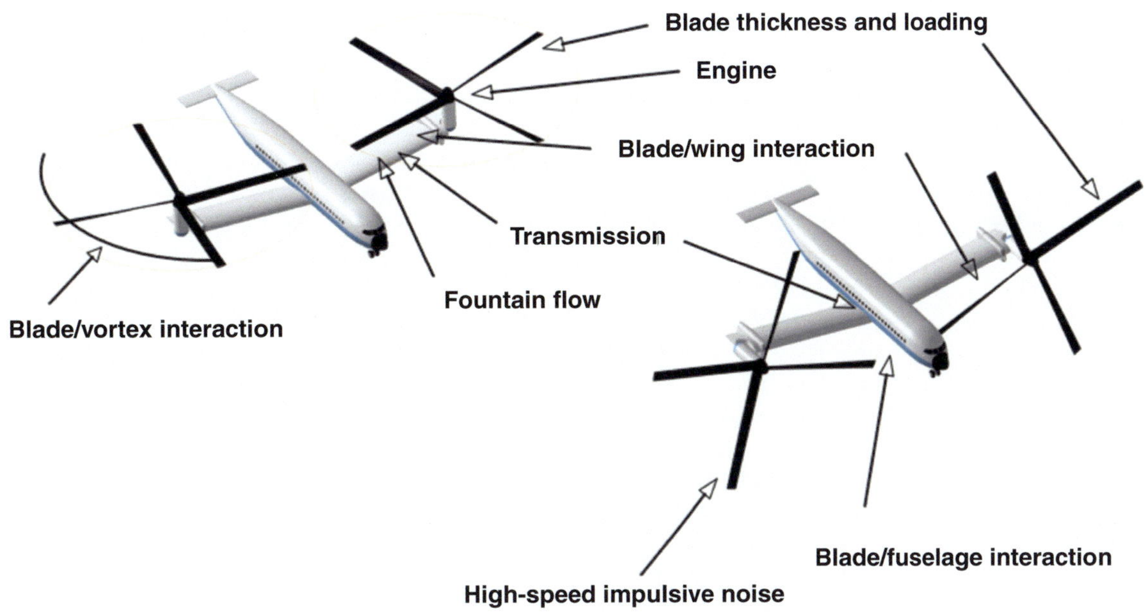

FIGURE 5-2 Breakdown of typical noise sources for a rotorcraft configuration. Source: Burley (2008).

Langley researchers have also used the Dutch anechoic wind tunnel to make helicopter noise measurements. Both the United States and Europe also have access to state-of-the-art flow measurement equipment (including particle imaging velocimetry) and modern phased microphone array systems. Most of these facilities have been described in great detail by Ahuja (1995).

U.S. NOISE REDUCTION PROGRAMS

The FAA and NASA have primary responsibility in the United States for R&D on aviation noise. The FAA focuses on the impacts of noise on communities, while NASA focuses on noise at its source—namely, aircraft engines and airframes. A recent congressionally requested report on aviation noise addresses (1) how well the FAA and NASA's R&D plans are aligned and (2) the likelihood that noise reduction goals will be met (FAA, 2008).

The FAA and NASA's R&D plans, aligned through partnerships and planning and coordinating mechanisms, include a wide range of projects for addressing aviation noise. The FAA sponsors aviation noise R&D in noise measurement, noise effects, interrelationships between noise and pollutant emissions, and flight procedures and technologies to mitigate

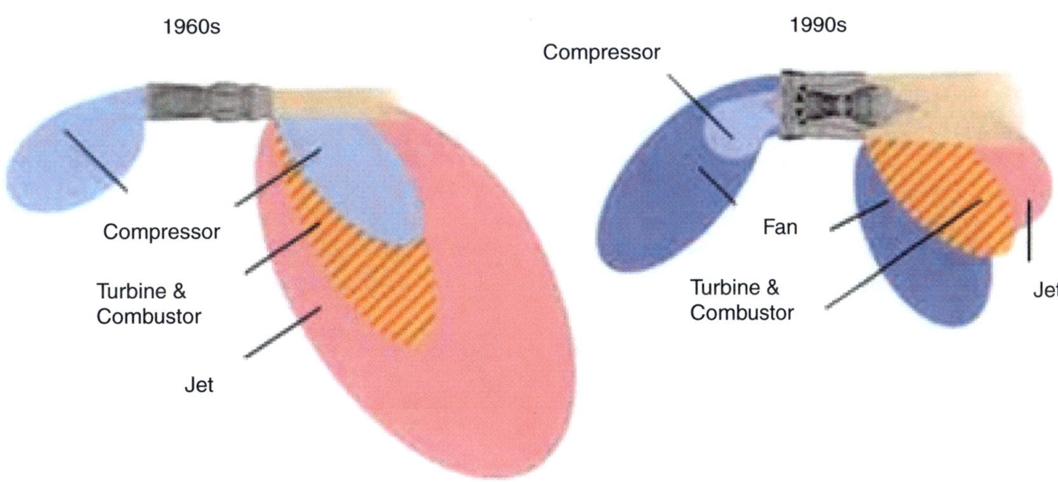

FIGURE 5-3 Noise sources for 1960s and 1990s jet engines. Source: Rolls-Royce (2005a).

the impact of noise on communities. Much of this R&D is funded through partnerships with universities, other federal agencies (including NASA), and industry. NASA's R&D can eventually lead to new technologies for substantially quieter aircraft. However, industry will have to integrate the research results into production-ready aircraft designs.

The FAA and NASA have worked with interagency planning and coordinating groups to establish objectives for the nation's aeronautical R&D and for specific research on the environmental impacts of next-generation aviation technologies. The strategic plans for the National Airspace System indicate how each agency's R&D will contribute to meeting noise reduction goals, which are designed to reduce public exposure to aviation noise primarily by reducing noise at its source (GAO, 2008).

In 1994, NASA initiated a seven-year program, the Advanced Subsonic Transport (AST) Noise Reduction Program, to develop technology to reduce jet transport noise by 10 dB relative to 1992 levels. Most of the goals of AST were met by 2001. However, because of an anticipated annual increase of 3 to 8 percent in passenger and cargo operations well into the twenty-first century and the slow introduction of new noise reduction technology into the fleet, the global impact of world aircraft noise is expected to remain essentially constant until 2020 (or perhaps 2030) and thereafter begin to increase. Therefore, NASA has begun planning with FAA, industry, universities, and environmental interest groups in the United States for a new noise reduction initiative.

One of the most important noise reduction technology programs in the United States is the so-called Quiet Technology Demonstrator (QTD1) Program, a partnership among Boeing, Rolls Royce, and American Airlines initiated in 2000 (Bartlett et al., 2004). A second phase, QTD2, a partnership among NASA, General Electric, Goodrich, and ANA, was begun in 2005. These programs have validated new, advanced noise reduction technologies, including nacelle inlet acoustical treatments and chevrons on engine exhaust ducts.

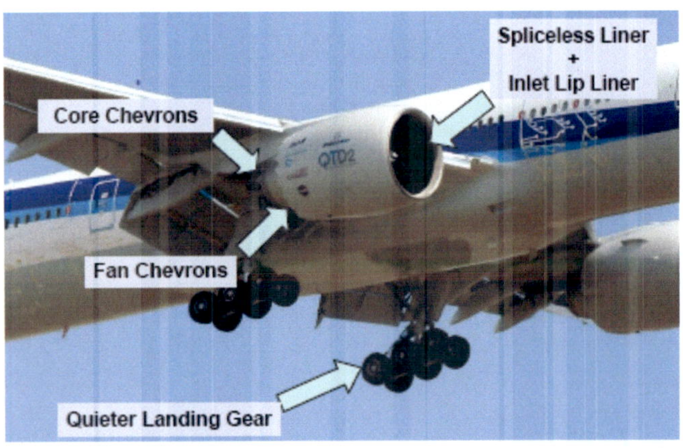

FIGURE 5-4 QTD2 noise reduction technologies. Source: Herkes (2006). Copyright Boeing. All rights reserved.

After rigorous testing, including measurements taken on the ground, in the passenger cabin, and on the airframe (Herkes, 2006), many noise reduction technologies, including nozzle chevrons, spliceless inlet linings, extended lining locations, and redesigned wing anti-icing systems (see Figure 5-4), as well as smooth fairings to reduce landing gear noise (see Figure 5-5), have been incorporated into existing airplanes and designs for future Boeing airplanes. Thus, Boeing's newer airplanes are significantly quieter for both passengers and airport communities (Herkes et al., 2006). A third phase, QTD3, is in the planning stages at Boeing.

Over the years the FAA has defined requirements for the reduction of aircraft noise *emissions* in terms of stages (1–4). The metric for describing the noise emissions is the effective perceived noise level in decibels (EPNdB), and well-defined microphone positions are used for the measurement. Note that this is quite different from the *immission* metric (DNL) used to describe the effects of aircraft noise on communities.

The goal of NASA's current Subsonic Fixed Wing Project is to reduce aircraft noise by 42 EPNdB cumulative below Stage 3 for conventional, small, tube-with-wing twin-jet aircraft, what NASA calls "$N + 1$ generation" aircraft, by 2012 to 2015 (Collier and Huff, 2007). An even more ambitious goal, set for the 2018 to 2020 period, is to reduce aircraft noise by 52 EPNdB cumulative below Stage 3 for $N + 2$ generation aircraft, which NASA envisions as an unconventional hybrid wing-body aircraft (see Figure 5-6). In addition to reducing noise, NASA expects dramatic improvements in the emission and performance of these aircraft.

EUROPEAN NOISE REDUCTION PROGRAMS

Driven by increasingly stringent noise requirements and strong competition from the United States, Europe has set

FIGURE 5-5 Toboggan landing gear fairings for reducing landing gear noise tested in QTD2. Source: Herkes (2006). Copyright Boeing. All rights reserved.

Corners of the Trade Space	N+1 Generation Conventional Tube w/Wing (2012-2015)	N+2 Generation Unconventional Hybrid Wing Body (2018-2020)
Noise (cumulative below Stage 3)	- 42 dB	-52 dB
LTO NOx Emissions (below CAEP 2)	-70%	-80%
Performance: Aircraft Fuel Burn (relative to B737/CFM56)	-33%	-50%*
Performance: Field Length (relative to B737/CFM56)	-33%	-50%

N+1 Conventional

N+2 Hybrid Wing/Body

Approach

- ***Enable Major Changes in Engine Cycle/Airframe Configurations***
- ***Reduce Uncertainty in Multi-Disciplinary Design and Analysis Tools and Processes***
- ***Develop/Test/ Analyze Advanced Multi-Discipline Based Concepts and Technologies***
- ***Conduct Discipline-based Foundational Research***

**** Estimate, still being evaluated***

FIGURE 5-6 Goals of the *N*+1 and *N*+2 generation aircraft. Source: Collier and Huff (2007).

very ambitious goals for reducing aircraft noise by 2020 (Collier and Huff, 2007). For example, as shown in Figure 5-7, the Advisory Council for Aeronautical Research in Europe (ACARE) has set a goal of a 50 percent reduction in noise annoyance (relative to their 2000 counterparts) for aircraft entering into service in 2020. This is equivalent to a 10-EPNdB reduction in the day-evening-night averaged sound level from fixed-wing airplanes. Along with the noise reduction, there must be a 50 percent reduction in specific fuel consumption (again relative to engines introduced into service in 2000).

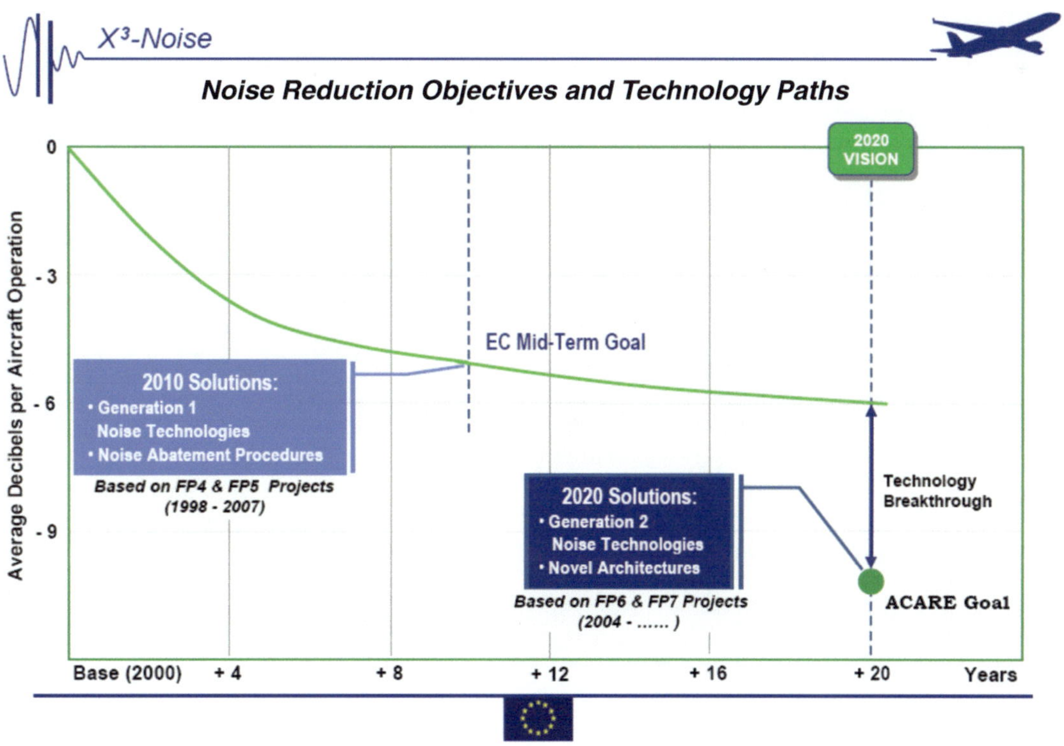

FIGURE 5-7 Noise reduction objectives and technology plans set by ACARE. Source: EU (2007).

Over the years a number of ambitious programs for reducing aircraft noise and, more recently, reducing aircraft emissions have been launched in Europe. Significant investments have been made under the so-called Framework Programs in which European Union (EU) industry and researchers in many countries work together to perform well-funded, coordinated R&D. A total of 20 aircraft noise R&D initiatives were launched in Europe between 1998 and 2006 with considerable participation by industry, small and medium enterprises (SMEs), research establishments, and government agencies (see Figure 5-8).

Four of the most noteworthy programs are (1) the Silence(R) Program, (2) the Silent Aircraft Initiative (SAX-40), (3) the EnVIronmenTALly (VITAL) Friendly Aero Engine, and (4) the EU Clean Sky Initiative. Each program is described in detail in a Society of British Aerospace Companies (SBAC) Aviation and Environment Briefing Paper (SBAC, 2009).[1] The summaries that follow are based on the SBAC briefing paper, a discussion at a Council of Academies of Engineering and Technical Sciences workshop in June 2008 (CAETS, 2008), and a presentation at the Workshop on Technologies for a Quieter America (Ahuja, 2008).[2]

The SILENCE(R) Program

SILENCE(R), the largest European aircraft noise research project ever undertaken, was a six-year program that began in 2001. Coordinated by Snecma, a French company, the €112 million program was a collaboration of 51 partners, including all major European airframe and engine manufacturers, major research institutes, and universities. The program addressed both engine noise (including jet noise, fan noise, compressor noise) (see Figure 5-9) and landing gear noise and airframe noise (see Figure 5-10).

Technologies for reducing jet noise included the ultra high bypass ratio fan; low-noise core and fan nozzles designed to improve the mixing of exhaust and bypass flows; internal and external exhaust plugs; and technologies to attenuate fan noise, including a zero-splice passive liner, active noise control technologies, and a negatively scarfed intake design to reflect fan noise away from the ground (see Figure 5-11). In flight tests the negatively scarfed fan was shown to reduce perceived noise by about 2.5 dB for an observer at a 60 degree angle to the engine (Rolls-Royce, 2005a).

Acoustical liners have traditionally been constructed from two or three pieces to facilitate manufacturing and maintenance. However, a continuous, zero-splice design greatly improved the absorption of fan noise, and the new technology is now being used in Rolls-Royce's Trent 900 engine on the Airbus A380. The change has resulted in a 4- to 7-dB reduction in fan-tone noise on takeoff and a 2-dB reduction in fan noise on approach (Coppinger, 2007). (A similar device was demonstrated by QTD2 in the United States.)

An active noise control system was also successfully demonstrated (SBAC, 2009). The system consisted of microphones mounted in the fan duct and actuators mounted on the stator vanes. The microphones measured fan noise and sent signals to the actuators, which generated "antinoise" (sound waves that were out of phase with the sound waves generated by the fan), canceling out the fan noise.

To reduce landing gear noise, some new low-noise designs for the nose and main landing gears were investigated. Ultimately, the noise was reduced by shielding the gears from each other and aligning them in the direction of the flow. Two aligned nose landing gears were demonstrated to be as much as 3 dB quieter than two independent gears (Coppinger, 2007).

Some of the noise technologies validated in SILENCE(R) are now in production engines. Others are either undergoing further work in R&D programs by individual manufacturers or have been carried over to other projects (e.g., VITAL, described below).

Silent Aircraft Initiative (SAX-40)

The Silent Aircraft Initiative (2006) (SAX-40) was a £2.3 million three-year research project run by Cambridge University and the Massachusetts Institute of Technology—with input from industry and government. SAX-40 culminated in a revolutionary concept design for a very quiet aircraft (see Figures 5-12 and 5-13). The concept design includes an airframe and engines designed specifically for a steep, low-speed climb and a low-noise approach that reduces both the amount of noise generated and the ground area of noise exposure. Some of the noise reduction technologies are listed below:

- a novel three-fan design that allows UHBR and hence lower jet noise
- low fan speeds that emit less noise
- extensive use of acoustic liners to absorb fan noise
- engines embedded in the fuselage, with intakes above the wings, to shield much of the engine noise from the ground
- variable area nozzles that allow engines to operate with low-speed, low-noise exhaust jets at takeoff and on ascent and then can be optimized for minimum fuel burn and carbon dioxide emissions at cruise
- elimination of flaps and slats
- low-noise fairing on the undercarriage

[1] SBAC is the Society of British Aerospace Companies. After a merger of three companies, it is now A|D|S, which is Aerospace|Defence|Security.

[2] Ahuja, K.K. 2008. Summary of the Aircraft Noise Day of the CAETS Workshop on Transportation Noise Sources in Europe, June 2–4, 2008, Southampton, United Kingdom. Presentation at the Workshop on New Technologies for a Quieter America, National Academy of Engineering, Washington, DC, June 11–12, 2008. Unpublished. A summary of the CAETS workshop is available online at *http://www.noisenewsinternational.net/docs/caets-2008.pdf*.

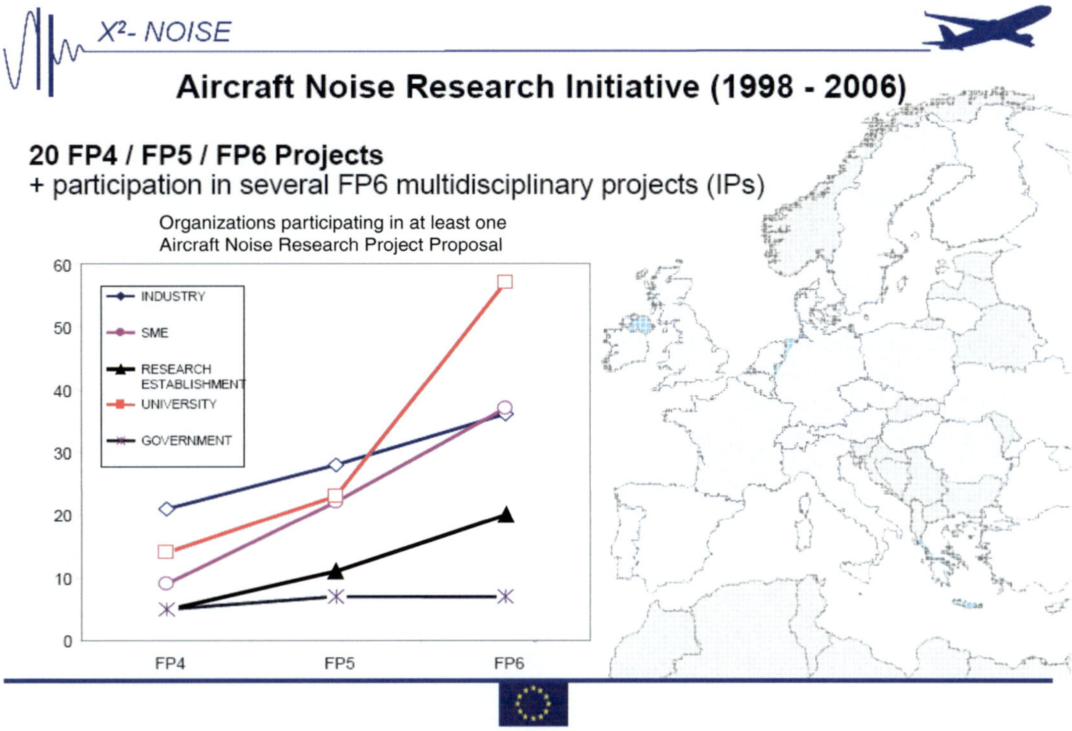

FIGURE 5-8 Aircraft noise research initiatives undertaken in Europe under the Framework Programs. Source: LEMA (2008).

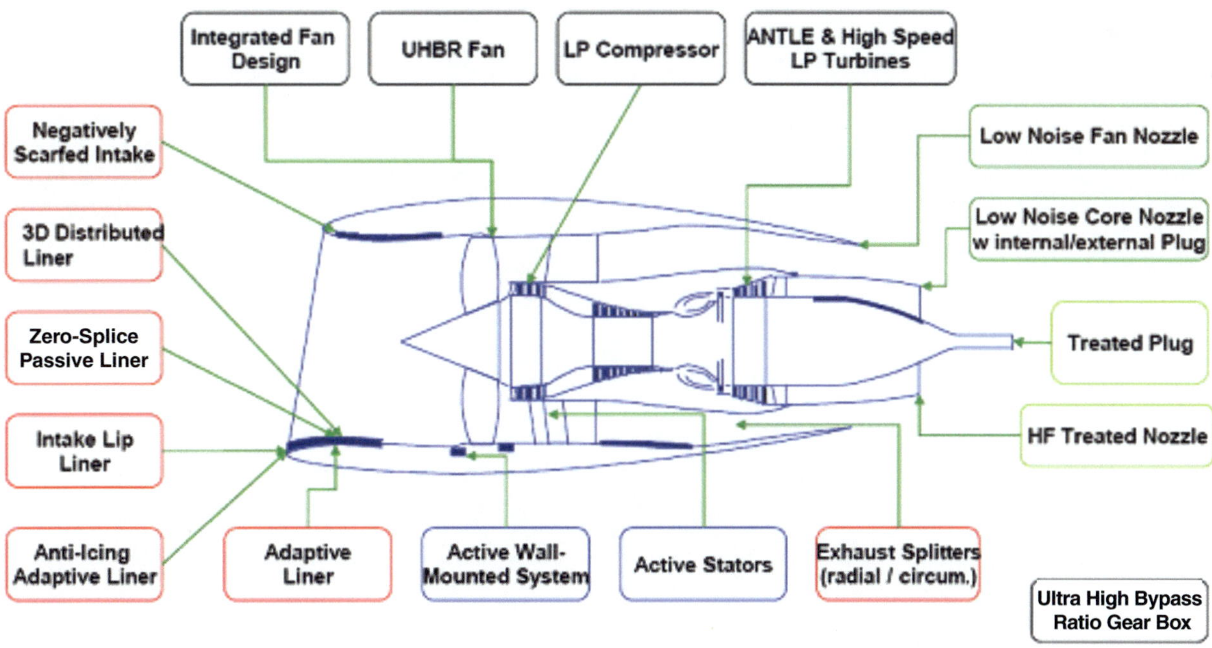

FIGURE 5-9 Engine/nacelle noise reduction technologies. UHBR = ultra high bypass ratio. Source: SBAC (2009).

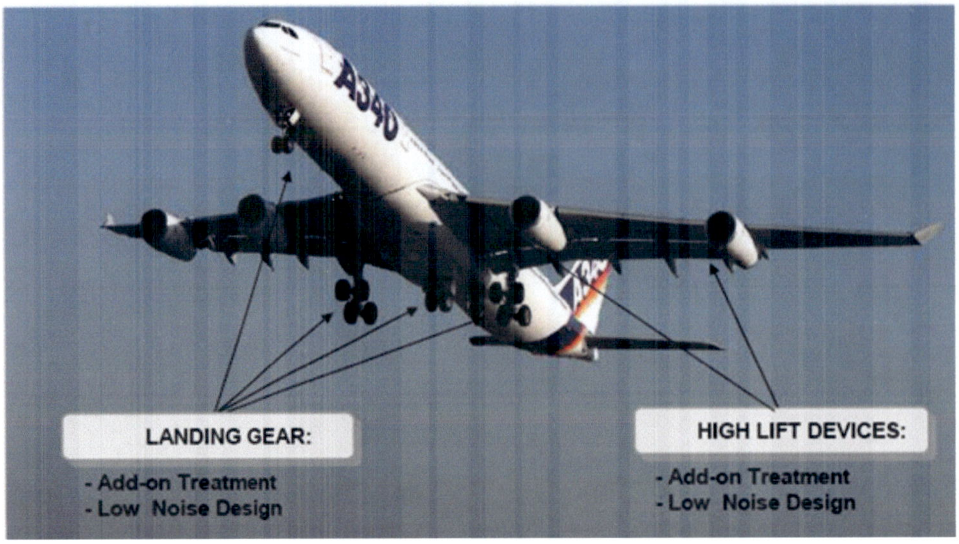

FIGURE 5-10 Aircraft noise reduction technologies. Source: SBAC (2009).

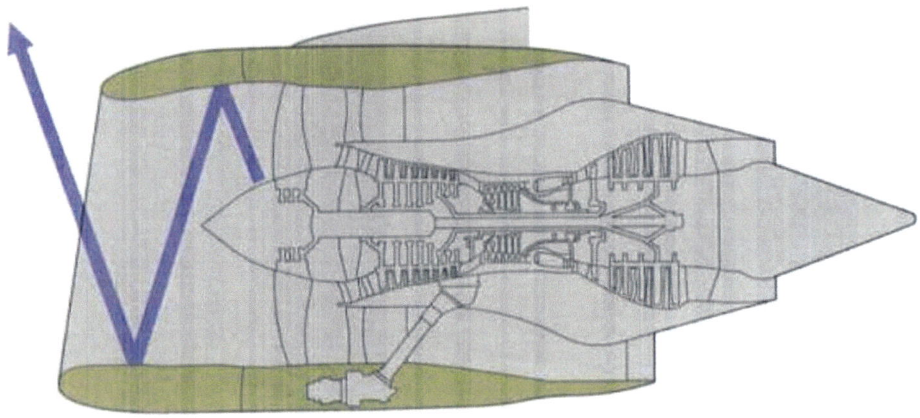

FIGURE 5-11 Negatively scarfed intake reflects fan noise away from the ground. Source: *The Jet Engine*, 2005. Reprinted with permission from Rolls Royce, 2005.

FIGURE 5-12 SAX-40 silent aircraft. Source: SBAC (2009). Copyright Silent Aircraft Initiative.

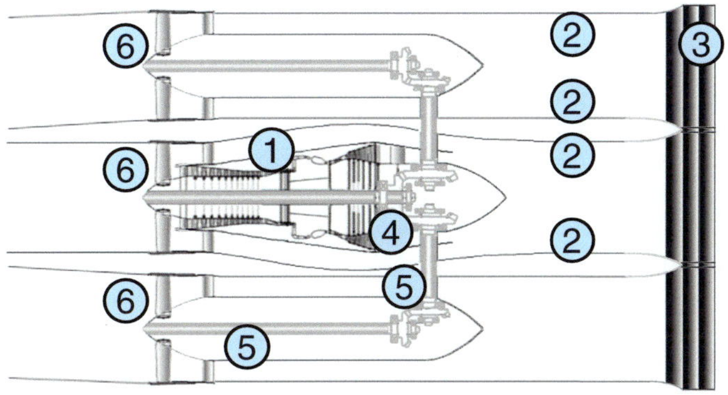

FIGURE 5-13 SAX-40 engine design. Source: SBAC (2009). Copyright Silent Aircraft Initiative.

SAX-40 is predicted to achieve a reduction in noise of 25 dB based on current standards and also a reduction in fuel consumption of about 25 percent for a typical flight. Although these results are impressive, the SAX-40 is a concept design only. Further work must be done to confirm the feasibility and develop and validate the novel technologies.

EnVIronmenTALly (VITAL) Friendly Aero Engine

The VITAL program is a four-year, €90 million, EU-wide R&D program that began in January 2005 and has 53 partners. The partners, major stakeholders in the European aviation industry, include all major engine manufacturers, Airbus, and equipment makers, as well as innovative small businesses, universities, and research centers.

The goal of this Snecma-led program is to integrate the results and benefits in noise reductions of the SILENCE(R) program with the emission reductions achieved in the Affordable Near Term Low Emissions and Component vaLidator for ENvironmentally friendly Aero Engine programs. By the end of VITAL, there should be a noise reduction of 8 dB per aircraft operation and an 18 percent reduction in carbon dioxide emissions, compared to engines in service prior to 2000.

To reduce engine noise, very high bypass ratio engines with novel low-noise, low-speed fan designs are being studied. One of these designs, the contrarotating turbo fan, is shown in Figure 5-14.

VITAL also plans to demonstrate a low-pressure compressor and turbine technologies designed for low noise and weight and compatible with the novel fan designs. An overview of the VITAL project was given by Bone (2009) at a European Engine Technology Workshop in Warsaw, Poland.

EU Clean Sky Initiative

The goal of the Clean Sky Initiative is to create a radically innovative air transport system with a reduced environmental impact based on less noise and gaseous emissions and better fuel economy. The specific objective is to reduce carbon dioxide emissions by about 40 percent, nitrogen oxide emissions by 60 percent, and noise by 50 percent in time for a major fleet renewal in 2015. The approach is to conduct an overall assessment of individual technologies at

FIGURE 5-14 Schematic drawing of contrarotating turbo fan design to be studied in VITAL. Source: EU (2007).

the fleet level to ensure the earliest possible deployment of research results.

The budget for Clean Sky is up to €800 million from the 7th Research Framework Program, which will be matched by funds from industry. The total budget could be as high as €1.6 billion. The research partners include all major aeronautical players in Europe, almost 100 organizations that are active in aeronautical R&D and many SMEs, research centers, and universities. The technical and geographical scope of a typical team is shown in Table 5-1.

The program is organized around six technical areas, called integrated technology demonstrators (ITDs), that will (1) perform preliminary studies, (2) select research areas, and (3) lead large-scale demonstrations either on the ground or in flight. The ITDs are "smart" fixed-wing aircraft, "green" regional aircraft, "green" rotorcraft, sustainable and "green" engine systems for "green" operations, and eco-design.

OVERALL OBJECTIVES OF ALL AERONAUTICS RESEARCH PROGRAMS

The goal of all of the programs described above is to have a "silent" aircraft in the future, that is, for the average sound pressure levels from all aircraft noise sources not to exceed sound pressure levels from other sources beyond airport boundaries during departure and arrival operations. In the next 20 years, newly designed aircraft are likely to be introduced at a rapid rate. These aircraft will likely be based on current aircraft but designed to achieve significant reductions in noise and carbon emissions. In the longer term (beyond 2025), further reductions in noise and carbon emissions are likely to require the development of entirely new aircraft and engine configurations.

The enabling technologies for both phases of development are becoming apparent. It appears that one version of the futurist aircraft, based on lessons learned from SILENCE(R) and SAX-40, will mimic a hybrid wing/body (HWB) configuration. As NASA continues to work toward the introduction of a new generation of highly fuel efficient large aircraft as early as 2020, it is already planning wind tunnel tests of low-noise HWB aircraft (Figure 5-15 shows a typical HWB). Convinced that the HWB is the only way it can meet its goals, NASA is providing funding for Boeing to study improvements to the configuration to further reduce noise and improve fuel burn.

NASA's subsonic *N*+2 research is now focused on a cargo version of the HWB, and if all goes well, an HWB freighter could be available by 2020, with a passenger version to follow within 10 years. According to a report in *Aviation Weekly* (2009), Boeing, with funding from NASA and the U.S. Air Force, will test two low-noise HWB configurations—N2A and N2B—in a wind tunnel in 2011. N2A has padded engines mounted above the aft fuselage. N2B has embedded engines and S-duct inlets for lower drag. Both designs incorporate hybrid laminar flow control to further reduce drag.

For NASA to achieve its goals of aircraft noise of 42 EPNdB cumulative below Stage 3 for the *N*+1 generation aircraft, considerable research will be needed in the following areas:

- target next-generation single aisle
- ultra-high-bypass engines
- noise reduction technologies for fans, landing gears, and propulsion airframe aeroacoustics
- lightweight acoustic treatment in multifunctional structures

TABLE 5-1 Team Members Available to Work on European Noise Reduction Programs

X-3 Team Partners	Country
Ain Shams University	Egypt
Alenia	Italy
ANOTEC	Spain
A2 Acoustics	Sweden
Budapest University of Technology and Economics (U.T.E.)	Hungary
COMOTI	Romania
Czech Technical University (T.U.)	Czech Republic
EADS CRC	Denmark
EPLF (Ecole Polytechnique Fédérale de Lausanne)	Switzerland
Federal University of Santa Catarina	Brazil
FFT (Free Field Technologies)	Belgium
Gediminas T.U. (Technical University)	Republic of Lithuania
INASCO	Greece
Institute of Aviation	Poland
Instituto Superior Tecnico	Portugal
ISVR	United Kingdom
National Aviation University	Ukraine
NLR	Holland
ONERA	France
Trinity College	Ireland

FIGURE 5-15 Hybrid wing/body aircraft with vertical tails on either side of the engines to shield jet noise. Source: NASA (2002).

To meet the goal of aircraft noise of 52 EPNdB cumulative below Stage 3 for the *N*+2 generation aircraft, considerable research will be needed in the following areas:

- noise reduction from wing shielding of engines
- drooped leading edge
- continuous-mold line flaps
- landing gear fairings
- long-duct, low-drag acoustic liners
- distortion-tolerant fans with active noise control

Objectives for the Air Transport System

Meeting NASA's Goals

If NASA can meet its targets for the next three generations of aircraft, successively quieter aircraft would enter into service by 2015, 2020–2025, and 2030–2035, respectively (GAO, 2002). The likelihood of meeting these targets depends on a number of factors. First, federal funding will have to be available not only for NASA's research but also for later-stage R&D, which NASA expects will be conducted by others.

Second, even if funding is available, the development of noise reduction technologies may be limited by concerns about global warming, because advances in noise reduction technologies could make it more difficult to achieve reductions in aircraft emissions of greenhouse gases. Third, manufacturers must be willing to integrate newly developed technologies into aircraft and engine designs. Finally, airlines must purchase new aircraft or retrofit existing aircraft with the new technologies in sufficient numbers to achieve targeted reductions in exposure to aviation noise.

If the FAA and NASA's noise reduction goals are not met, this could impede efforts to reduce congestion by expanding the capacity of the National Airspace System (FAA, 2007).

U.S. and European Visions of the Future

In 2002 the Federal Transportation Advisory Group published *Aeronautics Research and Technology for 2050: Assessing Visions and Goals*, which compares civil aeronautics in Europe and the United States. Although the United States recognizes that its national well-being depends on a national transportation system with a strong aviation element, there is no explicit goal to ensure the primacy of the U.S. aeronautics industry. On the contrary, competitiveness is central to the European vision, so much so that it appears in the title of the document that defines this vision: *European Aeronautics: A Vision for 2020—Meeting Society's Needs and Winning Global Leadership* (DG Energie et Transport and DG Recherche, 2001).

NASA's *Blueprint* (2002) and the *European Aeronautics* vision both specify that the ultimate goal in terms of operational impact is that aircraft noise be reduced to the point at which it is no longer a nuisance beyond airport boundaries and that airports be free of operational restrictions related to noise. The *European Aeronautics* vision highlights two areas not emphasized in any U.S. visions: (1) the quality and affordability of air transportation and (2) the global primacy of the aeronautics industry (FTAG, 2002a; NRC, 2002).

According to the GAO report, by including quality and affordability issues, the European vision acknowledges the importance of structuring R&D programs to focus on providing air transportation services that users want to buy and can afford. NASA's original goals issued in 1997 included reducing the cost of air travel by 50 percent in 20 years. However, this goal fell out of favor with Congress, which argued that meeting customer demands is an industry responsibility and not an appropriate goal for NASA's research. Congress then reduced NASA's aeronautics budget to eliminate research related to this goal (GAO, 2002).

The *European Aeronautics* document foresees the future in the following way:

> In 2020, European Aeronautics is the world's number one. Its companies are winning more than 50% shares of world markets for aircraft, engines, and equipment. The public sector plays an invaluable role in this success story. Crucially, they are coordinating a highly effective European framework for research cooperation, while funding programs that put the industry on more equal terms with its main rivals.

Future Operational Procedures

Limiting—on a yearly basis—the cumulative noise footprint in areas surrounding airports will effectively limit the capacity of the national aerospace system. Present departure and arrival procedures, which were developed when a limited range of navigational aids was available, are far from optimal from an environmental point of view. Therefore, in combination with new "silent" aircraft, the introduction of new approach, navigation, and flight management systems will make environmentally friendly procedures feasible.

FINDINGS AND RECOMMENDATIONS

A generation ago, "Higher, Farther, Faster" was the imperative for the future of air transport. Today, it is "More Affordable, Safer, Cleaner, and Quieter." This change reflects the new emphasis on combining cost effectiveness with safety and environmental objectives. Significant investment is being made on both sides of the Atlantic to meet the demands of the market as well as the needs of the community. In the United States much of this effort has been led by NASA; in Europe significant investments have been made under the Framework Programs, in which EU industry and researchers in many countries work together in well-funded, coordinated R&D programs.

The major challenge in the development of noise reduction technology for the future is that the design requirements for an aircraft with low emissions of chemical pollutants differ and sometimes conflict. Some design considerations are common to achieving both low noise and low pollution; for example, improved engine/aircraft aerodynamics result in lower noise as well as reduced fuel burn and thus reduced carbon dioxide emissions. Similarly, operating practices such as continuous descent approaches can reduce both pollution and noise (see SBAC Aviation and Environment Briefing Paper titled "Aircraft Traffic Management and Operations" at *www.sbac.co.uk* for more details). However, other design requirements are in direct conflict with each other, forcing engine and aircraft designers to make difficult compromises. One example of this is the requirement to increase the bypass ratio of the engine beyond the optimum to reduce fuel burn and reducing the fan speed to achieve a reduction in noise (particularly at takeoff), but at the cost of increasing fuel burn and chemical emissions at cruise speed. Future increases in air traffic, combined with the industry's desire to reduce contributions to global warming, will certainly necessitate even more such difficult decisions.

On the positive side, the "silent" aircraft concept, as envisioned under the SILENCE(R) program and the Silent Aircraft Initiative (SAX-40), promises reductions in both noise and pollution. The realization of a silent aircraft may be possible with known concepts, but bringing the enabling technologies to a suitable level of readiness constitutes a significant barrier and will require a significant investment in R&D.

Even if government investment in R&D in the United States and Europe remains strong, new quiet aircraft technologies may still not be available until 2020. Even then it will take many years for current airplanes to be phased out and for the full benefits of quiet aircraft to be realized.

The impact of increased number of airports and aircraft in service is likely to exceed the mitigating effects of near-term technological advances and operational improvements, and the number of people exposed to aircraft noise is likely to increase. In addition, the sensitivity of people to noise, or at least vocal objections to it, which often depends on attitudes and socioeconomic conditions, may also increase as people become more affluent. The implication for the aviation noise research community and government agencies that must support this community is that they cannot afford to be complacent.

Recommendation 5-1: The National Aeronautics and Space Administration (NASA) should continue to fund collaborative projects by engine, airframe, and aircraft systems manufacturers. Drawing on expert knowledge in research organizations and academic institutions, research should focus on the complex interrelationships between engine and airframe and the importance of reducing each constituent noise source to reduce the overall noise signature of aircraft. These projects should develop improved prediction tools, for example, for advanced propulsion designs; acoustic scattering and propagation models, including adequate weather and terrain models; models of the effects of interactions between engine installation and airframe configuration; and benchmark measurements necessary for the development and validation of these advanced tools.

Recommendation 5-2: The Federal Aviation Administration should continue to fund the development of novel operational and air traffic management procedures to minimize noise and should work with NASA and industry to make intelligent trade-offs between competing noise mitigation and chemical pollution goals.

NEW TECHNOLOGIES FOR REDUCING NOISE FROM ROAD TRAFFIC

DEFINING THE PROBLEM

Noise from motor vehicles is undoubtedly the most pervasive noise in our society (Bowlby, 1998; Sandberg, 2001). Individually, passenger cars, trucks, buses, and motorcycles emit relatively low levels of noise compared with aircraft and rail transit at equivalent distances. However, the sheer number of these vehicles in close proximity to sensitive receptors more than offsets their lower noise levels (Donavan and Schumacher, 2007).

Based on figures from the Environmental Protection Agency (EPA), approximately four times as many people are exposed to highway noise with DNL values of more than 65 dB than are exposed to aircraft noise and almost eight times as many are exposed to highway noise than are exposed to rail noise (Waitz et al., 2007). The high numbers of impacted people are, in part, institutionalized by federal policy, which uses as a threshold a worst-hour equivalent noise level (L_{eq}) "approaching" 67 dB for when noise abatement should be considered near new or expanding highways (23 CFR 772). Given a typical day/night urban freeway traffic distribution (Greene, 2002), the hourly level leads to DNL values of about 69 to 70 dB.

This DNL for road traffic noise is expected to "highly annoy" almost 30 percent of people exposed to it (Waitz et al., 2007) and is 15 dB higher than the 55-dB DNL criterion established by EPA as necessary to protect public health and welfare with an adequate margin of safety (EPA, 1974). A DNL value of 65 dB has also been identified by the EU as the threshold for negative health effects caused by noise.

Road traffic noise is the result of a combination of noise from several different vehicle types, each of which has its own characteristics. These include light vehicles (passenger cars, pickup trucks, sport utility vehicles, and passenger-size vans), medium trucks, heavy trucks, buses, and motorcycles. Of these, light vehicles and trucks tend to dominate traffic noise.

To support the modeling of traffic noise, an extensive database of vehicle pass-by noise emissions was collected in the mid-1990s by the Federal Highway Administration (FHWA) characterizing each vehicle type as a function of speed (see Figure 5-16). These data reveal a number of attributes. At speeds of more than 20 to 30 kilometers per hour, noise emissions from trucks and light vehicles increase rapidly with speed. At highway speeds, 80 kilometers per hour and above, noise from heavy trucks is about 10 dB higher than from light vehicles; medium trucks fall somewhere in between. For this reason, the level of traffic noise is strongly influenced by the mix of cars and trucks.

Contributions to overall noise from each vehicle type are typically considered in three categories: power train; tire/pavement interaction; and aerodynamic noise. Power train noise includes all sources associated with vehicle propulsion and strongly depends on engine speed. This source tends to dominate the overall noise emission at lower speeds at which speed has little effect on noise levels. At very high speeds, beyond legal speed limits in the United States, aerodynamic noise caused by flow over and under the vehicle becomes important. Noise emissions from this source are typically proportional to 60 times the logarithm of vehicle speed. Between these extremes, noise emissions from all three vehicle types are dominated by noise from tire/pavement interaction (as shown in Figure 5-17 for light vehicles).

Noise levels can also be visualized using acoustic beamforming technology, as shown for a light vehicle and heavy truck cruising at about 88 kilometers per hour in Figure 5-18. In this speed range, at which tire/pavement noise is the dominant source, vehicle noise emission levels increase at about 30 to 35 dB times the logarithm of speed (Sandberg and Ejsmont, 2002).

Power train noise, the dominant noise source at low vehicle speeds, will be greatly reduced as new hybrid and plug-in hybrid vehicles are introduced into the fleet. This is an example of noise being reduced not by noise control engineers but by the introduction of a new technology by manufacturers. Urban dwellers will benefit from reduced noise levels as these vehicles are introduced, but new problems are created. The sound of a vehicle in some cases serves as a warning signal, especially to children and sight-impaired persons, and consideration is being given to adding sound when vehicles are very quiet. This, however, creates opportunities for engineers interested in the product sound quality issues discussed later in this chapter.

NOISE BARRIERS

To address noise from motor vehicles, EPA has set a noise emission limit of 80 dB for new heavy and medium trucks, buses, and motorcycles. Although there is no federal limit for light vehicles, a sufficient number of state and local jurisdictions require that emissions be limited to a level of 80 dB, effectively making this a de facto national limit. The test procedures used to obtain these levels, which are conducted at low speed and under full-throttle acceleration, essentially deal with power train noise. Under these conditions, typically 40 percent or less of total noise emissions for light vehicles is due to tire noise (Donavan et al., 1998). As a result, these limits do little to address traffic noise under highway conditions (Sandberg, 2001). In Europe a limit on tire noise has been established for moderate speeds; however, there is no equivalent regulation in the United States.

Because there are no pertinent source emission regulations, road traffic noise is abated in the United States al-

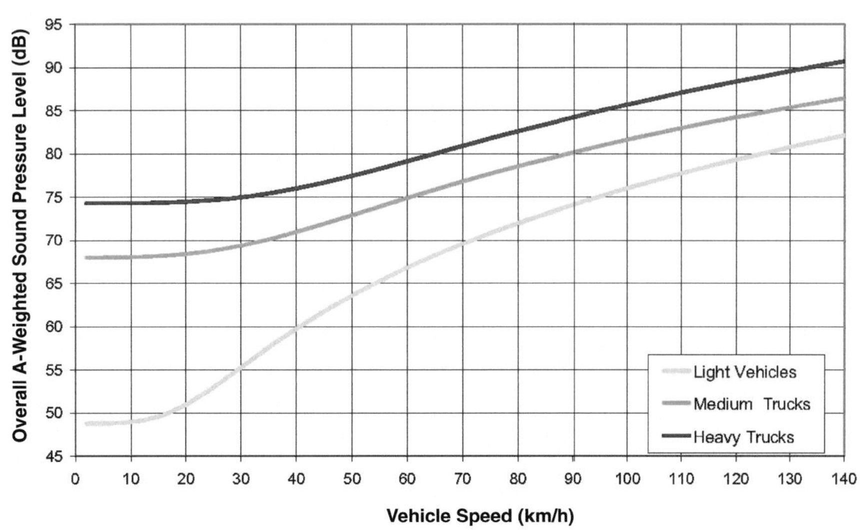

FIGURE 5-16 U.S. average pass-by noise levels under cruise conditions for light vehicles, medium trucks, and heavy trucks measured at a distance of 15 meters. Source: Fleming et al. (1996).

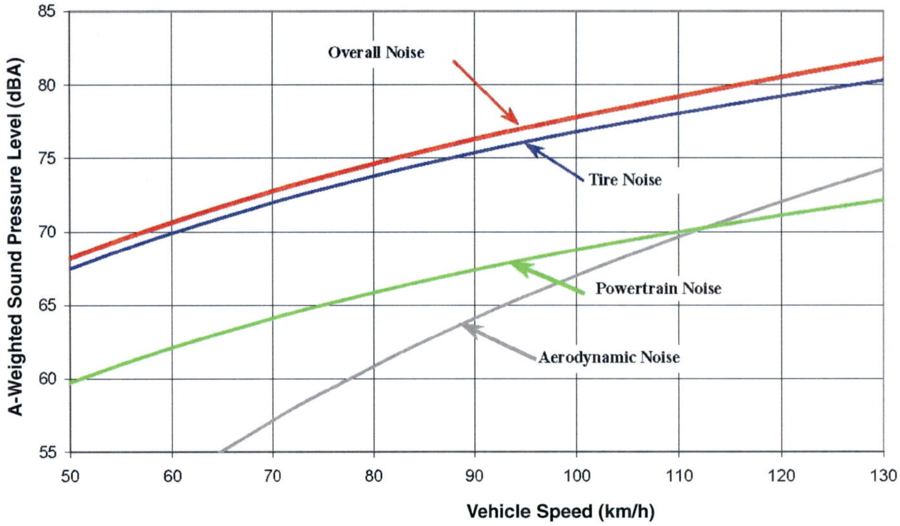

FIGURE 5-17 Typical levels for noise sources in light vehicles. Sources: Adapted from Donavan (1993); Donavan et al. (1998).

most exclusively by erecting noise barriers or sound walls alongside freeways. All but six states have used this method of noise abatement to some degree (Polcak, 2003). The extensive use of sound walls is driven largely by FHWA policies (23 CFR 772). For new highway projects or when the capacity of an existing highway is planned, and if federal funds are being used, federal policy allows only five types of noise abatement to be considered: traffic management, such as speed limits or vehicle restrictions; alterations of the highway alignment away from sensitive receptors; barriers in the form of sound walls or earthen berms; the creation of buffer zones along the highway; and sound insulation for some public buildings. For practical reasons, barriers are almost always selected.

Once an abatement method has been selected, it must pass tests in terms of "feasibility" and "reasonableness." *Feasible* in this context means: Will the barrier provide at least a 5-dB reduction in the predicted noise level for impacted residences? *Reasonable* has several dimensions, one of which is the cost of the barrier compared to the benefit received by the impacted residences. To implement the federal policy, each state develops its own policies and guidelines to define other parameters, such as the level at which noise abatement will be considered and the value of each impacted residence. If the state determines that a barrier or other form of noise mitigation is not feasible or reasonable, no abatement measures are required. Federal policy explicitly forbids the selection of pavement type for noise abatement, largely because of concerns that it will not maintain a given level of noise reduction performance over the life of the highway project.

QUIETER PAVEMENTS

Notwithstanding federal policy, because of the initial cost of highway barriers (see Chapter 7) and because they are not always feasible, both state and federal governments have an interest in investigating other possibilities for reducing road traffic noise. This interest has focused largely on source control at the tire/pavement interface. Although the two components of tire/pavement noise are inherently inseparable, when either the pavement or the tire remains constant, it appears that the greatest potential for noise reduction is in the pavement (if all pavement types are considered).

For most pavement types, the noise level from different

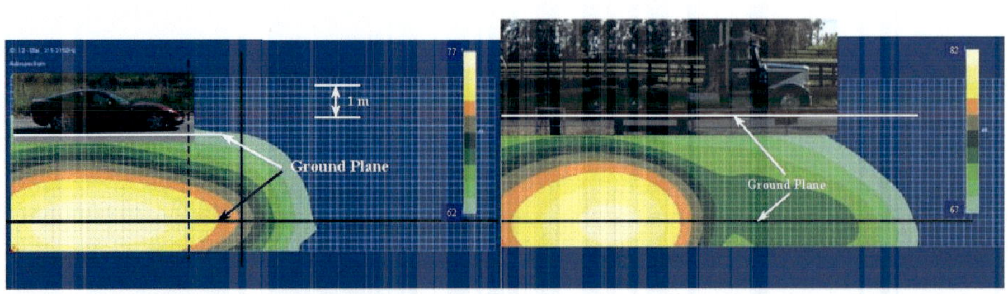

FIGURE 5-18 Acoustic images of typical noise source regions for light vehicles and heavy trucks obtained with acoustic beaming. Source: Adapted from Donavan and Rymer (2009).

tires has a range of about 3 dB (see Figure 5-19). If all-terrain tires, such as those that might be used on four-wheel-drive vehicles, are included, the noise range increases by about 1 dB. Even comparing typical treaded tires to blank-tread tires produces a range of 3 dB or less, depending on the pavement. However, pass-by noise levels at 97 kilometers per hour can easily demonstrate a 10-dB range (see Figure 5-20) on different pavements. Surveys of many highway pavements using measurements made on board the vehicle near the tire/pavement interface demonstrate that the range may be even greater than 13 dB (Donavan, 2006).

According to the FHWA Tire Pavement Noise home page, eight states are investigating and testing quieter pavements (FHWA, 2009), and others are considering such programs. In the longest-running program, by Caltrans in California, 9 kilometers of older dense-grade asphalt concrete (DGAC) was overlaid with 25 millimeters of new open-grade asphalt concrete (OGAC) on a six-lane portion of I-80 near Davis, California. Initially, this produced a reduction of about 6.5 dB in traffic noise levels measured alongside the freeway, and a level 6 dB lower than was predicted by the FHWA Traffic Noise Model (TNM). After 10 years, the performance

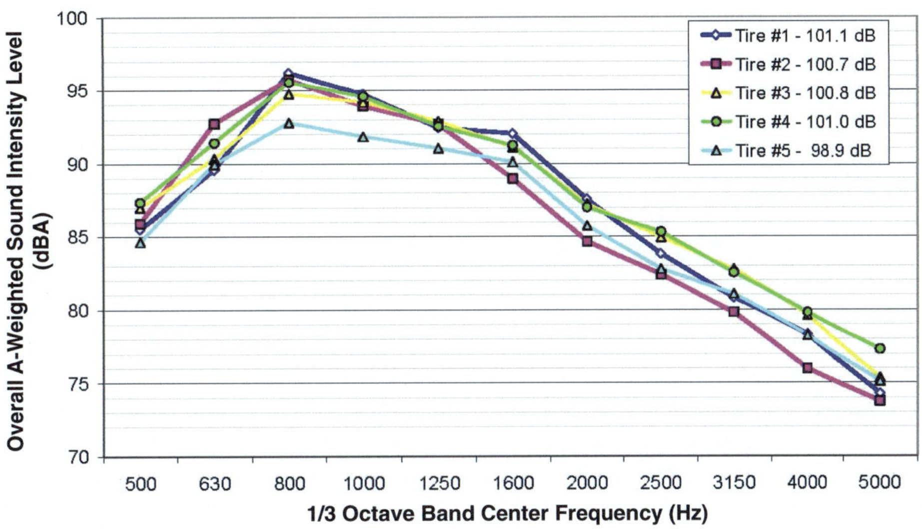

FIGURE 5-19 Range in one-third octave band sound intensity levels for tires measured at 97 kilometers per hour on a dense, graded, asphalt-concrete roadway. Source: Adapted from Donavan (2006).

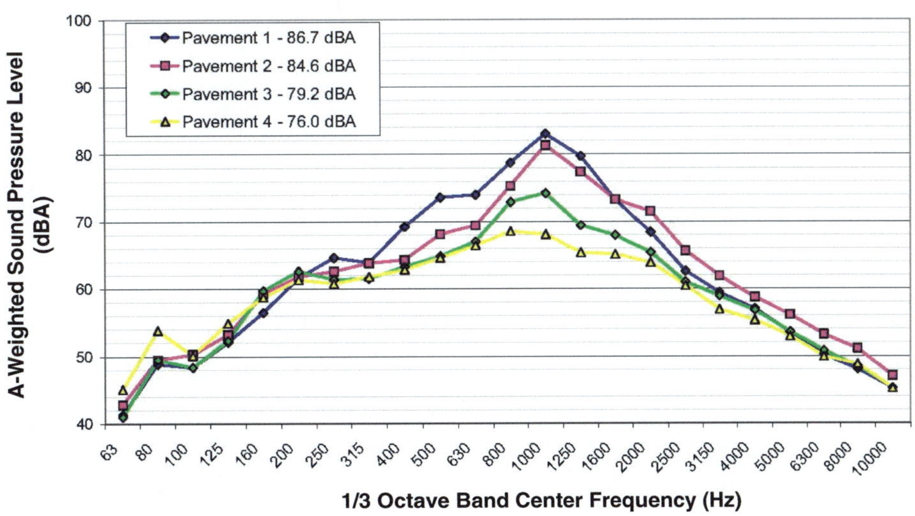

FIGURE 5-20 One-third octave band pass-by noise levels for the same car and tires operating on different pavements at 97 kilometers per hour. Source: Adapted from Donavan (2006).

of this pavement has deteriorated by a little more than 1.5 dB (Lodico and Reyff, 2009). In other projects, Caltrans has documented reductions of 3 to 6.5 dB by overlaying existing Portland cement concrete (PCC) with a rubberized open-graded asphalt concrete.

The largest application of quieter pavement in the United States was in the greater Phoenix area by the Arizona Department of the Transportation (ADOT). This Quiet Pavement Pilot Program (QPPP), conducted jointly with FHWA, allowed ADOT to take a 4-dB credit or reduction in the TNM predicted level attributable to a 25-millimeter overlay of rubberized asphalt friction course applied over new and existing PCC (Donavan, 2005; Reyff and Donovan, 2005). As measured at five wayside noise measurement research sites, this overlay reduced traffic noise levels by 6 to 12 dB measured 15 meters from the freeway. Using TNM predictions, this reduction was equivalent to a barrier height of 4 meters placed alongside the roadway. The QPPP also committed ADOT to continue to monitor the performance of the overlay for 10 years. After four years, the average tire/pavement noise reduction slipped from 8 to 6 dB, with most of the degradation in the first two years of the project.

The selection and modifications of PCC surface textures have also resulted in reduced tire/pavement and traffic noise levels (Donavan, 2005). As part of the Quiet Pavement Research Program completed by ADOT leading up to the QPPP, the agency investigated PCC texturing. Two types of rakelike tining textures applied perpendicular (transversely) to the direction of travel were compared to tining applied longitudinally to (with) the direction of travel. In the most extreme case, the longitudinal tining was found to reduce tire/pavement noise and the noise from light vehicle pass-bys by 7 dB compared to random transverse-tined PCC. The longitudinal tining also produced a level 5 dB lower than uniformly spaced transverse tining, which was the standard texture used by ADOT up to that time. Following the lead of California, Arizona—and several other states—have now adopted longitudinal tining as their design standard.

Modifying PCC surface texture by grinding has also been found to reduce tire/pavement and traffic noise. In an extreme example, a reduction of 10 dB was documented when Caltrans ground away an aggressively transverse bridge deck over the Sacramento River (Donavan, 2005). In less extreme cases and depending on the texture of the existing PCC, reductions of 2 to 9 dB have been reported (Donavan, 2005; Herman and Withers, 2005).

The Technology of Quieter Pavements

The examples of quieter pavements described above are primarily quieter versions of existing pavement designs that reduce road traffic noise. From research in Europe and more recently in the United States (Rasmussen et al., 2007), the two primary attributes of conventional pavement that dictate noise performance are surface texture and porosity. For asphalt concrete (AC), texture is largely determined by aggregate size; smaller sizes produce lower noise levels. Another factor is whether the texture is negative (embedded in the surface) or positive (protruding from the surface); negative texture is quieter.

Tining

For PCC, texture is more consciously designed on the surface by tining or dragging a material over the surface before it cures. As described above, the direction of tining is important; surfaces tined longitudinally are usually quieter than those tined transversely. Even in the longitudinal direction, however, tining can introduce some unwanted texture as the material is dug out of the surface. Surfaces that are ground or textured by dragging burlap over the uncured surface typically produce less textured surfaces and less resultant noise.

Porosity

Porosity, typically associated with AC pavements, is dictated by the percentage of void area in the pavement. To be most effective in reducing noise, voids must be interconnected so that the structure provides some degree of acoustic absorption. Porous pavements reduce tire/pavement noise both by absorbing sound propagated over the surface and by reducing the air-pumping mechanism created by air being trapped and ejected out of tire tread void areas.

OGAC mix designs generally achieve higher pavement void ratios. However, OGAC constructions do not always ensure a porous pavement. Other AC pavement designs, such as DGAC and stone matrix asphalt, typically have very low void ratios and hence are only quieter if smaller aggregate sizes are used (Donavan, 2006).

There are also porous PCC pavements, although these are very rarely used for highways. Porous PCC can produce tire/pavement sound levels that rival quieter AC pavements, provided that the texture is managed through grinding or other techniques.

As can be inferred from the discussion above, the quietest AC pavements measured to date fall into two categories: (1) highly porous, small aggregate OGAC pavements and (2) very fine textured, small aggregate DGAC pavements. The latter is rarely used in highway construction, and the former is not regularly achieved in practice. In fact, high porosity is often achieved with larger aggregate OGAC designs that tend to actually increase texture-generated noise.

In Europe there are examples of double-layer porous pavements (Figure 5-21), with a large aggregate lower layer and a small aggregate upper layer (2 millimeters layer over a 6 millimeters layer). These pavements have produced the lowest noise levels measured in either the United States or Europe to date. As a next step, research is being conducted in Sweden and Japan with poroelastic surfaces that have air void ratios of 20 to 40 percent and are constructed of resilient materials, such as recycled rubber from used tires.

FIGURE 5-21 Example of a double-layer porous asphalt pavement used in the Netherlands. Source: Donavan (2006).

Tire Noise

Many studies have been done in Europe on the effect of tire tread patterns on the generation of noise in tire/road interaction. Regulations have been promulgated to control the tire contribution, but these have been largely ineffective. The most important factors from a tire manufacturer's point of view are safety and performance. Next the manufacturer considers rolling resistance and durability. Noise is also "on the list" of considerations but is not likely to be a major factor in tire design, although there is a weak correlation between noise radiation and rolling resistance (Sandberg, 2001).

The demand for quieter tires comes primarily from vehicle manufacturers. For passenger vehicles in particular, *interior* noise from tires is a development issue as important as other noises, such as power train and wind noise. Vehicle manufacturers work closely with tire manufacturers to tune and develop original equipment tires to meet specific targets. However, this does not necessarily result in lower levels of *exterior* noise, because interior noise can often be controlled by structure-borne noise paths into the vehicle.

On occasion, exterior tire noise becomes an issue for meeting the regulated pass-by noise limits determined under the ISO 362 test procedure. Although this procedure is initiated at relatively low speed (50 kilometers per hour) and performed at wide-open throttle conditions, tire noise can contribute as much as 41 percent to the overall A-weighted pass-by noise level (Donavan, 2005). For vehicles that are over the limit in the vehicle development stage, selecting or developing quieter tires becomes a consideration.

Costs and Benefits of Reducing Noise from Tire/Road Interaction

The costs and benefits of reducing traffic noise have not been extensively studied in the United States, but there have been studies in Europe. Chapter 7 in this report includes a review of EPA and FAA cost-benefit activities and a summary of European activities. Because of the extensive use of noise barriers in the United States, the costs of barriers, although highly dependent on a number of factors, are reasonably well known. Several examples of the cost of low-noise road surfaces are also given. As the reader can tell, benefits have been described in terms of decibel reductions but not yet in terms usually used by economists (i.e., hedonic pricing, willingness to pay).

FINDINGS AND RECOMMENDATION

More people in the United States are impacted by road traffic noise than by aircraft noise. At highway speeds the main source of noise emission is interaction between vehicle tires and road surfaces. Considerable progress has been made in understanding this noise source, and development work in both the United States and Europe has shown that considerable reductions in noise emissions can be achieved by changing the design of the road surface. This technology requires further development both to increase noise reductions and to study other factors, such as durability of the road surface over time.

The primary defense against road traffic noise in the United States has been noise barriers. However, barriers are expensive and provide relief to a relatively small number of residents—they are most effective within about 200 feet of the highway. To allocate costs effectively, a cost-benefit analysis of the two alternatives (quiet pavements and noise barriers) should be undertaken.

Recommendation 5-3: Current activities of the Federal Highway Administration and several states to investigate noise reduction through new pavement design should be continued and expanded to speed up development and application of new technologies. Studies on the durability of pavement surfaces are essential, because durability has a direct effect on the life-cycle costs of applying quiet pavement technology, which has the potential to reduce noise where barriers are not feasible—for example, where homes are located on a hillside overlooking a busy highway.

RAIL NOISE

TYPES OF RAIL TRANSPORTATION

Urban Transit

Rail transit systems are ideal for densely populated urban and suburban areas. Typically, these systems include three types of transit: light rail transit (LRT), rail rapid transit (RRT), and commuter railroads (CRRs).

LRT systems operate in two forms: (1) exclusive rights-of-way in tunnels, at grade level, or on aerial structures and

(2) nonexclusive rights-of-way on streets or medians. LRT consists[3] tend to be one or two cars operated manually in urban areas with a maximum speed of 50 mph. Vehicles are generally powered by electricity supplied by overhead wires, and tracks are located close to buildings where people live, work, and shop. Stations are located every few blocks in downtown areas and less than a half-mile apart in outlying areas (AREMA, 2009).

RRT systems operate in fenced, exclusive rights-of-way in tunnels, at grade, or on aerial structures; no street crossings are permitted. RRT consists are typically made up of four or more cars operated either manually or automatically with a maximum speed of 80 mph. Vehicles are powered by electricity supplied by a third rail at trackside, and tracks are located close to buildings where people live, work, and shop. Stations are approximately one-half mile apart in downtown areas and 1 to 2 miles apart in outlying areas (AREMA, 2009).

CRRs, which carry passengers on intra- and intercity routes, transport people from the outer fringes and suburbs to center cities on a regular schedule and are concentrated heavily on morning and evening rush hours. Intercity railroads also carry people over longer distances with schedules distributed throughout the day and night. Passenger railroads in the United States are "conventional," that is, they have steel wheels and steel rails and generally operate at lower speeds than trains in other parts of the world. Only one Amtrak intercity operation—some sections of the Northeast Corridor—can be called "high-speed."

Passenger Railroads

Conventional passenger railroads operate on exclusive rights-of-way, primarily at grade, but road crossings are common. Railroad consists are typically made up of locomotive-hauled trains of six or more cars operated manually at speeds up to 80 mph. Locomotives are generally powered by diesel engines, with some exceptions in electrified areas around major cities. Tracks are located on shared rights-of-way with freight railroads and are not very close to buildings where people live, work, and shop. Stations separated by many miles are located in city or town centers.

There are no high-speed passenger railroads (HSRs) in the United States—with the exception of Amtrak's Northeast Corridor, where speeds of 150 mph are possible on limited sections of track in Massachusetts, Rhode Island, and New Jersey. According to EU Directives, the general definition of "high speed" requires operation at sustained speeds of 200 kilometers per hour (125 mph) or more. HSRs operate on exclusive, fenced, and grade-separated rights-of-way with as few road crossings as possible. Instead of traditional locomotives, HSR consists include one or more power cars driven by electrical motors that take the current from overhead wires. Because HSRs require considerable distances for acceleration and deceleration, stations are separated by many miles.

Two types of HSR are currently in operation: conventional trains with steel wheels on steel rails and magnetically levitated trains (maglevs) on special guideways. The speed record for a steel-wheel train, 357 mph, was set by a French TGV. The record for a maglev is 361 mph.

Freight Railroads

Freight railroads in the United States set the standard for efficiency and service and are the envy of the railroad world. Private companies operate the freight sector, with individual companies owning the rights-of-way over which their trains travel. Road crossings are common. Consists are made up of diesel locomotives pulling trains of freight cars, ranging from short lines and local freight trains that have one locomotive and 10 cars or less to typical coal trains in the Midwest that have six locomotives and more than 100 cars. Origins and destinations for freight trains are railroad yards or specialized facilities where cars are either unloaded for local distribution or switched to new trains for continuing transportation.

NOISE IMPACTS FOR EACH MODE OF RAIL TRANSPORT

Estimates have been made comparing rail transportation noise impacts with noise from other transportation modes. A study sponsored by EPA concluded that 4 million people in the United States were impacted by rail noise, 2 million by urban mass transit system noise and 2 million by railroad operations and yards (EPA, 1981; FRA, 2002). The current numbers are probably higher because of the expansion of public transit systems, especially LRT and CRRs, and the increase in freight operations in the past 30 years. Even more people are impacted by train horns at rail/highway grade crossings. In 2005 the Federal Railroad Administration (FRA) estimated that 6 million people were impacted by train horns (FRA, 2006). The EPA (1981) estimated that 2.5 million persons were impacted by rail noise. A reasonable estimate would be that at least 10 million people are impacted by rail transportation noise in the United States today.

Noise Sources

Wheel/Rail Interaction

With few exceptions, rail technology is associated with steel wheels rolling on steel rails.[4] Noise generated by wheels and rails can be categorized into three types:

[3] *Consist* is the term used to describe the makeup of trains (e.g., number of cars or locomotives and coaches).

[4] Guided-rail technology also includes maglev and rubber-tired subway systems.

- Rolling noise—roughness on the running surfaces of wheels and rails generates the ubiquitous rolling noise associated with moving trains. The condition of the running surfaces of wheels and rails ranges from smooth to rough, depending on how well they are maintained. Microscale roughness is associated with what appears to be smooth surfaces; macroscale defects, such as corrugations in the rail or skid flats on the wheel, are clearly visible. Noise is radiated by both wheel and rail vibrations, each with its own sound characteristics. Because rolling noise is a major source for rail systems, controlling it has been a focus of research, and reliable models have been developed to explain it.
- Impact noise—gaps or discontinuities in the rail running surface generate a distinctive impact noise, sometimes called the "clickety-clack of the railroad track." This noise occurs when steel wheels encounter joints (e.g., special track work associated with switches or crossings). Both wheels and rails radiate sound from the sudden application of force during the encounter.
- Squeal noise—sharp curves in the track cause steel wheels to radiate a piercing squeal. The noise is generated by a stick-slip mechanism as the wheel's running surface skids over the top of the rail on a curve. Because wheels are connected by solid axles and because the inside wheel has a shorter distance to travel than the outside wheel, one wheel has to slip in the curve. The slippage of wheel running surface and flange along the top and sides of the rail causes the wheel to resonate at its natural frequencies, some of which are in the most sensitive range of human hearing and are very annoying.

Propulsion and Equipment

Mechanical equipment associated with propulsion, braking, and air conditioning is a major source of noise:

- Traction motor noise—electric motors are used for propulsion in both diesel and electric locomotives, as well as electric transit systems. Although normally considered a minor noise source when compared with the powerful diesel engines in freight locomotives, electric motors generate considerable noise at certain rotational frequencies.
- Fan noise—cooling fans, which are required for all propulsion systems, can be a major noise source, especially at low speeds.
- Diesel engine noise—noise from the diesel engine emanates from both the exhaust and the engine casing. Exhaust noise generally dominates at all speeds.
- Compressor noise—air compressors are key components of all trains because braking systems rely on air pressure for their operation. Compressors are electrically driven pumps that generate noise as they work to fill the reservoirs.
- Air brake discharge noise—brakes on rail vehicles are released from wheels (tread brakes) or disks (disk brakes) by air pressure from the compressed air reservoirs. When air pressure is released, the brakes are engaged. Noise is generated as the air escapes the brake units. When the train comes to a complete stop, all of the air is "dumped" from the reservoir in the locomotive, causing a very loud sound.

Aerodynamics

HSTs generate noise from the sources listed above and one additional source—aerodynamic sound. The air surrounding the body of a moving train is pushed out of the way as the train moves through it. At low speeds the air moves away and closes back in without much disturbance. As the train speed increases, however, forces on the air also increase, causing turbulence at the boundary layer surface, vortex shedding at edges and appendages, and interactions with stationary objects beside the tracks. These aerodynamic phenomena generate noise levels that increase with speed faster than any of the noise from mechanical sources. At speeds of more than 150 mph, aerodynamic noise becomes the major noise source.

Warning Systems

Interactions between rail and highway systems invite disaster unless they are controlled by warning systems. Motorists and pedestrians must be warned of approaching trains, typically by the train horn, where roads and footpaths intersect railroad tracks (DOT, 2002). Unfortunately, these intersections often occur in residential areas where the warning sounds are a continuing source of annoyance.

- Horns—railroads that operate on the interstate railroad network are required to sound horns at all road crossings at grade level unless special conditions are met, in which case a waiver may be granted. Regulations specify a minimum sound level to ensure that sufficient warning is given, as defined by the Federal Railroad Administration (FRA). In a recently adopted rule, both minimum and maximum sound levels are specified (FRA, 2006). Very high sound levels are emitted from horns on top of the locomotive for one-quarter mile leading up to a grade crossing. In a residential area, one-quarter mile can cover many homes and generate a great deal of annoyance. Urban transit systems also use horns for emergencies to alert motorists and pedestrians at street crossings. Transit horns are unregulated, and a wide variety of sounds are used for warning purposes, depending on the policies and practices of the transit agency.

- Bells—stationary bells are used at busy street crossings of tracks, often in connection with gates to block traffic. Bells give pedestrians and motorists advance warning that a train is approaching and that the gate arms are about to descend. Sound is emitted in all directions from a crossing before, during, and after the passage of a train, disturbing neighbors with an annoying ringing sound for significant periods of time in the course of a day.

Structures

Heavy trains rolling on tracks on bridges and elevated structures cause the structures to vibrate, resulting in rumbling sounds like thunder emanating from steel and concrete beams. In urban areas this rumbling sound can be extremely annoying, especially when it occurs many times an hour throughout the day.

NOISE ABATEMENT

Noise from rail transportation is a major concern of nearby residents. Both federal agencies that oversee rail transportation, the Federal Transit Administration and FRA, have developed noise guidelines for new projects based on the "source-path-receiver" noise model (see Figure 4-1), with abatement approaches identified for each element in the model (FRA, 2005; FTA, 2006).

Treatment of Noise Sources

Control at the source is generally the most cost-effective way to approach a noise problem. In the case of rail systems, the owners of rail systems are usually in control of the trains operating on their systems and therefore can be deemed responsible for their noise. Diagnostic techniques and applications of new technologies are focused on rail vehicles themselves and on their interaction with tracks.

Treatment of the Sound Path

Interrupting the path between the source and the receiver is a traditional way of handling a noise problem, especially in locations where the source is not under the control of the mitigating agency. Typically, noise barriers are used to screen the noise source. Although these can be effective, they are permanent structures—blank walls that are present whether or not the noise source is present. The cost of erecting a noise barrier must always be balanced against its effectiveness in reducing noise as well as the number of persons who benefit from the reduced noise level.

Treatment of the Receiver

Treatments for individual residences that receive unwanted sounds from rail systems are the least cost-effective approach. However, if sources cannot be controlled (e.g., horns at grade crossings) and path treatments are not feasible (e.g., roadway site lines prohibit noise walls), the only noise abatement approach may be to treat the windows, doors, and walls of homes exposed to the noise.

NEW TECHNOLOGIES

The history of efforts to control rail noise in the United States dates back to the early 1960s. The San Francisco Bay Area Rapid Transit District was the first new transit system to identify noise abatement as an important part of the design of rail systems. In the 50 years since then, transit systems, railroads, and public agencies have all addressed the problem of noise as a public concern. Methods of analyzing noise and noise control treatments have proliferated. Some new technologies for reducing rail system noise are described in this section.

Diagnostic Tools

The track wheel interaction noise system model, based on seminal research at Bolt Beranek and Newman, was developed in the mid-1970s and is used throughout the world to estimate the mechanisms in noise generation. In 1983, Remington provided the basis for development in Europe of a noise model to define the causes of wheel/rail noise.

Modern, high-speed computers have enabled the development of microphone arrays that can pinpoint the sources of noise on trains as they pass by on a track. This technology is especially valuable for locating sources of aerodynamic noise on HSTs.

Wheel/Rail Treatments

Research has shown that roughness on the running surfaces of wheels and rails, even on the microscale, is the root cause of rolling noise. Improved wheel truing and rail grinding practices are used to reduce this source of noise. The radiation of noise from steel wheels, especially on tight curves where squeal is a problem, can be reduced by a variety of damping devices attached to the webs of the wheels. Radiated noise from rails is a major problem on tangent (straight) track. Damping devices attached to the rails have been demonstrated to reduce this noise (Remington, 1983).

Friction Modifiers

Train wheels traversing tight curves emit a high-pitched squeal sound via a stick-slip phenomenon at the wheel/rail interface that vibrates the wheels at their natural frequencies, similar to the way a bow generates vibrations in a violin string. Changing the friction coefficient between the wheel and rail has been found to eliminate this phenomenon and eliminate squeal. Various fluids, including plain water, have been used to modify friction by lubrication.

Warning Systems

Because sounds from locomotive horns are generally omnidirectional, the sounding of horns creates noise along the sides of the track ahead of a train as it approaches a highway grade crossing. Directional horns focus the sound forward where the warning is needed and eliminate much of the wayside noise. An even more effective technique is to install horns at the grade crossing pointing toward oncoming traffic. This technology eliminates the use of locomotive horns entirely, thereby eliminating horn noise in residential areas on both sides of the tracks.

Engine Noise

Research supported by FRA provided proof-of-principle that loudspeakers placed around the exhaust stack of a diesel locomotive can be tuned to cancel out the sound of idling locomotives (Remington et al., 2005). Current research (Johnson et al., 2009) related to the design of next-generation locomotive cabs, also supported by FRA, has shown that low-frequency noise from a diesel engine can be drastically reduced by active noise cancellation techniques. Low-frequency noise is the main cause of crew fatigue on long-distance runs.

Structures

Applying damping materials to steel beams and isolating tracks are two proven ways of reducing the low-frequency rumbling noise that emanates from bridges and elevated structures as trains roll over them.

Aerodynamic Noise

Maglev trains are supported above their guideways by magnetic forces, thereby reducing noise from rolling contact with tracks. However, because these systems are capable of traveling at speeds of more than 300 mph, they must be designed to reduce aerodynamic noise. At high speeds the noise is generated by aerodynamic forces interacting with the vehicle body. Consequently, maglevs are designed to be extremely smooth in all configurations. Designs are carefully shaped in wind tunnels to minimize air resistance and aerodynamic noise-generating mechanisms. Full-scale testing of the designs is conducted on dedicated test tracks where large microphone arrays are used to diagnose noise sources.

Special conventional HSTs can be built that are capable of speeds of more than 300 mph. These trains must also be designed to reduce aerodynamic noise.

FINDINGS AND RECOMMENDATION

It has been estimated that 10 million people are impacted by rail transportation noise. As in the case of road traffic noise, interaction between two surfaces—in this case a steel wheel and, typically, a steel rail—is a major source of rail noise. To ensure that technology to control this noise and other noise sources is applied to the design of new systems, there must be careful planning, and manufacturers must understand what noise *emission* levels must be to control noise *immission* in a particular environment. Available planning tools can be used on the local level for developing and planning new projects.

In Europe there are major problems with noise from freight wagons that the United States does not have. However, considerable progress has been made in controlling rail noise that could be applied to problems in the United States. Both Europe and Japan have considerable experience with controlling noise from high-speed trains that would be useful if new high-speed lines are developed in the United States. Noise from warning horns must also continue to be addressed. Horn blowing at highway grade crossings is the dominant noise source for many railroad corridors, not wheel/rail interaction or diesel noise. Horns are essentially omnidirectional, even though they appear to be pointing in the forward direction.

Recommendation 5-4: Planning tools available from modal agencies of the U.S. Department of Transportation, such as the Federal Railroad Administration and the Federal Transit Administration, should be used in planning new rail transportation systems, and supplemental metrics should be developed and used to estimate the effects of noise on people. The public would benefit if warning horns were made more directional; research and development related to warning horn directivity should be undertaken to better understand the effects on safety and benefits to the public.

NOISE CONTROL IN BUILDINGS

Noise in and around buildings affects 100 percent of the population all or most of the time. Although acoustics are a factor in many aspects of building design and construction, the best technologies are not always used in buildings for a variety of reasons. This section provides an overview of significant demands for the building design community, particularly for acoustics in health care facilities, multifamily dwellings, classrooms, and entertainment structures.

Because of the sheer size of the building industry, widespread educational efforts are needed and design standards must be readily understood if satisfactory acoustical conditions are to be achieved in buildings. There is a need for better information on the acoustical characteristics of building materials and for the will to implement noise control measures—sometimes at added cost. At one time the National Bureau of Standards (now the National Institute of Standards and Technology) had an active program in building acoustics, but that program no longer exists. One result of the program was publication of *Quieting: A Practical Guide to Noise Control* by Berendt et al. (1976). As mentioned later in this chapter, there is an active program on building acoustics at the National Research Council of Canada, and design guidelines are being produced. Examples of currently available information are Cavanaugh and Wilkes (1998) and Harris (1994). Two documents related to noise in buildings published by the U.S. Department of Housing and Urban Development (HUD, undated; HUD, 1967) badly need updating. HUD noise policy can be found in 24 CFR 51.

To make building acoustics more accessible to the design community, two fundamental principles must be understood: (1) the source-path-receiver model (see Figure 4-1) and (2) the classification of building acoustics into six interrelated areas: exterior noise, interior room finishes, interior room noise levels, sound isolation between rooms, alarms and other electroacoustical systems, and building vibration (Cavanaugh et al., 2010). Although understanding these principles does not eliminate the need for special expertise, it does encourage the inclusion of someone with acoustical expertise on the design team.

HEALTH CARE FACILITIES

> Unnecessary noise, or noise that creates an expectation in the mind, is that which hurts a patient. It is rarely the loudness of the noise, the effect upon the organ of the ear itself, which appears to affect the sick. Of one thing you can be certain, that anything which wakes a patient suddenly out of his sleep will invariably put him into a state of greater excitement, do him more serious, aye, and lasting mischief, than any continuous noise, however loud.
>
> . . . .
>
> I have often been surprised at the thoughtlessness, (resulting in cruelty, quite unintentional), of friend or of doctor who will hold a long conversation just in the room or passage adjoining to the room of the patient, who is either every moment expecting them to come in, or who has just seen them, and knows they are talking about him. . . . If it is a whispered conversation in the same room, then it is absolutely cruel; for it is impossible that the patient's attention should not be involuntarily strained to hear.
>
> Florence Nightingale
> *Notes on Nursing: What It Is, and What It Is Not (1860)*

Florence Nightingale's words are echoed by many, perhaps most, current hospital patients. Noise is cited as the first or second source of complaints by patients and staff in hospitals today (Anjali and Ulrich, 2007). This level of concern has led to growing interest in the acoustics of health care facilities.

In "Are Acoustical Materials a Menace in Hospitals?" a presentation by Charles Neergaard at the third meeting of the Acoustical Society of America in 1930, he referred, at least obliquely, to noise in hospitals as a design issue that requires control of reverberations (Neergaard, 1930). Neergaard's studies of noise in hospitals were later described in another presentation, "What Can the Hospital Do About Noise?" at the twenty-sixth meeting of the Acoustical Society of America in 1941 (Neergaard, 1941). In a study of nine hospitals in Berlin, de Camp (1979) found noise levels that could interfere with patient well-being.

The use of modern instrumentation to study noise in hospitals was described as early as 1958 (Taylor, 1958). More recently, MacLeod et al. (2007) demonstrated a firm connection at Johns Hopkins hospital between treating surfaces to control sound reverberation and patient and staff satisfaction with improvements in the acoustical environment. In a recent paper, Pelton et al. (2009) demonstrate the need for noise isolation between rooms, especially when painful procedures are being performed. Unfortunately, the Institute of Medicine did not include a discussion of the adverse effects of noise in hospitals in its influential report, *To Err Is Human* (IOM, 2000). This may be a good subject for a future study. Almost every aspect of hospital design and operation is in dire need of noise control.

The Facility Guidelines Institute (FGI) Health Guidelines Revision Committee, in conjunction with the American Society for Healthcare Engineering (ASHE), has adopted recommendations from the *Interim Sound and Vibration Design Guidelines for Hospital and Healthcare Facilities* (FGI, 2009). The *Interim Guidelines* provide comprehensive recommendations for incorporating sound and vibration controls into the 2010 edition of the *FGI/ASHE Guidelines for the Design and Construction of Health Care Facilities*, which is issued every four to five years (ASHE, 2010) and is the primary reference for all aspects of hospital design in the United States.

Among the several needs is the development of a broader understanding of the implication of the use of sound-absorptive materials in nosocomial infection in health care settings. This includes not only whether sound absorptive materials pose a health risk, but also how that risk is measured and what the terms "clean-ability" and "microbial resistant" mean and how they can be measured.

Other materials that should be in design guidelines include controlling noise from air distribution systems, especially in operating rooms that require air curtains to control infection, and the use of pagers in hospitals and alarm systems. Many hospitals are considering eliminating areawide pagers in favor of silent pagers. In most hospitals today a gaggle of instrumentation is used to monitor patient health, each with its own alarm and of varying importance to patients' health, comfort, and safety; these instruments can produce a bewildering cacophony that places the patient at risk. Inaudible warning signals and signals with easily identifiable tonal qualities are being studied, and some are already available (McNeer et al., 2007).

Finally, some less obvious methods of reducing sound in hospitals have been found to have some success. For example, dark colors in corridors seem to encourage staff and visitors to lower their voices. Music can also be beneficial, as Florence Nightingale (1860) noted:

> The effect of music upon the sick has been scarcely at all noticed. In fact, its expensiveness, as it is now, makes any general application of it quite out of the question. I will only remark here, that wind instruments, including the human voice, and stringed instruments, capable of continuous sound, have generally a beneficent effect.

As a result of the promulgation of privacy requirements applicable to all health care facilities under the Health Insurance Portability and Accountability Act (HIPAA) of 1996, hospitals are now under added scrutiny by regulators (Sykes and Tocci, 2008). The American National Standards Institute Committee S12, Noise, Working Group 44 Speech Privacy, is currently developing a draft standard on speech privacy in health care facilities in response to the HIPAA legislation as a service to both regulators and health care facilities. The need is urgent as hospitals have been cited by the Joint Commission on Accreditation of Healthcare Organizations for permitting conditions of insufficient speech privacy, even though there are no speech privacy guidelines for health care environments.

MULTIFAMILY DWELLINGS

The desire for access to public transportation, close proximity to workplaces, and lifestyle changes have led to an increase in the construction of condominium units in many urban areas. However, perceived, or actual, inadequate sound isolation between units is a chief complaint of many residents. Putting aside complaints that arise from adjacent residents with incompatible lifestyles (e.g., a young couple with a new baby next to a young man who parties with his friends late into the night), there may be legitimate problems in the design and construction of these buildings.

Residential condominiums are predominantly of two types: (1) wood frame and (2) concrete deck/steel post and beam construction. Both are lightweight, and both have some advantages. Wood-frame buildings tend to be found in suburban areas where they blend in with existing wood-frame houses and have height limitations of three to four stories. Concrete and steel buildings can be taller and often fit better in urban environments. There are no hard-and-fast rules about which type of building is constructed, but concrete and steel buildings generally have a clear advantage in isolating sound.

The three most common sound isolation problems in multifamily buildings are floor/ceiling sound isolation, common-wall sound isolation, and plumbing sound isolation. Separate stud frames, doubling layers of gypsum wallboard, and glass fiber in common-wall cavities nearly always provide satisfactory sound isolation. However, in wood buildings, common walls are often used as shear walls for seismic restraint, which sometimes precludes double- or staggered-stud frames, and a single-stud frame must be constructed with plywood attached to both sides. In these cases, resilient channels do not work, so a separate sound isolation wall must be constructed on at least one side of the common wall to augment sound isolation.

Wood-frame buildings almost always need some concrete cover for structural decks to ensure the isolation of impact sounds. The thicker the concrete, the better the sound isolation, but also the more seismic restraint required in the building structure. Much work has been done with 1.5-inch concrete thickness on wood decks (Warnock and Birta, 2000), but current design conventions, motivated by cost and desires to maximize headroom, have reduced the thickness to 0.75 inches or less, which reduces the sound isolation of the floor/ceiling interface.

To meet the most common building code requirement of a minimum impact insulation class (IIC) rating of 50 requires multiple layers of gypsum wallboard on a resilient ceiling suspension system, as well as sound-absorptive material added to the floor/ceiling cavity and either carpeting or a hard finish in a resilient interlayer for the floor finish. The most common resilient suspension system is the resilient channel, of which only a single-leg type with long slots has been shown to provide reliable sound isolation (Lilly, 2002). However, newer elastomeric suspension elements used in lieu of resilient channels are now known to provide better and more consistent sound isolation than most available resilient channels (Kinetics, 2009; PAC International, 2009).

Resilient underlayers generally provide the much-needed increase in impact sound isolation when nearly all hard-finished floor systems are used in either wood or concrete buildings. Many manufacturers seized the need for resilient underlayers as an opportunity for using their products that were designed for other purposes. For example, foundation drainage and soil erosion control mats are being used as resilient underlayers in buildings. Even though this is effective in most cases, the multifamily housing industry would benefit greatly from a more systematic development of floor/ceiling assemblies and products that act together by design.

Sound isolation performances stipulated in building codes often fall short of occupants' expectations. Although building design teams often realize this and set higher design goals, their principal sources of guidance are collections of sound transmission class (STC) and IIC test reports, which are often conducted under early standards with products of unknown performance characteristics. The result is a wide range of measured ratings for ostensibly identical sound isolation assemblies.

This problem has been largely remedied for sound isolation wall assemblies by a large body of test reports published by the National Research Council (NRC) of Canada, which has also published floor/ceiling STC and IIC ratings for a variety of assemblies. For a summary of the NRC of Canada activities related to sound transmission in buildings, see Quirt (2009). There is more variation in floor/ceiling assemblies than common-wall assemblies, and it would be helpful if the NRC of Canada provided data on the wide range of floor/ceiling assemblies in the built environment.

The housing industry would also benefit from the development of theory and testing to characterize improvements to the design of acoustical materials such as the dynamic stiffness of resilient underlayers and how this information can be used to evaluate the IIC rating of floor/ceiling assemblies. Recent advances in the incorporation of damping into panelized building materials such as drywall should also be rated to provide a better understanding of how damping works in building sound isolation systems; this would also encourage product development. New material concepts should also be explored, such as distributed absorbers composed of heavy lumped masses embedded in a lossy sheet binder, which has been shown to improve the sound isolation of low-frequency airborne noise, and nanogels that offer high sound absorption and partial translucency.

CLASSROOMS

American National Standard S12.60, Acoustical Performance, Criteria, Design Requirements, and Guidelines for Schools is the first widely used standard for acoustical conditions in classrooms (available online at *http://asastore.aip.org*). This standard establishes limits for sound isolation between spaces; background sound produced by mechanical, electrical, and plumbing equipment and systems; and reverberation time.

For the most part, sound isolation and reverberation control methods and materials are well known. However, this is not true for in-room unitary HVAC (heating, ventilation, and air-conditioning) units or classroom ventilation units. Currently, sound produced by these units exceeds the American National Standards Institute (ANSI) recommended maximum sound pressure level of 35 dB(A), thus requiring the use of central air distribution using air handlers, air heating and cooling methods, and air distribution terminal units. The cost of these systems, according to manufacturers, school building owners, and designers, is considerably higher than the cost of typical classroom ventilation units.

Arguments by classroom equipment manufacturers to exclude or significantly raise permitted sound levels in order to permit the use of noisier conventional units have not persuaded the standards and education communities; the standard has not been modified. Nevertheless, the cost of school buildings and the need for flexibility are important issues. Hence, quiet design concepts for classroom ventilation units should be investigated. So far, manufacturers have had only limited success in developing units that are comparable in cost to more conventional central system equipment.

Certain manufacturers of electroacoustical products (microphones, loudspeakers, etc.) have argued that their systems can be used in place of more expensive architectural solutions to background sound, sound isolation, and reverberation. Most of these are one-way systems; the teacher speaks into a microphone and students wear hearing assistance devices. These systems generally do not work for student-to-student communication or student-to-teacher communication. The use of electroacoustical solutions to architectural acoustics problems is hotly disputed in the architectural acoustics profession. However, there may be a place for electroacoustical devices in classrooms, particularly for hearing-disabled students or those who have different learning styles.

GREEN ACOUSTICAL DESIGN

The importance of Leadership in Energy and Environmental Design (LEED) certification[5] for newly constructed buildings and for the reuse/rehabilitation of existing buildings is rapidly becoming a focal point of building design. Whereas only two years ago little attention was paid to green design, including acoustics in the green design of buildings, it is now being addressed in some cases, notably in classrooms and hospitals. Up to now, acoustics has played a minor role in the LEED rating of a building, although significant contributions to LEED ratings have been possible through high-recycled-content products, such as acoustical ceilings, duct silencer fill, and the use of acoustical products produced near project sites. It is expected that the availability of green acoustical products will increase over time.

Green factors affect all building systems, which in turn affect the acoustics of a building. In a post-occupancy survey of building acoustics (see Muehleisen, 2009), it was found that bad acoustics was at the top of the list of undesirable features (acoustics, thermal comfort, air quality, lighting, etc.) for *all* buildings and was considered even *more* undesirable for green buildings. Some features of green buildings that are considered important for reasons other than acoustics include more use of natural lighting, natural ventilating systems, use of hard interior surfaces, maximum use of windows (especially when they must open), and the lack of conventional (porous) acoustical materials. All of these features tend to degrade the acoustical quality of workspaces.

Some green features include lower partial-height partitions, which may be used to extend natural light farther into an open-plan building space. However, this can reduce speech privacy between workstations. Another feature is the use of green materials that, in many cases, absorb less sound than conventional materials. However, this can result in an excessively reverberant environment and reflections from the ceiling can compromise speech privacy in open-plan offices. Natural ventilating systems are considered to be desirable in green buildings, but they can transmit noise throughout a building. The ability to open windows is considered desirable but can result in transportation noise entering a building and being transmitted through the ventilation system. Lack of conventional acoustical materials in buildings can affect

[5] LEED is an initiative of the U.S. Green Building Council.

speech privacy, as mentioned above, and can also affect speech intelligibility in conference rooms.

Electronic sound masking, widely used in open- and closed-plan offices since the 1960s, is now necessary as a means of maximizing speech privacy. But, although electronic sound masking can go a long way toward ensuring acceptable speech privacy, it is usually not a sufficient solution. Green solutions to office workstation partition height and sound absorption will have to be developed. The requirement for more natural ventilation, including opening of windows, just adds to the challenge.

Razavi (2009) has reported on some acoustical improvements in green building ventilation systems, but the noise control engineering and architectural acoustics community face a major challenge in integrating good acoustical conditions into green buildings. The *Green Guide for Health Care* and the *Green Guide for Schools* establish design objectives for acoustical building characteristics, including reverberation, sound isolation, and ambient sound in building spaces (*http://www.gghc.org*; *http://www.buildgreenschools.org*). LEED points[6] are added if these objectives are met using green materials and methods.

AIR DISTRIBUTION SYSTEMS

The American Society of Heating, Refrigerating and Air-Conditioning Engineers (ASHRAE) has been the pioneer and sole standard bearer in the development of standards and estimation methods for sound produced by air distribution systems. Much to the organization's credit, it has funded most of the research that is now the exclusive basis for estimating and evaluating sound produced by building ventilation systems.

Designers of building mechanical systems rely on the *ASHRAE Guide* (ASHRAE, 2007), a handbook updated every five years that covers all aspects of the design of building mechanical systems (thermal and ventilation). The *ASHRAE Guide* includes a chapter on design goals for sound produced by building mechanical systems. The design goals are divided into noise criteria, room criteria ratings, and A-weighted sound pressure levels. The algorithms for estimating sound in building spaces are based on work by Reynolds and Bledsoe (1989).

Little progress has been made since 1989, when the algorithms were published, despite efforts by TC 2.6 (the ASHRAE committee on sound and vibration) to improve the situation. In fact, it has been generally agreed that the previously used general method of estimating the sound power level of ventilation fans should be dropped from the *Guide* because of its unreliability.

It would be beneficial if industry and academia formed a partnership to study the acoustical literature, produce some additional theory and testing, and include new information in the *ASHRAE Guide*. This would reinforce the tools used by the mechanical engineering profession to address sound produced by new green mechanical systems, such as numerous small fans operating in parallel in lieu of a single large fan, new concepts in passive induction units that replace fan-powered terminal units, and the development of new, quiet classroom ventilators (discussed above).

ENTERTAINMENT VENUES

The rapid increase in multifamily urban dwellings is likely to increase demand for public entertainment venues, both inside and outside buildings, particularly small venues that can nurture a sense of community. Small venues can provide opportunities for the performing arts in intimate, attractive performance spaces. Entertainment in the broadest sense includes music, cinema, and theater but also dining and parks.

The proliferation of small entertainment venues would open the door to commercial opportunities in lighting, sound system equipment, and computer-controlled software, all of which have been addressed in the marketplace. However, the proximity of entertainment venues to living spaces, and community annoyance from sound that sometimes results, can be a significant challenge. Rather than prohibiting such proximity, communities and developers should be guided by planning guidelines and codes that protect residences with only minimal compromises in performance or entertainment. Conflicts that arise between performers and the public were discussed and resolved in a decision by the U.S. Supreme Court in 1989 (Ward, 1989).

FINDINGS AND RECOMMENDATIONS

More people are probably affected by noise inside buildings—such as sound transmission in multifamily buildings, noise (and reverberation) in classrooms, noise in residences from road, rail, and air traffic, and noise in hospitals—than in any other environment. Clearly, trade associations and professional societies will play important roles in the design and construction of quieter interior spaces.

Recommendation 5-5: The acoustics and noise control communities should actively promote the inclusion of noise criteria in requirements for Leadership in Energy and Environmental Design (LEED) certification of buildings, not only to improve the noise environment but also to ensure that the acoustical environment is not degraded. Design standards (e.g., building codes) must be improved to ensure that good acoustical practices are followed in the construction of buildings.

Recommendation 5-6: The National Institutes of Health and/or the Facilities Guidelines Institute should fund the de-

[6]LEED certification involves awarding points for various aspects of green designs.

velopment of improved materials for hospital environments, where traditionally used materials may harbor and promote the growth of bacteria and other harmful biological agents.

MODELING, SIMULATION, AND DATA MANAGEMENT

Perhaps the greatest change in technologies for noise reduction has occurred because of increased computational power, which has changed the way products are designed, tested, and analyzed. We now have tools for defining and manipulating structures and mechanisms, for modeling and simulation, for laboratory measurements on prototypes, and for processing and interpreting voluminous amounts of data.

Mechanism analysis programs compute the motions and forces of gears, cams and followers, cranks, and sliders that are the sources of audible energy in many products. It is said that a sewing machine contains more interesting mechanisms per dollar than any other product. The forces that these mechanisms place on their supports lead to vibrations in the product structure that are analyzed using finite element analysis. These vibrations in turn cause radiated sound, analyzed by boundary element analysis.

The computer also has just as important a role in the experimental testing that is part of the product engineering process. Accelerometer arrays allow the measurement and display of the natural modes of structural vibration, and postprocessing using modal analysis programs is used to test the validity of both the measured modes and those computed using finite element analysis. Microphone arrays allow the quantification and display of the radiation of sound from the product using software for acoustic intensity and acoustical holography.

As discussed below, the existence of these technologies does not mean that product companies are able to take advantage of them. Cost—in terms of the acquisition of the software/ hardware and the commitment to the training and retention of specialized personnel—can be a problem, particularly to smaller companies. Making these new methods more affordable and available to companies is a challenge to be met.

MODELING AND SIMULATION

Traditional modeling for sound has been based on "canonical problems" representing different aspects of a sound source. Simple examples include radiation from bending waves on a plate to estimate sound from machine or equipment housing and a simple monopole source of sound to represent the radiation of sound from the unsilenced inlet of a compressor. These models can be useful aids to understanding, but they cannot deal with all aspects of design.

Some modeling procedures are oriented toward describing and analyzing mechanisms. These models, which can compute motions and forces attributable to cams, followers, and other components, enable computation of forces at supporting points, combined with structural finite element analysis, to predict vibrations of the structure. Other software packages can use information about structure and vibration to compute radiated sound. Although at one time these capabilities were available only in distinct packages, software companies today offer them as an integrated package.

In some products, airflow and heat transfer, accompanied by noise from fans and airflow, represent a different kind of interaction between mechanisms, product geometry, and sound production. Progress toward an integrated procedure is not as advanced as in the example cited above, but there is little doubt that integration will be achieved in the near future.

DATA MANAGEMENT AND ANALYSIS

Microphone arrays are now commonly used to characterize radiation from a structure. The analysis of these data can take the form of acoustic intensity or acoustical near-field holography. Data rates are typically 50 kilobits per second for 24-bit words; thus, a 10-second recording is 1.2 megabits of data and a 100-microphone array will generate 120 megabits of data per experiment. This kind of data collection can be done with modern (even ordinary) computers, but keeping track of all of these data for later processing can be a challenge. Generating the intensity and/or hologram graphs for N channels of data may require as many as $N(N-1)/2$ cross spectra for these data records of a simple 10-second experiment that will be repeated many times.

Similar issues arise in collecting and processing vibration data to correlate with acoustical data. Accelerometers are the most widely used sensors, but new scanning, three-axis laser vibrometers are increasingly being used. The latter have signal processing, in the form of cross spectra between channels, "built in" to the system. A laser vibrometer channel is much more expensive than an accelerometer channel, but in some situations being able to analyze data without physical contact between the sensor and the structure or the airflow can be valuable.

CONSUMER PRODUCTS

MAKING PRODUCTS QUIETER AND SELLING QUIET

U.S.-made white goods (major household appliances such as refrigerators, dishwashers, and cookers), health care devices, personal care products, and other products are mostly sold on the domestic market; the export sector is relatively small. In addition, foreign competitors are moving into U.S. markets and challenging U.S. companies abroad. The sound and sound quality of products is important for market acceptance, and technology for improving sound and/or producing quieter products is important for maintaining U.S. competitiveness. (See Chapter 6.)

Although the current economic situation may slow reductions in product noise, there is little doubt that consumers have concluded that quieter products are better built and have "real quality" and not just better "sound quality." On the other hand, the market has not favored developments that result in increased prices. Thus many consumer products become commodities with different manufacturers meeting the same price points and offering very similar products.

In some product sectors, however, consumers are willing to pay more for products with extra features or materials. For example, new countertop cookers and refrigerators with brushed steel exteriors and countertops made of granite have become status symbols and statements of achievement. Kitchens are becoming gathering places where these products are displayed. These upscale products (made both in the United States and abroad) are generally quieter and have profit margins sufficient to support extra engineering and manufacturing costs. But these products, although growing, remain a smaller part of the market. There is still a need to make the technology for better noise control more available in the manufacturing environment where cost constraints are very important.

PRODUCT SOUND QUALITY

Metrics for product sound are important for controlling noise exposure, measuring customer satisfaction, and guiding design. The acceptability of the sound of a product is influenced by user expectations, context, and signal content or information. Unfortunately, noise control professionals have labeled product sound as "product noise," implying that any sound from a product is undesirable. Perhaps as a reaction to the notion of product sound as product noise, the most attention has been paid to metrics, such as A-weighted sound pressure level, that measure noise exposure, annoyance, and hearing impairment and reflect negative reactions to sound.

However, hearing scientists (psychologists and engineers) and product designers are aware that A-weighted sound level is an imperfect measure for predicting product sound acceptability. Recent work has focused on defining physical metrics that can select out certain sound signal features that are separately audible and are likely to be associated with positive or negative reactions to sound.

In some cases the link between metrics and design is very strong. Product engineers in the automotive industry can sit at a workstation, manipulate signals by filtering and other means, and decide that certain signal features (tones, modulation, and transients) should be changed to achieve a more desirable sound for the driver and passengers in a car. The "sound quality" programs used allow them to modify signals and process the resulting signals to determine changes in 20 to 30 physical metrics. The changes in values are an indication of how the sound should be evaluated as design changes are made. In this case the first evaluation is made by an engineer or a product designer.

Jury (listening panel) studies are a useful mechanism for designing for better sound quality. Listeners are presented with a group of sounds from real or virtual products and asked to rate them in terms of acceptability. The number of sounds, their order, the number of listeners, and the scaling of responses are all part of the experimental design. In a sense the jury is a measuring instrument, the output of which is a measure of sound quality. But to anticipate the effect of future design changes on sound quality, either the jury study must be repeated or a correlation must be found between physical metrics and the jury's response.

Historically, acousticians have associated perceptual aspects of sound with individual physical metrics. Thus, the perception of loudness correlates well with the physical metric of "loudness." A similar correlation between the perception of annoyance and the metric "noisiness" was developed for jet aircraft and later applied to other noise sources. But as the perceptions become more complex, involving expected, informative, and hedonistic dimensions, the correlation between any single physical metric and perception breaks down, and one is required to look for patterns of acceptability or sound quality of a product, and that correlation will be different for each product. This has been expressed as "a good lawnmower does not sound like a good washing machine."

Physical metrics in use include tonality (the presence of tones in the signal), spectral balance (high-frequency versus low-frequency content), fluctuation strength (presence of modulation), and roughness (nonharmonic dissonant components) as well as loudness and noisiness. One sound quality program evaluates nearly 20 such physical metrics to form a profile of values to correlate with jury judgments of product sounds. Products for which such metrics profiles have been used to correlate with jury study judgments of sound quality include washing machines, dishwashers, vacuum cleaners, cookers, and room air cleaners.

The metrics profile that best correlates with good sound quality (or most acceptable) will be different for different products, but there are certain features of the sound that are undesirable for any product. Loudness, noisiness, tonality, and fluctuation strength are all undesirable if too strong. Modulation is an interesting example because it is very desirable in music as vibrato or tremolo but undesirable in a product sound. The reason seems to be that modulation captures our attention—desirable in music, undesirable in a product.

There is little cost to generating a profile of 20 or more metrics since this only requires running the same sound samples that are to be presented to a jury through the signal processing algorithm for each metric. Using a larger set of physical metrics can give some reassurance that nothing has been missed, but making sense of the profiles can be difficult. If the metrics profile for each sound is labeled with the jury evaluation for that sound, it is possible to combine the metrics into a smaller set of variables using the method of principal component analysis.

Manufacturers would like to have a single metric such as A-weighted sound level that would enable them to claim their products have better sound than their competitors and can also be used in product development. Organizations such as Consumers Union that routinely evaluate products for sound would also like such a metric. Unfortunately, the correlation between any single metric and sound quality and the outcome of jury studies has not been generally accepted by the acoustics community, so claims that one product has "better sound" than another cannot be supported by physical metrics, even though improvement in the sound quality of a particular product in a particular organization is possible. For more information on product sound quality, see Lyon (2000, 2004) and Lyon and Bowen (2007).

R&D IN SUPPORT OF QUIETER PRODUCTS

Sound is very important for some products (e.g., automobiles), and companies spend heavily in terms of facilities and personnel to make these products quiet and pleasing. But in the past 40 years or so, the price of an automobile has risen by a factor of more than 10, while the price of a dishwasher has risen by a factor of 3 to 4. One result is that while the automobile companies have developed large staffs and good facilities for sound, most appliance companies have not (with one notable exception). In typical appliance and health care products companies, engineers are "jacks of all trades," working one day on problems of airflow or heat transfer and the next on product sound. Also, these engineers may have significant motivation to move around in a company where the path upward is through management and not technical expertise.

Another factor that affects nonautomotive producers is the pace of model changes. Appliances, health care, and personal care products go through much more frequent changes, so consumers will replace older products or choose to buy a newer product because of a desired feature. The effect of this is to compress development schedules and to limit the transfer of a new development (e.g., a quieter way to support a small motor) into the new model.

It would appear that simpler products such as a sleep apnea device should have noise issues that are simpler. But this product has a couple of brushless DC motors, a fan, an air pump, and valves, each of which produces audible sound in a device that is in someone's bedroom at night. In addition, cost and utility constraints mean the enclosure is lightweight and stiff, a perfect construction for the efficient radiation of sound. The manufacturer probably buys the motors from a manufacturer in China and finds it impossible to convince his supplier to do the engineering to make the motor quieter.

There are other trends that are not helpful in terms of product sound. Design for manufacturing has a cachet that is attractive to industry because of lower assembly costs and easier model changes. One such method is "layering," in which an assembly is achieved by placing components into the supporting structure in a sequence that minimizes the need for reconfiguring the assembly. When this method was applied to a popular electric mixer, its noisiness was significantly increased because of the increased tolerances in the drive train gearing inherent in this method of assembly.

The basic message is that issues of product sound are very complex and do not become simpler and easier to handle because a product is simpler and less costly. Indeed, the situation may be quite the opposite. But there are good tools for meeting the need. The question is: are they being used and, if not, why not?

TOOLS FOR QUIET PRODUCT DESIGN AND TESTING

Most companies now use computer-aided design (CAD) software to visualize their product designs and to anticipate problems of parts interference and fit before a prototype is built. These CAD programs can be interfaced with certain computer-aided engineering programs like finite element analysis for structural analysis (stiffness, resonant modes, mass distribution) or dynamic analysis for mechanism forces. But these programs (discussed above) while useful, are limited in their assistance in designing for quiet function.

For example, a fan can be analyzed using a computer fluid dynamics (CFD) program, which most likely does not reflect the actual flow environment of a typical product. Also, these programs are very expensive to run, and considerable expertise is needed to run them. Most consumer products companies will not make the investment in personnel or funds to have their products analyzed in this way. Some CFD providers will work with manufacturers on a consulting basis to provide such analyses, but the process remains expensive and the idealized calculations may not provide the information needed for design decisions.

Manufacturers are more likely to invest in experimental facilities than software for analysis for several reasons. First, the cost of experimental equipment has been coming down and its capabilities are increasing. Multichannel systems of microphones and vibration sensors (accelerometers) involving dozens of sensors are now commonplace, and the software to analyze the patterns of sound and vibration, such as acoustical near-field holography and modal analysis, is widely available. Also, experimental work is generally more relied on in product development than is analysis. The ability to keep engineers in place long enough to become proficient in the use of both hardware and software remains an issue but seems to be much less of an issue than for the analytic software.

WHAT'S NEXT?

Although the current economic situation may slow sound improvements, there seems little doubt that consumers have become convinced that quiet products are better built and have "real quality" and not just better "sound quality." So the issue of better sound as a marketing feature will not go away, and the need to support the industry in its attempts to meet this marketing and technical challenge will not go away. Thus the technology for better sound must be more available in the manufacturing environment where cost constraints are very important.

ACTIVE NOISE CONTROL

The most efficient and cost-effective way of reducing noise is to design equipment to produce less noise. If this strategy has been fully implemented and additional noise reduction is needed, add-on measures must be applied. Active noise control is one of these measures.

Most noise sources produce noise in a wide frequency range. Passive noise control measures (such as silencers, acoustic enclosures, wrappings, barriers, etc.) usually provide sufficient noise reduction at middle and high frequencies (approximately 200 Hz and above), and they are robust, reliable, and cost effective. However, they are ineffective at low frequencies (below about 200 Hz). At these low frequencies, active control becomes an alternative; it may be the only solution for frequencies below 100 Hz.

Noise sources such as gas turbines and large reciprocating compressors produce high levels of low-frequency noise. Almost without exception, the noise control of such sources requires a combination of both passive and active measures. The passive measures attenuate the mid and high frequencies, and the active measure attenuates the low frequencies.

There are four major active noise control strategies:

1. Reducing the sound radiation efficiency of the sound source by placing a secondary source (loudspeaker in an enclosure) in its immediate vicinity and driving it with an electric signal that produces the same magnitude but opposite phase fluctuating volume flow as the primary noise source. In this case the air volume pushed out of the primary source during the positive cycle fills the void generated by the receding volume of the secondary source and, conversely, the receding volume flow of the primary source is supplied by the outflow from the secondary source. This strategy, which reduces the radiation efficiency of the original source and effectively reduces the noise level at all locations, is sometimes referred to as "global" noise reduction.
2. Creating a limited "zone of silence" in the vicinity of the receiver (the person to be protected) by sensing the local sound pressures, driving the loudspeaker with an electric signal (located as near to the receiver as practicable) that produces a sound pressure of the same magnitude and opposite in phase as the primary signal. This is the only situation where "noise cancellation" is appropriate. This active noise control strategy, in almost all cases, is inferior to the first strategy because its effectiveness is limited to a single area. Because this strategy does not affect the sound power output of the primary source and creates a secondary source, the overall noise level is increased in locations where cancellation does not occur. A good practical application of this strategy are noise-canceling headphones, such as those manufactured by the Bose Corporation that achieve a significant reduction in sound pressure level in the ear canal.
3. Increasing the low-frequency sound attenuation of tuned dissipative silencers by placing actuators (loudspeakers) in the cavity behind the thin porous lining as described by Vér (2000). The sound pressure is sensed behind the porous lining by a microphone and entered into a control system that feeds the loudspeaker with a signal so that for a wide frequency band it produces (nearly) zero sound pressure immediately behind the porous liner. This condition maximizes the sound pressure gradient across the liner and consequently its ability to absorb sound. In a passive silencer this condition occurs only at single frequencies where the depth of the airspace is one-quarter the acoustic wavelength and at odd multiples of that frequency (frequency, f, and wavelength, λ, are related by $f = c / \lambda$, where c is the speed of sound).
4. When the noise is produced by the sound radiation of a structure exited to vibration by localized dynamic forces (such as the attachment points of the wing of an airplane to a ring frame), the most efficient way to obtain global noise reduction is to mount a shaker at the attachment point and feed it by a control system to produce nearly zero vibration (i.e., render nearly zero power input to the structure). Here, again, the noise that is attributable to the vibration force is reduced at all locations.

One early example of active control was the electronic sound absorber (Olson and May, 1953), which was a microphone, phase inverter, and loudspeaker that could be used to create a "zone of silence" around the head of a factory worker. At that time all of the circuits were analog, and phase shift through the system was critical. It was not until digital signal processing became feasible that applications began to be developed.

Active control of sound is effective only when the wavelength of the sound is long compared with the dimensions of the volume in which cancellation is desired. For example, the most successful application of the technology is in active headsets where cancellation of sound in the (small-volume) ear canal is desired. Another example is cancellation in the cabin of a turboprop commuter airplane, which requires a large number of microphones and loudspeakers and is only effective at low frequencies.

This limitation of cancellation to low frequencies also

has implications for sound perception, sound quality, and hazard to hearing. A listener may perceive the sound as lacking in low frequencies. Hence, it may sound "hissy." The A-frequency weighting network already attenuates low-frequency sound, and therefore additional attenuation through active control may not produce a significant decrease in the A-weighted sound level. According to current standards, a small decrease in A-weighted sound level produces only a small decrease in the hazard to hearing.

APPLICATIONS OF ACTIVE CONTROL

Despite the complexity of active control and the above limitations, the technology has been applied in a number of cases. Some examples are given below. Active headsets provide noise reduction and both comfort and protection from hazardous noise for the user. The Federal Railroad Administration has demonstrated both active control in locomotive cabs and proof of principle for active control of exhaust stack noise from idling locomotives. Hansen (2005) developed an active control system to control sound propagation in the exhaust stack of a spray dryer unit in a dairy factory. Scheuren (2005) discussed a number of engineering applications of active control, including wind tunnel buffering, control of combustion burners, noise control in gas turbines, and modification of sound in the cabin of automobiles. Cancellation of the blade passage tone in a small axial flow fan was achieved by Sommerfeldt and Gee (2003) by using four small cancellation loudspeakers placed around the fan. There are a number of applications of active control in the aerospace industry; these have been described by Maier (2009). Gorman et al. (2004) produced noise reduction on the flight deck of an airplane, and Cabell et al. (2004) have shown how active control can be used to control chevrons and produce noise reduction of a jet engine exhaust. Finally, Fuller et al. (2009) reduced noise from a portable generator set by using active control.

Impediments to Commercial Development

Despite the long history of the development of active control technology and digital processing systems, there are few devices (except for active headsets) on the market today. Some of the barriers to commercial development are expense and reliability as well as the materials used and characteristics of transducers, amplifiers, and materials.

Active control systems are expensive to implement because of the required microphones (or accelerometers), loudspeakers (or force transducers), and electronic control systems. If a universal control system were to be developed, it would have to be versatile because the control algorithm will depend on the type of noise being canceled (e.g., a single-frequency tone, a tone in noise, or broadband noise). Reliability is also an issue in complex systems.

For high-intensity noise sources, high-powered amplifiers and special loudspeakers may be required. There is also the problem that the materials used for transducers (microphones, accelerometers, loudspeakers, force transducers) must, in many cases, withstand hostile environments. Examples are hot exhaust gases and turbulent flow.

There is a rich literature on active control of sound and vibration. This includes books (Hansen and Snyder, 1997; Nelson and Elliott, 1993), technical articles (Nelson and Elliott, 1993; Tichy, 1996), and conference proceedings papers (ACTIVE, 2009; Fuller, 2002).

Recommendation 5-7: Research agencies should fund university research on active noise control to address situations where the use of traditional noise-control materials is problematic or where they are not suitable for attenuating noise in the appropriate frequency range. Investigations into hybrid active-passive and adaptive-passive noise control systems and the development of low-cost microphones and loudspeakers that can be used in hostile environments should also be funded.

SUMMARY

Active controls of sound and vibration have been under development for many years, but few products on the market have incorporated them, and many barriers must still be overcome.

In this chapter, technologies for controlling noise from a large variety of sources have been described. Clearly, aircraft noise control technology is much more advanced than technologies for addressing other noise sources, and the funds expended to reduce the noise of airplanes themselves as well as mitigation measures around airports are far greater than for other noise sources. Road traffic noise has been controlled mostly by constructing noise barriers, but work is being done on promising technologies for reducing noise generated by tire/road interaction. Technologies are available for reducing noise from rail-guided vehicles, and these will become more important as the nation develops light rail systems and high-speed trains. Technologies for the built environment will also become more important as building construction is driven by LEED certification and "green" principles.

REFERENCES

23 CFR 772. Procedures for Abatement of Highway Traffic Noise and Construction Noise. Available online at *http://www.fhwa.dot.gov/hep/23cfr772.htm*.

ACTIVE. 2009. Collected papers from the ACTIVE series of international symposia on active control of sound and vibration: 1995, 1997, 1999, 2002, 2004, 2006, and 2009. Available online at *http://www.atlasbooks.com/marktplc/00726.htm*.

Ahuja, K.K. 1995. Aeroacoustic Performance of Open-Jet Wind Tunnels with Particular Reference to Jet/Collector Interactions. Final Report AEDC-SBIR-94-02. Reston, VA: American Institute of Aeronautics and Astronautics.

Anjali, J., and R. Ulrich. 2007. Sound Control for Improved Outcomes in

Healthcare Settings. Issue Paper 4. Concord, CA: Center for Health Design.

AREMA (American Railway Engineering and Maintenance-of-Way Association). 2009. Rail Transit. Chapter 12, Table 12-2-6, in Manual for Railway Engineering. Lanham, MD: AREMA. Available online at *http://www.arema.org/eseries/scriptcontent/custom/e_arema/pubs/pubs.cfm?activesection=pubs*.

ASHE (American Society for Healthcare Engineering). 2010. 2010 Guidelines for the Design and Construction of Health Care Facilities. Dallas, TX: Facilities Guidelines Institute. Available online at *http://www.fgiguidelines.org/index.html*.

ASHRAE (American Society of Heating, Refrigerating and Air-Conditioning Engineers). 2007. ASHRAE Handbook—HVAC Applications. Atlanta, GA: ASHRAE.

Aviation Weekly. 2009. NASA Pushes Blended Wing/Body. January 13. Available online at *http://www.aviationweek.com/aw/generic/story.jsp?id=news/Body011309.xml&headline=NASA%20Pushes%20Blended%20Wing/Body%20&channel=space*.

Bartlett. P., N. Humphreys, N., and P. Phillipson. 2004. The Joint Rolls-Royce/Boeing Quiet Technology Demonstrator Programme. American Institute of Aeronautics and Astronautics (AIAA) 2004-2869. Presented at the 10th AIAA/CEAS Aeroacoustics Conference, Manchester, United Kingdom, May 10–12.

Berendt, R.D., E.L.R. Corliss, and M.S. Ojalvo. 1976. Quieting: A Practical Guide to Noise Control. National Bureau of Standards Handbook 119. Washington, DC: U.S. Department of Commerce.

Bone, D. 2009. VITAL Project Overview. Presentation at the European Engine Technology Workshop, Warsaw, Poland, June 2–3. Available online at *http://www.newac.eu/69.1.html*.

Bowlby, W. 1998. Highway Noise Prediction and Control. Pp. 48.1–48.23 in Handbook of Acoustical Measurements and Noise Control, edited by C. Harris. Melville, NY: Acoustical Society of America.

Cabell, R.H., N. Schiller, J.H. Mabe, R.T. Ruggeri, and G.W. Butler. 2004. Feedback Control of a Morphing Chevron for Takeoff and Cruise Noise Reduction. Proceedings of ACTIVE 04, The 2004 International Symposium on Active Control on Sound and Vibration, Williamsburg, VA, September 20–22. Available online on the ACTIVE 2009 CD at *http://www.bookmasters.com/marktplc/00726.htm*.

CAETS (International Council of Academies of Engineering and Technological Sciences). 2008. The Design of Low-Noise Vehicles for Road, Rail, and Air Transportation. Sponsored by the Royal Swedish Academy of Engineering Sciences and the Royal Academy of Engineering (U.K.), Institute of Sound and Vibration Research, Southampton, U.K., June 2–4. Available online at *http://www.noisenewsinternational.net/docs/caets-2008.pdf*.

Cavanaugh, W.J., and J.A. Wilkes. 1998. Architectural Acoustics: Principles and Practice. New York: John Wiley & Sons.

Cavanaugh, W.J., G.C. Tocci, and J.A. Wilkes, eds. 2010. Architectural Acoustics: Principles and Practice, 2nd ed. Hoboken, NJ: John Wiley & Sons.

Collier, F.S., and D.L. Huff. 2007. NASA's Subsonic Fixed Wing Project. Presented at Revolutionary Aircraft for Quiet Communities Workshop, Hampton, VA, July 24–26. Available online at *http://www.nianet.org/workshops/docs/QA/presentations/KFS/Huff.pdf*.

Coppinger, R. 2007. Quietly does it. Flight International 172(5104):11–17.

De Camp, U. 1979. Measurement of sound levels in hospitals. Noise Control Engineering Journal 13(1):24–27.

DG Energie et Transport and DG Recherche. 2001. European Aeronautics: A Vision for 2020—Meeting Society's Needs and Winning Global Leadership. Luxembourg: Office for Official Publications of the European Communities. Available online at *http://www.eurosfaire.prd.fr/7pc/bibliotheque/consulter.php?id=1130*.

Donavan, P. 1993. Tire/Pavement Interaction Noise Under Vehicle Operating Conditions of Cruise and Acceleration. Document No. 931276. Proceedings of the Noise & Vibration Conference and Exposition, Traverse City, MI, May. Warrendale, PA: SAE International.

Donavan, P. 2005. Quieting of Portland Cement Concrete Highway Surfaces with Texture Modifications. Proceedings of NOISE-CON 05, The 2005 National Conference on Noise Control Engineering, Minneapolis, MN, October 17–19. Available online at *http://www.bookmasters.com/marktplc/00726.htm*.

Donavan, P. 2006. Comparative Measurements of Tire/Pavement Noise in Europe and the United States—NITE Study. Report No. FHWA/CA/MI-2006/09. Sacramento, CA: California Department of Transportation.

Donavan, P., and B. Rymer. 2009. Measurements of the Vertical Distribution of Truck Noise Sources During Highway Cruise Pass-bys Using Acoustic Beam-Forming. Compendium of Papers, Transportation Research Board 88th Annual Meeting, Washington, DC, January.

Donavan, P., and R. Schumacher. 2007. Exterior Noise of Vehicles—Traffic Noise Prediction and Control. Chapter 120 in Handbook of Noise and Vibration Control, edited by M. Crocker. Hoboken, NJ: John Wiley & Sons.

Donavan, P.R., R.F. Schumacher, and J.R. Stott. 1998. Assessment of tire/pavement interaction noise under vehicle passby test conditions using sound intensity measurement methods (A). Journal of the Acoustical Society of America 103(5): 2919.

DOT (U.S. Department of Transportation). 2002. Final Environmental Impact Statement: Final Rule for the Use of Locomotive Horns at Highway-Rail Grade Crossings. Washington, DC: Office of Railroad Development, Federal Railroad Administration.

EPA (U.S. Environmental Protection Agency). 1974. Information on Levels of Environmental Noise Requisite to Protect Public Health and Welfare with an Adequate Margin of Safety. Report No. 550/9-74-004. Arlington, VA: EPA Office of Noise Abatement and Control.

EPA. 1981. Noise in America: The Extent of the Noise Problem. EPA/ONAC Report No. 550/9-81-101. Washington, DC: EPA.

EU (European Union). 2007. "Silence(R)" Aircraft Noise Reduction Project Holds Final Meeting. Available online at *http://ec.europa.eu/research/transport/news/article_5637_en.html*.

FAA (Federal Aviation Administration). 2004. Report to Congress—Non-military Helicopter Urban Noise Study. Available online at *http://www.faa.gov/regulations_policies/policy_guidance/envir_policy/media/04Nov-30-RTC.pdf*.

FAA. 2007. Capacity Needs in the National Airspace System. Available online at *http://www.faa.gov/airports/resources/publications/reports/media/fact_2.pdf*.

FAA. 2008. Aviation and the Environment: FAA's and NASA's Research and Development Plans for Noise Reduction Are Aligned But the Prospects of Achieving Noise Reduction Goals Are Uncertain. GAO-08-384. Available online at *http://www.gao.gov/new.items/d08384.pdf*.

FGI (Facilities Guidelines Institute). 2009. Interim Sound and Vibration Design Guidelines for Hospital and Healthcare Facilities. Draft, November 1, 2006. Available online at *http://www.fgiguidelines.org/pdfs/Interim%20Sound%20and%20Vibration%20Design%20Guidelines%20for%20Hosp%20and%20Healthcare%20Fac.PD1-Watermark.pdf*.

FHWA (Federal Highway Administration). 2009. Tire Pavement Noise. Available online at *http://www.fhwa.dot.gov/environment/noise/tire-pavement_noise/*.

Fleming, G., A. Rapoza, and C. Lee. 1996. Development of National Reference Energy Mean Emission Levels for the FHWA Traffic Noise Model (FHEA TNM), Version 1.0. Report No. DOT-VNTSC-FHWA-96-2. Washington, DC: U.S. Department of Transportation.

FRA (Federal Railroad Administration). 2002. Final Environmental Impact Statement: Final Rule for the Use of Locomotive Horns at Highway-Rail Grade Crossings. Washington, DC: FRA.

FRA. 2005. High-Speed Ground Transportation Noise and Vibration Impact Assessment. HMMH Report No. 293. October. Washington, DC: FRA.

FRA. 2006. Use of Locomotive Horns at Highway-Rail Grade Crossings; Final Rule. 49 CFR § 222 and 229 (FR 71, 159, August 17, 2006). Washington, DC: FRA.

FTA (Federal Transit Administration). 2006. Transit Noise and Vibration Impact Assessment. FTA-VA-90-1003-06. Washington, DC: FTA. Available online at *http://www.hmmh.com/cmsdocuments/FTA_cover_sec01.pdf*.

FTAG (Federal Transportation Advisory Group). 2002a. Vision 2050: An Integrated Transportation System. Available online at *http://scitech.dot.gov/polplan/vision2050/index.html*.

FTAG. 2002b. Aeronautics Research and Technology for 2050: Assessing Visions and Goals. Available online at *http://scitech.dot.gov/polplan/vision2050/index.html*.

Fuller, C.R. 2002. Active Control of Sound Radiation from Structures, Progress and Future Directions. Paper P236. Proceedings of ACTIVE 02, The 2002 International Symposium on Active Control of Sound and Vibration, Southampton, United Kingdom, July 15–17. Available online at *http://www.bookmasters.com/marktplc/00726.htm*.

Fuller, C.R., C. Papenfuss, and T.D. Saux. 2009. Active Control of Portable Generator Set Noise: Heuristic Verses Design. Proceedings of ACTIVE 09, The 2009 International Symposium on Active Control of Sound and Vibration, Ottawa, Canada, August 20–22. Available online at *http://www.bookmasters.com/marktplc/00726.htm*.

GAO (U.S. Government Accountability Office). 2002. Aviation and the Environment: Impact of Aviation Noise on Communities Presents Challenges for Airport Operations and Future Growth of the National Airspace System. Testimony before the Subcommittee on Aviation, Committee on Transportation and Infrastructure, U.S. House of Representatives. GAO-08-216T. Available online at *http://www.gao.gov/new.items/d08216t.pdf*.

GAO. 2008. Aviation and the Environment: FAA's and NASA's Research and Development Plans for Noise Reduction Are Aligned But the Prospects of Achieving Noise Reduction Goals Are Uncertain. GAO-08-384. Washington, DC: U.S. Government Printing Office.

Gorman, J., R. Hinchliffe, and I. Stothers. 2004. Active Sound Control on the Flight Deck of a C130 Hercules. Proceedings of ACTIVE 04, The 2004 International Symposium on Active Control on Sound and Vibration, Williamsburg, VA, September 20–22.

Greene, M. 2002. Typical Diurnal Traffic Noise Patterns for a Variety of Roadway Types. Proceedings of INTER-NOISE 02, The 2002 International Congress and Exposition on Noise Control Engineering, Dearborn, MI, August 19–21. Available online at *http://www.bookmasters.com/marktplc/00726.htm*.

Hansen, C.H. 2005. Current and future industrial applications of active noise control. Noise Control Engineering Journal 53(5):181–196.

Hansen, C.H., and S.D. Snyder. 1997. Active Control of Noise and Vibration. Boca Raton, FL: Taylor & Francis.

Harris, C.M. 1994. Noise Control in Buildings: A Practical Guide for Architects and Engineers. New York: McGraw Hill. Available online at *http://www.atlasbooks.com/marktplc/00726.htm*.

Herkes, B. 2006. Quiet Technology Demonstrator 2 Flight Test. Presented at the Aviation Noise & Air Quality Symposium, Palm Springs, CA, March 7. Available online at *http://www.techtransfer.berkeley.edu/aviation06downloads/herkes.pdf*.

Herkes, W.H., R.F. Olsen, and S. Uellenberg. 2006. The Quiet Technology Demonstrator Program: Flight Validation of Airplane Noise-Reduction Concepts. AIAA 2006-2720. Presented at the 12th AIAA/CEAS Aeroacoustics Conference (27th AIAA Aeroacoustics Conference), Cambridge, MA, May 8–10.

Herman, L., and J. Withers. 2005. Effectiveness of Tire/Road Noise Abatement Through Surface Retexturing by Diamond Grinding. Report No. FHWA/OH-2005/009. Columbus: Ohio Department of Transportation.

HUD (U.S. Department of Housing and Urban Development). 1967. Airborne, Impact, and Structure Borne Noise—Noise Control in Multifamily Dwellings. Washington, DC: HUD. Available online at *http://www.eric.ed.gov/ERICWebPortal/custom/portlets/recordDetails/detailmini.jsp?_nfpb=true&_&ERICExtSearch_SearchValue_0=ED024212&ERICExtSearch_SearchType_0=no&accno=ED024212*.

HUD. Undated. The Noise Guidebook. Washington, DC: HUD. Available online at *http://www.hud.gov/offices/cpd/environment/training/guidebooks/noise/index.cfm*.

IOM (Institute of Medicine). 2000. To Err Is Human: Building a Safer Health System. Washington, DC: National Academy Press.

Johnson, T., et al. 2009. Development of Passive and Active Noise Control for Next-Generation Locomotive Cabs. Proceedings of INTER-NOISE 09, The 2009 International Congress on Noise Control Engineering, Ottawa, Canada, August 23–26.

Kinetics (Kinetics Noise Control). 2009. Kinetics Model IsoMax: Resilient Sound Isolation Wall and Ceiling Clip. Patent No. 7,093, 814. Available online at *http://www.kineticsnoise.com/arch/isomax/index.aspx*.

LEMA (Laboratory of Electromagnetics and Acoustics). 2008. Overview of EC Funded Aircraft Noise Research: Networks and Research Projects. Available online at *http://lema.epfl.ch/images/stories/LEMA/AC/X3_noise_workshop/presentation/080527_x3noise%20workshop_01_collin.pdf*.

Lilly, J. 2002. Resilient Channel Update (December). Available online at *http://www.jglacoustics.com/acoustics-rc_1.html*.

Lodico, D., and J. Reyff. 2009. Long-term noise performance of open graded asphalt concrete (OGAC)—Results of 10-year long study. Noise Control Engineering Journal 56(2):84–93.

Lyon, R.H. 2000. Designing for Product Sound Quality. New York: Marcel Dekker.

Lyon, R.H. 2004. Product Sound Quality—From Design to Perception. Proceedings of INTER-NOISE 04 207(1):4417–4420. Available online at *http://www.noisenewsinternational.net/docs/lyon-2004.pdf*.

Lyon, R.H., and D.L. Bowen. 2007. Designing quiet products. The Bridge 37(3):13–17.

MacLeod, M., J. Dunn, I.J. Busch-Vishniac, J.E. West, and A. Reedy. 2007. Quieting Weinberg 5C: A case study in hospital noise control. Journal of the Acoustical Society of America 121(6):3501–3508.

Maier, T. 2009. Challenges and Applications for Active Noise and Vibration Control in Aerospace. Proceedings of ACTIVE 09, The 2009 International Symposium on Active Control of Sound and Vibration, Ottawa, Canada, August 20–22. Available online at *http://www.bookmasters.com/marktplc/00726.htm*.

McNeer, R.R., J. Bohorquez, O. Ozcan, A.J. Varon, and P. Barach. 2007. A new paradigm for the design of audible alarms that convey urgency information. Journal of Clinical Monitoring and Computing 21(6):353–363.

Muehleisen, R.T. 2009. Noise Problems and Opportunities in "Green Buildings." Presentation at the Second CAETS Forum on Worldwide Noise Sources, Ottawa, Canada, August 25. Available online at *http://www.noisenewsinternational.net/docs/caets-2009.pdf*.

NASA (National Aeronautics and Space Administration). 2002. Aeronautics Blueprint, 2002. Washington, DC: NASA. Available online at *http://www.nasa.gov/pdf/55403main_20%20AT.pdf*.

Neergaard, C.F. 1930. Are Acoustical Materials a Menace in Hospitals? Third Meeting of the Acoustical Society of America, New York, May 9–10.

Neergaard, C.F. 1941. What Can the Hospital Do About Noise? Twenty-Sixth Meeting of the Acoustical Society of America, New York, October 24–25.

Nelson, P.A., and S.J. Elliott. 1993. Active Control of Sound. San Diego, CA: Academic Press.

Nightingale, F. 1860. Noise. Chapter 4 in Notes on Nursing: What It Is, and What It Is Not. New York: D. Appelton. Available online at *http://digital.library.upenn.edu/women/nightingale/nursing/nursing.html*.

NRC (National Research Council). 2002. Aeronautics Research and Technology for 2050: Assessing Visions and Goals—Letter Report. Washington, DC: National Academies Press.

Olson, H.F., and E.G. May. 1953. Electronic sound absorber. Journal of the Acoustical Society of America 25(6):1130–1136.

PAC International. 2009. Resilient Sound Isolation Clip (RSIC-1™). Available online at *http://www.pac-intl.com/products.html*.

Pelton, H.K., E. Ryherd, and M. Martin. 2009. Acoustical design of a burn acute care unit for enhanced patient comfort. Noise Control Engineering Journal 57(1):32–41.

Polcak, K. 2003. Highway traffic noise barriers in the U.S.—construction trends and cost analysis. Noise/News International 11(3):96–108. Available online at *http://www.noisenewsinternational.net/archives_idx.htm*.

Quirt, J.D. 2009. Controlling Air-Borne and Structure-Borne Sound in Buildings. Proceedings of INTER-NOISE 2009, The 2009 International Congress on Noise Control Engineering, Ottawa, Canada, August 23–26. Available online at *http://www.bookmasters.com/marktplc/00726.htm*.

Rasmussen, R., R. Bernhard, U. Sandberg, and E. Mun. 2007. The Little Book of Quieter Pavements. Report No. FHWA-IF-08-004. Washington, DC: Federal Highway Administration. Available online at *http://www.tcpsc.com/LittleBookQuieterPavements.pdf*.

Razavi, Z. 2009. Acoustical Improvements with Natural Air Ventilations in the Liu Institute at the University of British Columbia. Proceedings of INTER-NOISE 2009, The 2009 International Congress on Noise Control Engineering, Ottawa, Canada, August 23–26. Available online at *http://www.bookmasters.com/marktplc/00726.htm*.

Remington, P.J. 1983. Control of Wheel/Rail Noise and Vibration. Final Report No. TOT-TSC-UMTA-82-57. Washington, D.C: Urban Mass Transit Administration, U.S. Department of Transportation.

Remington, P.J., S.J. Knight, D. Hanna, and C. Rowley. 2005. A hybrid active/passive exhaust noise control system for locomotives. Journal of the Acoustical Society of America 117(1):68–78.

Reyff, J.A., and P.R. Donavan. 2005. Reduction of Traffic Noise and Tire/Pavement Noise: First Year Results of the Arizona Quiet Pavement Program. Proceedings of NOISE-CON 05, The 2005 National Conference on Noise Control Engineering, Minneapolis, Minnesota, October 17–19. Available online at *http://www.bookmasters.com/marktplc/00726.htm*.

Reynolds, D.R., and J.M. Bledsoe. 1989. Algorithms for HVAC Engineers. Atlanta, GA: ASHRAE.

Rolls-Royce. 2005a. The Jet Engine. Stamford, U.K.: Key Publishing Ltd.

Rolls-Royce. 2005b. Noise Facts Leaflet. Stamford, U.K.: Key Publishing Ltd.

Sandberg, U. 2001. Noise Emissions of Road Vehicles: Effectiveness of Regulations. Final Report from the Working Party on Noise Emissions of Road Vehicles. International Institute of Noise Control Engineering. Available online at *http://www.i-ince.org/data/iince011.pdf*.

Sandberg, U., and J. Ejsmont. 2002. Tire/Road Noise Sources and Generation Mechanisms. Pp. 127–138 and 147 in Tire/Road Noise Reference Book. Kisa, Sweden: Informex.

SBAC (Society of British Aerospace Companies). 2009. SBAC Aviation and Environment Briefing Papers: 1. Aircraft Noise. Available online at *www.sbac.co.uk/community/dms/download.asp?txtFilePK=5185*.

Scheuren, J. 2005. Engineering applications of active sound and vibration control. Noise Control Engineering Journal 53(5):197–210.

Silent Aircraft Initiative. 2006. Silent Aircraft Creeps Closer to Reality. Available online at *http://silentaircraft.org/news/article/default.aspx?objid=1602*.

Sommerfeldt, S.D., and K.L. Gee. 2003. A compact active control implementation for axial cooling fan noise. Noise Control Engineering Journal 51:325–344.

Sykes, D.M., and G.C. Tocci. 2008. Speech privacy: Momentum grows in healthcare. Acoustics Today 4(4):30–33.

Taylor, F.B. 1958. Noise control in a research hospital. Noise Control 4(5):9–62.

Tichy, J. 1996. Applications for active control of sound and vibration. Noise/News International 4: 73–86. Available online at *http://www.noisenewsinternational.net/docs/tichy-1996.pdf*.

Vér, I.L. 2000. Active Muffler, United States Patent 6,160,892. Issued December 12.

Waitz, I., R.J. Bernhard, and C.E. Hanson. 2007. Challenges and promises in mitigating transportation noise. The Bridge 37(3):25–32.

Ward. 1989. Ward v. Rock Against Racism, 491 U.S. 781 (1989). Available online at *http://www.oyez.org/cases/1980-1989/1988/1988_88_226*.

Warnock, A.C.C., and J.A. Birta. 2000. Detailed Report for Consortium on Fire Resistance and Sound Insulation of Floors: Sound Transmission and Impact Insulation Data in 1/3 Octave Bands. Internal Report IRC IR-811. Ottawa, Canada: National Research Council of Canada.

6

Standards and Regulations for Product Noise Emissions

This chapter examines national, regional, and international standards setting, regulation, and compliance testing with regard to product noise emissions and their implications for U.S. manufacturers. The industrial sectors of interest include consumer products/home appliances; computers, printers, and other information technology (IT) products; portable power generation equipment; air compressor equipment; air-conditioning and ventilation equipment; yard care equipment; small engine manufacturers; and construction equipment. These are sectors for which there are significant variations in national and regional standards and regulation of noise emissions. Airplanes and road vehicles are not addressed in this chapter, since there are international bodies—the International Civil Aviation Organization for airplanes and Working Party 29 of the United Nations Economic Commission for Europe (UNECE) for road vehicles—that deal with the harmonization of noise emission requirements for these vehicles worldwide.

Over the past two decades Europe has been particularly active in the development of product noise emission standards (e.g., voluntary limits that have been agreed upon by a nongovernmental body), regulations (e.g., noise measurements that must be complied with and certified), and efforts to increase the amount of information provided to consumers with respect to product noise emissions, such as voluntary and mandatory product labeling requirements. During this time European standards organizations have exercised considerable leadership in international standards bodies, thereby making the International Organization for Standardization (ISO) and, to a lesser extent, the International Electrotechnical Commission (IEC) worldwide leaders in the product noise emissions standards community.

In contrast, since the early 1980s, the U.S. government's interest in regulating product noise emissions based on evolving U.S. or international standards has advanced very little. Moreover, the participation and influence of U.S. standards organizations in the ISO and IEC in the area of product emission standards has been circumscribed by the structure and funding of U.S. standards bodies and the nature of ISO/IEC governance. America has only a single vote in ISO and IEC working groups and in the approval of standards—the same as every member country in the European Union.

European noise emission regulations are more stringent and more closely aligned with those of international standards bodies than their American counterparts, and European regulations based on these standards are more extensive than regulations in the United States. ISO standards committees have superseded many American-based standards committees and organizations that U.S. manufacturers have relied on in the past. To sell in global markets it has become increasingly important that U.S. manufacturers comply with European and ISO standards.

Different product noise emission regulations in foreign markets can drive up costs for a U.S. manufacturer seeking to sell in those markets by making compliance and certification more difficult. Adding to U.S. manufacturer's challenges are costs not only for additional testing and documentation but also for the need to carry multiple "silencer" packages and parts inventories needed to meet the demands of multiple foreign regulations.[1] If a market is too small to be worth the additional design and manufacturing costs, a company may decide not to compete there. The point is that the effect of national or regional differences in regulations can be to shut U.S. competitors out of markets.

At the same time it is important to recognize that, although more stringent noise requirements can sometimes be a burden for U.S. manufacturers, they can also encourage innovation. A U.S. manufacturer's desire to design a low-noise machine for sale in European or world markets is a positive force that could lead to the introduction of "quiet" products into American markets and provide an incentive for manufacturers and purchasers to cooperate in "buy quiet" programs.

---

[1] DeVries, L. 2007. Presentation at NAE workshop on Impact of Noise on Competitiveness of U.S. Products, Washington, D.C., June.

The remainder of this chapter provides more detailed information about international noise emission requirements, standards for noise emissions, noise emission labeling, accreditation and certification requirements, and the role of the Office of the U.S. Trade Representative in compliance and enforcement issues.[2]

IMMISSION VERSUS EMISSION

To understand how noise standards and regulations affect the ability of manufacturers to compete in national and international markets, it is important to distinguish between noise *emission* and noise *immission*.

Standards for noise *emission*—the sound emitted by a product independent of its location—allow a manufacturer to make a measurement of a specific piece of equipment under specified operating conditions and report the noise level, usually in the form of a "guaranteed level." Usually, but not always, noise emission information is reported as the A-weighted sound power level. Appendix A is a primer on quantities used in noise control and acoustics.

Requirements related to noise *immission*—the sound pressure level at a listener's ear—have been promulgated to address community noise worldwide. These requirements have been summarized by the International Institute of Noise Control Engineering (I-INCE, 2009).[3]

DETERMINING PRODUCT NOISE EMISSIONS

A wide variety of policies, regulations, and standards on noise emissions—local, national, regional, and international—have been published, and most countries have national standards organizations. In the United States the American National Standards Institute (ANSI) is the major nongovernmental organization that deals with product noise standards. In past decades the U.S. Environmental Protection Agency (EPA) was responsible for regulating some product noise emissions at the national level. For the purposes of this report, the most significant regional standards organizations and regulatory body for product noise emissions outside the United States are in Europe.

There are three major European nongovernmental standards organizations involved with product noise emission standards setting: the European Committee for Standardization (CEN), the European Committee for Electrotechnical Standardization (CENELEC), and Central European Initiative (CEI). The European Commission (EC) is responsible for regulating product noise emissions throughout the European Union (EU) using standards developed by CEN, CENELEC, and CEI. The international counterparts to CEN and CENELEC are the International Organization for Standardization (ISO; *http://www.iso.org*) and the International Electrotechnical Commission (*http://www.iec.ch*), which set product noise emission standards at the international level. In this section standards-setting activities and associated regulations are reviewed as they relate to noise emissions of machinery and equipment.

Product Noise Emission Standards and Regulations in the United States

American National Standards Institute

According to its website, "The American National Standards Institute (ANSI) is a private, non-profit organization that administers and coordinates the U.S. voluntary standardization and conformity assessment system." ANSI's mission is "to enhance both the global competitiveness of U.S. business and the U.S. quality of life by promoting and facilitating voluntary consensus standards and conformity assessment systems, and safeguarding their integrity" (ANSI, 2009). ANSI represents the United States in the ISO and IEC.

ANSI neither develops standards nor funds the U.S. standards system. Rather it accredits and audits standards-setting committees that are funded and administered by engineering and scientific professional societies, industry associations, and other nongovernmental organizations. ANSI's activities are supported by fees from these organizations (ASA, 2009b).

> When ANSI allows a standard to be called an "ANSI Standard," it is not making a technical judgment on the standard but stating that the standard was developed in accordance with operating procedures that facilitate openness, balance, and due process, and that the standard represents a consensus among those substantially concerned with its scope and provisions. Consensus is established when, in the judgment of the ANSI Board of Standards Review, substantial agreement has been reached by directly and materially affected interests. Substantial agreement means much more than a simple majority, but not necessarily unanimity. Consensus requires that all views and objections be considered and that a concerted effort be made toward their resolution. ANSI's approval represents approval of the process, not the content.

The most important of these organizations are the four standards committees of the Acoustical Society of America (ASA) on noise, acoustics, mechanical vibration and shock, and bioacoustics. ANSI-accredited standards committees related to noise are listed in Appendix C, Part A.

Even though ANSI standards reflect a consensus and are not mandatory, the procedures or criteria in those standards may be required by law, regulation, building code, or contract in specific situations. Thus, many federal regulations reference ANSI standards.

[2]Vehicle noise emissions are not covered in this chapter. The World Forum for Harmonization of Vehicle Regulations, a group in the UNECE, deals with vehicle standards (UNECE, 2009).

[3]Noise *immission* requirements in the workplace are discussed in Chapter 4.

Adoption of International Standards

When a standard relates to international commerce (e.g., standards for product noise emissions), international standards may be adopted. In these instances it is important that the American standard be identical (or nearly identical) to the international standard for a given product. If there is an ISO or IEC standard suitable for use in the United States and recommended by a U.S. technical advisory group (see Appendix C, p. 150), the ASA standards committees may adopt the standard as written (or with minor changes) as an American National Standard (ASA, 2009a). American standards can also be used as the basis for international standards (i.e., early versions of the sound power standards).

U.S. Regulation of Product Noise Emissions

U.S. regulation of product noise emissions is relatively limited and outdated. Following enactment of the Noise Control Act (NCA) of 1972 (codifed in 49 U.S. 4901-4918), EPA's newly established Office of Noise Abatement and Control (ONAC) was given the authority to undertake a range of activities to reduce noise pollution. These included "identifying sources of noise for regulation, promulgating noise emission standards, coordinating federal noise research and noise abatement, working with industry and international, state and local regulators to develop consensus standards, disseminating information and educational materials, . . . [and] sponsoring research concerning the effects of noise and the methods by which it can be abated." With the passage of the Quiet Communities Act of 1978, ONAC's mandate was expanded to include provision of grants to state and local governments for noise abatement. During ONAC's brief existence, from 1972 to 1982, when it was defunded by Congress at the request of the Reagan administration, the office promulgated only four product and six transportation noise standards and was unable to implement product labeling or the Low-Noise Emission Product Program (Shapiro, 1991).

While Congress has repeatedly refused to restore funding to EPA for its noise abatement activities, the NCA and the authority it gives to EPA to regulate noise remain in effect. Without resources to implement its mandate, however, EPA has been unable to promulgate any further product noise emission standards; and the four product noise standards it promulgated during the 1970s have not been subjected to critical evaluation since, despite advances in relevant science and technology and improved understanding of the effects of noise on people. Since 1982, EPA has also lacked the resources to participate in private standards-setting efforts or to provide technical assistance to state and local governments. (An exception is the efforts to improve the standard on the performance of hearing protective devices described in Chapter 4.) By retaining its authority under the NCA without the funding to execute it, EPA has effectively preempted state and local governments from adopting updated noise emission and labeling standards of their own for the sources and products that EPA has already regulated (Shapiro, 1991).

Product Noise Emission Standards and Regulations in the European Union

European Standards Organizations

There are several important differences between the organizations and structures of standards-setting processes in Europe and the United States. In contrast to the decentralized nature of standards bodies in the United States, European standards bodies at the national and regional (EU) levels are centralized in structure. European standards activities are organized by nation and region, whereas in the United States they are organized by sector. Standards-setting organizations are largely publicly funded in Europe, whereas they are mostly privately funded in the United States. Finally, membership in national and regional standards organizations in Europe is restricted to European entities or those that have a business interest or manufacturing presence in Europe (with the exception of the European Telecommunications Standards Institute, where participation is open to other nationals). In the United States, membership in most full-consensus standards-developing bodies is unrestricted, and in many instances membership on U.S. technical committees can be international in composition.

Similar to ANSI standards, standards of European regional and national standards bodies reflect a general consensus and are not mandatory, and the procedures or criteria in European standards may be required by European and/or national law, by regulation, by building code, or by contract in specific situations. Unlike in the United States, however, European regulation of product noise emissions based on standards developed by regional and international standards bodies has been very active and expansive in recent decades.

European Regulation of Product Noise Emissions

The 1996 Green Paper (EC, 1996), which stated the intent to extend the existing six directives on noise source emissions to cover more than 60 types of equipment and to require the reporting of guaranteed noise emission levels of machinery and equipment, signaled a significant change in EU noise policy (EC, 1996). One direct result of the Green Paper was the publication in 2000 of the outdoor equipment directive, 2000/14/EC (EC, 2000), and its amendment, 2005/88/EC (EC, 2005). These directives set noise emission limits on a wide variety of equipment used outdoors, such as compaction machines, tracked vehicles, wheeled vehicles, concrete breakers, cranes, welding and power generators, compressors, lawn mowers (Figure 6-1), and lawn trimmers/lawn edge trimmers. Noise emission is expressed as an A-weighted sound power level, and limits guarantee the noise emission levels of these products.

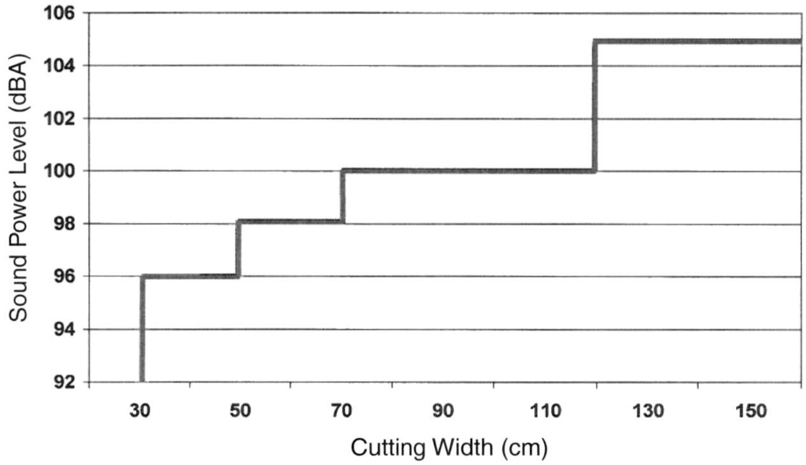

FIGURE 6-1 Permissible sound power levels (dB(A)) for lawn mowers, based on width of cut. Source: Directive 2000/14/EC of the European Parliament (EU, 2000).

The directives also set limits and require labeling of guaranteed sound power levels for a variety of other kinds of equipment, such as building-site band saw machines, chain saws, concrete and mortar mixers, conveyer belts, drill rigs, and hedge trimmers. A more detailed list and further explanation of the directives have been published by TÜV-SUD America (TÜV, 2009a). The EC also maintains a database of noise emission levels for equipment covered by the directives (EC, 2006a).

Since the 1996 Green Paper, the EC has adopted a new version of the machinery directive, 2006/42/EC, which sets standards for the safety of machinery (EC, 2006b). These standards include noise emissions, and reporting of noise emissions is required under certain circumstances:

- The A-weighted emission sound pressure level must be reported at workstations when the value exceeds 70 dB(A).
- The peak C-weighted instantaneous sound pressure value must be reported at workstations when the value exceeds 63 Pa (130 dB re 20 μPa).
- The A-weighted sound power level emitted by machinery must be reported wherever the A-weighted emission sound pressure level at workstations exceeds 80 dB(A).

Alternative test conditions are allowed under certain circumstances; this means that manufacturers must know at least the emission sound pressure level and peak instantaneous level of machinery and equipment.

The EC physical agents (noise) directive sets noise *immission* limits for workplaces and may indirectly influence the selection of low-noise machinery in manufacturing facilities (EC, 2003). The EU has also issued a directive for noise from household appliances (86/594/EC) that allows member states to label the level of noise emissions from household appliances and establishes the A-weighted sound power level as a measure of noise emission (EC, 1986).

Noise Emission Limits in Other Countries

China has set noise emission limits (A-weighted sound pressure level) according to GB/T 7725-2004 and noise limits for room air conditioners and heat pumps. In addition, noise limits have been set on household and similar electrical appliances according to GB 19606-2004. In India, immission limits are spelled out in "Air Quality Standard in Respect of Noise." Korea has also set noise emission limits (A-weighted sound pressure level) according to KS C9036. Japan, too, has set noise emission limits (A-weighted sound pressure level) for package air conditioners according to JIS B8612.[4] Canada has a standard (CSA-Z107.58-02) on declaration of noise from machinery, but it does not set noise limits.

Sweden's noise standard (Statskontoret 26:6) spells out noise emission requirements in terms of guaranteed sound power levels for a wide variety of IT equipment, including equipment in data-processing areas, servers, printers and imagers, laptops, data projectors, and other desktop devices. The limits and test methods specified are suitable for inclusion in purchase specifications (Statskontoret, 2004).

[4]Mézache, M. 2007. Presentation at the NAE Workshop on the Impact of Noise on Competitiveness of U.S. Products. Washington, D.C., June 20–21.

INTERNATIONAL ORGANIZATION FOR STANDARDIZATION

The ISO has an International Classification for Standards (ICS). Noise emission standards fall into the following category:

17. Metrology and measurement. Physical phenomena
 17.140 Acoustics and acoustic measurements
 17.140.20 Noise emitted by machines and equipment

A list of standards in ICS 17.140.20 can be found at *http://www.iso.org/iso/iso_catalogue/catalogue_ics/catalogue_ics_browse.htm?ICS1=17&ICS2=140&ICS3=20.*

Many ISO technical committees are involved with setting noise emission standards. One committee, ISO TC43/SC1 (noise), develops, among other standards, generic standards for the measurement of noise emissions. In the United States, ASA manages the technical advisory group for this technical committee. The ISO TC43/SC1 Secretariat in Denmark coordinates standards activities.

For the purposes of this report, the two most important series of standards are ISO 3740 and ISO 11200. The former describes the methods of measuring noise emissions from machinery in terms of sound power level, both A-weighted and in frequency bands, such as octave bands. The latter describes the measurement of emission sound pressure level. Most standards written by other ISO technical committees to determine sound power levels are similar to the 3740 series, and standards from this series (2000/14/EC and 2005/88 EC) are used by the EU in its directives on noise emissions from outdoor equipment.

The ISO 11200 series describes methods of measuring emission sound pressure levels (i.e., the level at the operator or bystander's position measured in a controlled acoustical environment). These measurements are important in determining compliance with EU Directive 2006/42/EC, which sets emission sound pressure level requirements for machinery and equipment and, under some conditions, A-weighted sound power level according to the ISO 3740 series.

Other standards related to noise emissions have been issued by the ISO technical committee (TC) 43/SC1. These include methods of declaring and verifying noise emission values for machinery and equipment (ISO 4871) and statistical methods of determining and verifying stated noise emission values for machinery and equipment (ISO 7574, parts 1–4). In addition, there are standards for determining sound power through sound intensity (ISO 9614, parts 1–3), noise emission in the IT industry (ISO 7779, ISO 9295), and noise from rotating machinery (ISO 1680). A complete list of standards under the jurisdiction of ISO TC43/SC1 (including some in ICS 13.140, Noise with respect to human beings, and other classifications) can be found at *http://www.iso.org/iso/iso_catalogue/catalogue_tc/catalogue_tc_browse.htm?commid=48474.*

ISO TC43/SC1 also maintains relationships with a number of other ISO and IEC TCs. These include TC72/SC8—Textiles (Working Group 2 on noise), TC 117—Industrial Fans (Working Group 2 on fan noise testing), and TC 118—Compressors and Pneumatic Tools, Machines and Equipment/Air Compressors and Compressed Air Systems.

ISO TC43/SC1 is just one of the ISO TCs that issue standards related to noise emissions. Many of the committees that cover work on standards for specific types of machines have working groups related to noise standards. One benefit of this arrangement is that the committees can specify realistic operating conditions in which noise measurements should be taken and can anticipate special situations that should be accommodated. Nevertheless, the proliferation of committees greatly complicates the harmonization of measurement standards for noise emissions. A partial list of ISO TCs with an interest in noise or sound can be found in Walters[5] (2007) and in this report in Appendix C.

The EU and the European standards organizations have made ISO a leader in the global standards community. Given the one country/one vote governance of ISO activities and European governments' financial support for their national nongovernmental standards committees in ISO activities, the member states of the EU exercise considerable influence on ISO working groups. With only one vote, no public support, and only limited private-sector support for the participation of U.S. standards committees in ISO, U.S. manufacturers' influence on ISO working groups is much less than that of its collective European counterparts.

As was pointed out in a presentation at the National Academy of Engineering workshop in June 2007, most Western countries that are members of ISO provide funding for a central standards office; national dues to ISO and IEC; funding for staff, including ISO or IEC committee secretariats and ISO working group secretariats; and funding or subsidies for travel for members of ISO working groups.[6] In sharp contrast, ANSI receives no federal funding and charges its accredited standards-setting nongovernmental organizations fees to support a central standards office; national dues to ISO and IEC; salaries for staff, including ISO or IEC committee secretariats and working group secretariats; and charges for IEC working group members. This means that the United States depends on nongovernmental organizations to raise funds to support U.S. participation in international standards activities.

To influence ISO draft standards, individuals from interested countries must be present at working group meetings when decisions are made. Most product-specific noise standards rely on "basic" or "fundamental" standards developed by ISO TC43 and TC43/SC1. U.S. companies do not fund

[5]Walters, J. 2007. Presentation at the NAE Workshop on Impact of Noise on Competitiveness of U.S. Products, Washington, D.C., June 20–21.

[6]Schomer, P. 2007. Impact of Noise on Competitiveness of U.S. Products—the Role of Standards. Presentation at the NAE Workshop on Impact of Noise on Competitiveness of U.S. Products, Washington, D.C., June 20–21.

fundamental or basic standards related to a by-product (i.e., noise) the way they fund applied standards related to a product. Thus, the United States is at a significant disadvantage in terms of representation on ISO committees involved in basic or fundamental noise standards.

INTERNATIONAL ELECTROTECHNICAL COMMISSION

The IEC develops and publishes international standards for electrical, electronic, and related technologies. Like ISO's standards, IEC standards are developed by TCs with international representation. TC29, Electroacoustics, develops standards for microphones, filters, sound-level meters, hearing aids, and other electroacoustical devices. The standards produced by this committee are used in making measurements, and some modern instruments are described in Appendix E. The work of this committee is vital for the measurement of noise but will not be emphasized in this chapter. In addition, a number of other TCs have developed noise standards for specific areas, such as consumer products. Examples of IEC TCs that develop standards related to noise are listed in Appendix C, Part E. In general, IEC TCs and ISO TCs with common interests have liaison programs with varying degrees of effectiveness.

TC59 deals with standards for many products that use a common descriptor for noise emission, the A-weighted sound power level, which is not widely used in the United States. Nevertheless, if international efforts to develop a common noise label for consumer products proceed, it is likely that the sound power level descriptor will be used.

ACCREDITATION AND CERTIFICATION OF NOISE EMISSIONS

The EU Outdoor Equipment Directive (2000/14/EC) is the prime example of a noise emission regulation that requires certification of product noise levels. This directive applies to more than 50 types of equipment used outdoors, and manufacturers are responsible for initiating and completing the certification process. According to the directive, three options are available to a manufacturer that wants to document noise emission in compliance with the directive:

1. Internal control of production with assessment. The manufacturer takes full responsibility for initial certification, documentation, and ongoing monitoring of production units. A "notified body" must be contracted to verify the manufacturer's documentation and noise-level conformance on a regular basis.
2. Unit verification. The manufacturer submits an application to a notified body, which is then contracted to examine the equipment and carry out the certification and documentation process.
3. Full quality assurance procedure. The manufacturer takes full responsibility for initial certification, documentation, and ongoing monitoring of production units. If the manufacturer has a certified quality assurance system in place, only periodic audits by a notified body are required.

A manufacturer incurs significant direct and indirect costs with each of these options. All notified bodies are based in EU countries and are approved by the EC to carry out their responsibilities; thus, U.S. manufacturers incur travel costs. These costs can be avoided, but only if the manufacturer takes on the cost of maintaining a quality assurance system, as well as noise measurement systems to monitor noise-level variances in production.

U.S. ACCREDITATION

National Institute of Standards and Technology

The National Institute of Standards and Technology (NIST) has four programs that can contribute to the development of technology for a quieter America. The National Voluntary Laboratory Accreditation Program plays an important role in the accreditation of laboratories for noise emission measurements to meet national and international standards.

The Global Standards and Information Group provides technical information to the federal government and industry. Although there are no known activities related to noise emission accreditation, this group could become important when foreign countries seek U.S. accreditation. The National Center for Standards and Certification Information could play a role in informing American manufacturers about noise emission standards and requirements. The Calibration Laboratory for Microphones is essential for accurate noise emission measurements and noise measurements in general, and microphone calibration must be traceable to NIST.

National Voluntary Laboratory Accreditation Program

The National Voluntary Laboratory Accreditation Program (NVLAP) accredits laboratories in the United States and other countries for measurements according to an accepted standard. The program is established under 15 CFR 285.

The Acoustical Testing Services, one of a wide variety of available accreditation programs, includes laboratories that perform a variety of acoustical tests—mainly according to the American Society for Testing and Materials International, ANSI, and ISO standards (NIST, 2009a). These include evaluating hearing protective devices, the properties of sound-absorptive materials, sound transmission loss, and noise emissions from many sources. As of November 2008, 26 laboratories were accredited to perform measurements according to one or more acoustical standards, and, of these, 14 were accredited to perform tests according to one or more standards for noise emissions (NIST, 2009b). These include independent testing laboratories, corporate laboratories, and one government laboratory. In addition, one laboratory is accredited in Canada and one in Japan. No list of standards has been issued for which accreditation is available. Accreditation is available for *any* standard—presumably from a recognized national or international standards development organization.

The procedure for accreditation includes submission of an application and payment of a fee to NVLAP, an on-site inspection by an independent technical expert, resolution of problems, and, if all problems are resolved, issuance of a certificate.

This program is not a certification of test data. It is designed to determine if a specific laboratory is qualified to perform measurements according to a specific standard or set of standards. Thus, it differs from the procedure followed by notified bodies that review data for the EU. Notified bodies examine test data for a specific product, which is (or is not) certified. Evaluations by notified bodies are based on the following international standards:

- ISO/IEC 17025, general requirements for the competence of calibration and testing laboratories
- ISO/IEC 17011, conformity assessment—general requirements for accreditation bodies accrediting conformity assessment bodies

Global Standards and Information Group

The Global Standards and Information Group (*http://ts.nist.gov/Standards/Global/ contact.cfm*) is involved in international conformity and assessment activities. Although there is no known current activity related to noise emission, the mission of the GSIG is such that it could play a role in determining if noise requirements in standards or regulations are fulfilled.

National Center for Standards Certification Information

The National Center for Standards Certification Information (NCSI) provides technical information related to standards activities. Although NCSI is not involved in any activities related to noise emission, American manufacturers would benefit from a database of information on national and international standards and requirements related to noise emission.

U.S. Trade Office

The Office of the U.S. Trade Representative (USTR) is a cabinet-level agency with more than 200 professionals on its staff whose role is to facilitate and expand trade with foreign countries through trade agreements, trade policy, and trade dispute resolutions. The USTR reports to the president of the United States and is the president's principal advisor, negotiator, and spokesperson on issues related to trade.

The study committee investigated how USTR might help U.S.-based companies compete more effectively in regions with noise-related regulations. Within that framework, the committee asked two questions: Can the USTR ensure that the opinions of manufacturers have an impact on the writing of regulations/standards? Can the USTR help mediate disputes over the application of regulations and standards when U.S.-based manufacturers believe they are being used to prevent them from competing in a market?

In answer to the first question, the committee found that since 1974 the USTR has had private-sector advisory committees to provide expertise in their areas (USTR, 2009). However, because noise is a by-product (usually unwanted) of the equipment being regulated, and because the measurement and reporting of noise is a complicated technical issue that applies to multiple sectors, it is difficult to present a uniform opinion that can be acted on in negotiations for trade agreements.

In fact, regulations and standards that apply to noise are developed and implemented separately from the general trade negotiations conducted by USTR. Therefore, U.S. manufacturers must be present and committed to participating in trade organizations and standards-making bodies that develop the regulations and standards for product noise levels (Schomer et al., 2008).

In answer to the second question, which may involve mediating disputes, the USTR and NIST may be in a better position to become directly involved. NIST can provide technical support to document testing and certification and can forward complaints/inquiries to USTR for notification. In addition, if a manufacturer believes that a regulation is being misapplied or is being used solely as a barrier to trade, the manufacturer can contact USTR for assistance and dispute resolution. USTR also has many interagency connections (e.g., in the U.S. Department of Commerce International Trade Administration) that can be called on for support.

INTERNATIONAL ACCREDITATION

Two international organizations accredit laboratories: the International Laboratory Accreditation Cooperation (ILAC, 2009) and the International Accreditation Forum (*http://www.iaf.nu*). There are also regional accreditation organizations for the Asia-Pacific region (*http://www.aplac.org*), the Inter-American region (*http://www.iaac.org.mx*), and Europe (*http://www.european-accreditation.org/content/home/home.htm*).

Role of Notified Bodies

The EC has defined a notified body in the following terms: "Notification is an act whereby a Member State informs the Commission and the other Member States that a body, which fulfils the relevant requirements, has been designated to carry out conformity assessment according to a directive" (EC, 2009). The role of a notified body is to certify that the requirements of a particular directive have been met. If they have been met, CE (*Conformité Européenne*) marking can be put on the product.[7]

Of the European directives listed above (2000/14/EC, 2006/42/EC, 2003/10/EC, 86/594/EC), only the 2000/14/EC outdoor equipment directive requires that sound testing be assessed by a notified body. If an equipment manufacturer is ISO certified, it can perform sound power testing independently; the notified body audits the testing and certifies the results. Manufacturers that are not ISO certified have two options. They can have a notified body perform the required sound tests and write the reports and declaration of conformity. Or the manufacturer can perform the sound tests and have the notified body approve the resulting reports and declaration of conformity before selling the product in Europe.

Outdoor products covered by European Directive 2000/14/EC require a label of "Guaranteed Sound Power Level." When the 2000/14/EC directive was published, TÜV SÜD America published an article, "The Father of All Noise Directives," on the implication of this document for manufacturers" (TÜV, 2009b).

LABELING OF NOISE EMISSIONS

The term *noise label* can be defined as information on product noise emissions provided to final customers. The information may be on a label affixed to the product or on the packaging, in a product brochure or user's manual, or on a manufacturer's website. Some noise-labeling programs are mandatory, but most are voluntary. If uniform labeling appears on all products, it can be a benefit to consumers. If it is not uniform, it can create confusion and be an unfair competitive advantage or disadvantage. This section describes noise labeling in the United States, trade associations, and other organizations in the EU and other countries—including "eco-labels," which indicate "environmental friendliness."

Mandatory Labeling in the United States

In the late 1970s, EPA established a noise-labeling program for products (*http://www.epa.gov/history/topics/nca/01.htm*). However, since funding for the EPA Office of Noise Control was cut in 1981, no labels have been required for stationary noise-emitting products, with the exception of portable air compressors, which must have a label certifying compliance with the relevant EPA noise limit. Unlike other areas of the world, the United States has no other mandatory requirements for reporting noise emission values of stationary products.

Voluntary Labeling in the United States

The Centers for Disease Control and Prevention has a database of noise information for hand-powered tools that have been tested by the National Institute for Occupational Safety and Health (NIOSH). NIOSH is conducting ongoing research to "fill" the database.

Sears has requirements on some but not all types of appliances; some labels include sound power values.[8] Consumers Union (CU) uses a five-level pictograph scale to rate product noise for some types of products (e.g., vacuum cleaners). Details of CU testing and methods of rating are not known to manufacturers, nor are actual product emissions available.

The Institute of Noise Control Engineering has a technical committee on product noise emissions and is working to develop a simple, easy-to-understand format for noise labels. One proposal under consideration is a noise label similar to the EU energy label, which has simple graphic comparisons that enable consumers to make a quick judgment; they also provide simple numerical values for consumers who want more details.

Trade Associations and Industry-Specific Voluntary Labels

To meet growing customer demand for standardized, comparable product environmental information for IT and communications technology and consumer electronics, in 2006 IT Företagen and Ecma International harmonized their separate eco-declarations into ECMA-370 "The Eco Declaration—TED." ECMA-370 does not include criteria, but the document enables reporting of environmental attributes, including product noise emissions. All claims in TED are subject to verification. As of 2006, more than 6,000 eco-declarations had been issued by the predecessor organizations. The declarations are available on company websites.

The Home Ventilation Institute has administered a sound certification program for more than 35 years using a simple noise value on packaging of ventilator fans. The Air Movement and Control Association and Air Conditioning and Refrigeration Institute have "certification programs" that include published noise emission levels.

Mandatory Labeling in Europe

Several European directives and their amendments require that product noise emission values be included on product labels or in product literature. The three primary directives are 92/75/EEC—Energy Labeling for Household Appliances, 2006/42/EC—Machinery Safety Directive, and 2000/14/EC—Outdoor Equipment. The provisions of the household appliance noise directive are intended to provide consumers with information on noise in their homes, whereas the provisions in the machinery noise directive are intended to provide information on machinery that may cause hearing damage in

[7]For a list of notified bodies, see *http://ec.europa.eu/enterprise/newapproach/nando/index.cfm?fuseaction=country.main*.

[8]Vukorpa, V. 2007. Presentation at the NAE Workshop on Impact of Noise on Competitiveness of U.S. Products, Washington, D.C., June 20–21.

the workplace. The outdoor equipment directive is intended primarily to reduce environmental noise and provide noise information to purchasers of outdoor equipment.

The EU Energy Label for household appliances is required to include sound power level values in addition to energy consumption information. Noise measurements are according to the IEC 60704 series for most appliances. The label must be prominently displayed on the product in stores and on packaging. Products with this label include refrigerators, freezers, washing machines, dryers, dishwashers, and air conditioners.

The Machinery Safety Directive requires the publication, in user documentation, of the A-weighted sound pressure level at the workstation, if the level is greater than 70 dB(A), and the A-weighted sound power level if the level at the workstation is greater than 85 dB(A). The machinery safety directive does not apply to products covered by the low-voltage directive (2006/95/EC), which does not include requirements for information on noise emissions of products, either on labels or in user information. Office and home computer products, including personal computers and printers, are not required to report noise emission values in Europe. The outdoor equipment directive (2000/14/EC) requires a simple label with the declared sound power level.

Mandatory Labeling in Other Countries

China requires noise information, either on a label or in the user's manual, for some domestic appliances. Experience has shown that no manufacturers put this information on a label on the product. Since 2006, Argentina has required a label with noise information on some appliances. Brazil has no labeling requirements on some small appliances but requires certifiable noise values on product packaging.

The German Equipment and Product Safety Law requires publication of noise emission values for all products, including IT products—even if they are not included in the EU Machinery Safety Directive. However, because the intent of the machinery directive is to prevent hearing loss, the only requirement for most IT products is a statement that the sound pressure level emissions do not exceed 70 dB(A); this information does not describe the noise emissions of IT products used in businesses, offices, and homes.

In Germany the "GS Mark" indicates that a product complies with the minimum requirements of the German Equipment and Product Safety Act (GPSG). The GS Mark is a licensed mark of the German government and may only be issued by an accredited testing and certification agency (e.g., TÜV). Products in Germany routinely carry a GS mark indicating that they are "safe." However, in some instances, test houses that certify GS marks require additional provisions that are not included in GPSG, and this can cause problems for manufacturers. For example, GS test houses require voltage output to personal computer and notebook computer headsets—requirements that are the same as for personal portable music systems (Walkmans, iPods, and MP3 players) without considering the differences in risk of hearing loss due to different exposure times and preferred listening levels. This unique GS requirement can act as a barrier to trade.

Voluntary Eco-Labels

Voluntary environmental labels, or "eco-labels," signify the "environmental acceptability" of a product. Eco-labels, which are popular in many countries, include noise emission information. Although labeling or reporting product noise to customers is not required, meeting the acoustical criteria and displaying the eco-label symbol on products and in advertising implies acoustical acceptability (and possibly superiority) of the product. Some eco-label programs are the German Blue Angel (since 1977), the Nordic White Swan (since 1989), the Dutch Milieukeur, the Swedish TCO, and the EU Flower. Products with eco-labels with noise criteria include personal computers, printers, copiers, projectors, chain saws, garden tools, and construction machinery. The same issues that have been raised for other labels about uniformity of testing and verification also apply to eco-labels.

In contrast to eco-labels in other countries, the popular U.S. Energy Star program has no product noise emission criteria. EPA does, however, have the authority to label the noise emissions of products that emit noise capable of adversely affecting public health and welfare (42 USC 65, Section 4907).

Two different product groups have different ways of treating product noise emissions in the same eco-label program. During the development of the EU Flower criteria for personal computers and notebook computers, no consideration was given to noise levels that are acceptable or "green" in homes and offices. The primary consideration was an arbitrary decision that 25 percent of existing products be required to meet the new criteria. Similarly, the German Blue Angel noise criteria for personal computers are the same as for notebook computers. No consideration was given to differences in product functionality, costs of compliance, or customer expectations. At the same time, the Blue Angel noise criteria for construction equipment require only that products meet the limits set in the EU outdoor equipment directive, 2000/14/EC.

Issues and Concerns

The study committee is in favor of a uniform system for labeling the noise emissions of products. This is reflected in Recommendation 6-1 below. However, there are issues with noise labeling that need to be resolved. The major concerns about noise labels are consistency of labeling requirements and test standards (one test worldwide) and verification or consistency of testing by manufacturers. Many manufacturers have expressed concerns about favoritism and inappropriate labeling by other manufacturers, especially those in nearby countries. The lack of consistency from one product group to another can cause confusion for customers. The lack of consistency from one country to another for the same types of products can cause problems for manufacturers who are required to perform different tests or provide different information for different countries.

Information about the availability of noise values to the public is also an issue. The public may not be aware that Web-based information, such as eco-declarations, is available. Noise information that is available only in user's manuals or other product documentation is of no help to consumers making purchasing decisions.

The noise emission values of appliances and outdoor equipment are readily available in Europe, as required by law. However, they are not available (or not easily available) in the United States for the same products. Noise emission values for some IT products used in homes and offices are available from some, but not all, manufacturers, and they may not be readily available to potential customers.

FINDINGS AND RECOMMENDATIONS

Countries in the EU have recognized the importance of standards and taken the lead, making the ISO a leader in the standards community. ISO standards committees have superseded many American-based standards committees and organizations that U.S. manufacturers relied on in the past. America's voice on the ISO standard committees is weakened by the lack of U.S. manufacturers' leadership in ISO working groups. America has only a single vote, the same as every member country in the EU. The EU has been a leader in the development of noise regulations based on ISO standards. Because these regulations are more extensive than those that exist in the United States, European manufacturers have gained a competitive advantage over their U.S. counterparts in meeting consumer demand for low-noise machinery and other products worldwide.

At the time of purchase, consumers rank noise as one of the top five characteristics when comparing product performance. Other concerns are energy efficiency, cost, reliability, and serviceability. Noise levels for U.S. products are often buried in product literature and are reported using a variety of noise metrics, making it difficult for consumers to compare noise levels at the time of purchase. Thus, consumers are unable to make informed decisions about the noise emission of a product. This problem could be corrected if product noise levels were prominently displayed and manufacturers adopted a system of self-enforcement.

American manufacturers have the ingenuity to design quiet products. However, manufacturers and trade associations, as well as the voluntary standards community, have been unable to agree on a uniform standard for measuring and labeling product noise.

Recommendation 6-1: The Environmental Protection Agency should encourage and fund the development of a uniform system of labeling product noise. The system should be self-enforced by manufacturers but should have strict rules and penalties if products are deliberately mislabeled. The rules should specify standard methodologies for measuring product noise. Uncertainties in noise emission values should be acknowledged. Product noise labels should be prominently displayed so that consumers can make informed purchasing decisions. In a world with proliferating eco-labels and different requirements, international cooperation to develop one label recognized worldwide would be of great benefit to American manufacturers and consumers everywhere.

Recommendation 6-2: Government, trade associations, and industry should fund the participation of U.S. technical experts on standards bodies that develop international standards for determining product noise emissions.

Recommendation 6-3: The National Institute of Standards and Technology should take the lead in providing assistance to American manufacturers with noise regulation compliance by establishing a database of information on U.S. and international product noise emission standards and requirements.

Recommendation 6-4: To establish their credibility, organizations that determine noise emission data according to a certain standard as part of a voluntary labeling program should be accredited to test products. Managers at the National Institute of Standards and Technology and its National Voluntary Laboratory Accreditation Program should promote their accreditation program, especially in industrial laboratories.

REFERENCES

ANSI (American National Standards Institute). 2009. Mission Statement. Available online at *www.ansi.org*.

ASA (Acoustical Society of America) Standards Secretariat. 2009a. American National Standards Committees and US Technical Advisory Groups (U.S. TAGs). Available online at *http://www.acosoc.org/standards/*.

ASA. 2009b. Guide to Participation in the ASA Standards Program. Fourth Edition. Available online at *http://www.acosoc.org/standards/Guide%20to% 20Participation%20-%20Updated%20June%202009.pdf*.

EC (European Commission). 1986. Council Directive 86/594/EEC of 1 December 1986 on Airborne Noise Emitted by Household Appliances. Available online at *http://eur-lex.europa.eu/LexUriServ/LexUriServ.do?uri=CELEX:31986L0594:EN:HTML*.

EC. 1996. Future Noise Policy. European Commission Green Paper. Available online at *http://ec.europa.eu/environment/noise/pdf/com_96_540.pdf*.

EC. 2000. Directive 2000/14/EC of the European Parliament and of the Council of 8 May 2000 on the Approximation of the Laws of the Member States Relating to the Noise Emission in the Environment by Equipment for Use Outdoors. Available online at *http://ec.europa.eu/environment/noise/pdf/d0014_en.pdf*.

EC. 2003. Directive 2003/10/EC of the European Parliament and of the Council of 6 February 2003 on the Minimum Health and Safety Requirements Regarding the Exposure of Workers to the Risks Arising from Physical Agents (Noise). Available online at *http://eur-lex.europa.eu/LexUriServ/site/en/oj/2003/l_042/l_04220030215en00380044.pdf.*

EC. 2005. 2005/88/EC of the European Parliament and of the Council of 14 December 2005 Amending Directive 2000/14/EC on the Approximation of the Laws of the Member States Relating to the Noise Emission in the Environment by Equipment for Use Outdoors. Available online at *http://eur-lex.europa.eu/LexUriServ/LexUriServ.do?uri=OJ:L:2005:344:0044: 0046:EN:PDF.*

EC. 2006a. Enterprise and Industry: Noise 1.5. Available online at *http://ec.europa.eu/enterprise/mechan_equipment/noise/citizen/app/.*

EC. 2006b. Directive 2006/42/EC of the European Parliament and of the Council of 17 May 2006 on Machinery, and Amending Directive 95/16/EC (Recast). Available online at *http://eur-lex.europa.eu/LexUriServ/LexUriServ.do?uri=OJ:L:2006:157:0024:0086: EN:PDF.*

EC. 2009. NANDO (New Approach Notified and Designated Organisations) Information System. Available online at *http://ec.europa.eu/enterprise/newapproach/nando/.*

EU (European Union). 2000. Permissible Sound Power Levels (dB(A)) for Lawn Mowers, Based on Width of Cut. From Directive 2000/14/EC of the European Parliament

I-INCE. 2009. Survey of Legislation, Regulations, and Guidelines for the Control of Community Noise. Final report of the I-INCE Technical Group on Noise Policies and Regulations (TSG 3). TSG3 Report 09-01. Available online at *http://www.i-ince.org/data/iince091.pdf.*

ILAC (International Laboratory Accreditation Cooperation). 2009. Welcome to ILAC. Available online at *http://www.ilac.org.*

NIST (National Institute of Standards and Technology). 2009a. Overview of the National Voluntary Laboratory Accreditation Program (NVLAP). Available online at *http://ts.nist.gov/standards/accreditation/index.cfm.*

NIST. 2009b. Directory of Accredited Laboratories: Acoustical Testing Services. Available online at *http://ts.nist.gov/standards/scopes/acots.htm.*

Schomer, P.D., G.F. Stanley, and W. Chang. 2008. Visitor Perception of Park Soundscapes: A Research Plan. Proceedings of INTER-NOISE 08, The 2008 International Congress and Exposition on Noise Control Engineering, Shanghai, China, October 26–29. Available online at *http://www.internoise2008.org/ProceedingsforSale.htm* or *http://scitation.aip.org/journals/doc/INCEDL-home/cp/.*

Shapiro, S.A. 1991. The Dormant Noise Control Act and Options to Abate Noise Pollution. Report to the Administrative Conference of the United States. Available online at *http://www.nonoise.org/library/shapiro/shapiro.htm.*

Statskontoret. 2004. Technical Standard 26:6: Acoustical Noise Emission of Information Technology Equipment. Available online at *http://www.statskontoret.se/upload/2619/TN26-6.pdf.*

TÜV (TÜV SÜD America, Inc.). 2009a. Industries: Automotive: Noise Testing Services. Available online at *http://www.tuvamerica.com/industry/automotive/noise.cfm.* (See also CIMA Newsletter, June 2000, The New EC Noise Directive Becomes Effective on July 3, 2000. Available online at *http://www.tuvamerica.com/tuvnews/articles/cima.cfm.*)

TÜV. 2009b. Description of the Activities of TÜV SÜD America. Available online at *http://tuvamerica.com/newhome.cfm.*

UNECE (United Nations Economic Commission for Europe). 2009. World Forum for Harmonization of Vehicle Regulations, WP.29. Available online at *http://www.unece.org/trans/main/wp29/wp29wgs/wp29gen/wp29age.html.*

USTR (Office of the U.S. Trade Representative). 2009. Mission of the USTR. Available online at *http://www.ustr.gov/Who_We_Are/Mission_of_the_USTR.html* (as posted on January 14, 2008).

7

Cost-Benefit Analysis for Noise Control

The Federal Aviation Administration (FAA) is developing methods of using cost-benefit analyses to assess noise around airports. The costs and benefits of reducing highway noise have received less attention in the United States and have been emphasized in this chapter. Highway noise barriers are an effective means of noise reduction because they interrupt the propagation path between the noise sources and nearby homes. At highway speeds most of the noise is generated by the interaction between vehicle tires and the road surface. The generation of sound by this interaction is complicated and involves "air pumping" as the tread alternately engages and releases from the road surface, vibration of the sidewalls, and other mechanisms. A layman's discussion of the various sources can be found in *The Little Book of Quieter Pavements*, published by the Federal Highway Administration (FHWA; Rasmussen et al., 2007). A great deal of research has been done on the characteristics of pavements that result in lower noise levels. What is needed is a cost-benefit analysis of highway noise reduction to ensure that the best mitigation methods are being applied.

The committee decided to focus on surface transportation noise for several reasons. In a 1981 report the U.S. Environmental Protection Agency (EPA) estimated that 24 million people were exposed to high day-night average sound levels (DNLs; greater than 65 dB) of surface transportation noise (19.3 million for highway noise and 4.7 million for rail noise); in the same report the number of people exposed to air transportation noise was estimated at 2.5 million people. Although no studies have been conducted to determine surface transportation exposures since then, it is likely that population growth, increased residential development near highways, and increased traffic volume have also increased exposures to highway traffic noise.

However, in the past 30 years, air transportation has led the way in technological developments and operational improvements to reduce noise and more recently to reduce environmental impacts; in addition, economic analysis tools have been developed for determining the costs and benefits of these improvements to the environment. In fact, a 2007 study showed that the number of people exposed to high levels of air transport noise in the United States had decreased to approximately 500,000 (Waitz et al., 2007).

As the numbers above reflect, reducing air transportation noise has been the focus of intense efforts by the public and by policy makers since the advent of the jet age. As Figure 7-1 shows, advances in technology and airport management have resulted in significant reductions in airport noise contours (e.g., the perimeter around airports where DNLs exceed 65 dB). As a result, the number of people exposed to noise in excess of 65 dB has decreased dramatically over the past several decades, although there are still many serious noise problems around airports.

Technologies to reduce highway noise generated by surface vehicles have also been investigated. Studies have shown that the most significant source of noise at highway speeds is interaction between tires and highway surfaces. Attempts to modify tires have had limited success because the primary concerns in tire design are safety and performance. However, highway surfaces can be modified to reduce overall noise without compromising safety.

The current approach to addressing noise levels along proposed highways is to construct sound barrier walls in residential areas to protect occupants from excessive noise levels as measured at the property line nearest the highway. However, noise barriers are expensive, and residents often consider them an eyesore because they obstruct views and are sometimes subject to graffiti. In addition, they provide significant noise reduction for only the first one or two rows of houses behind the barrier (FHWA, 2009a).

One question of interest to the committee was whether a greater number of residents would benefit if "quiet" pavement technology were used instead of barriers to reduce the noise level at the source. To make that assessment, the committee believes that cost-benefit analysis tools developed by EPA and the FAA should be used to identify variables and measurable characteristics and relate them to one another in commensu-

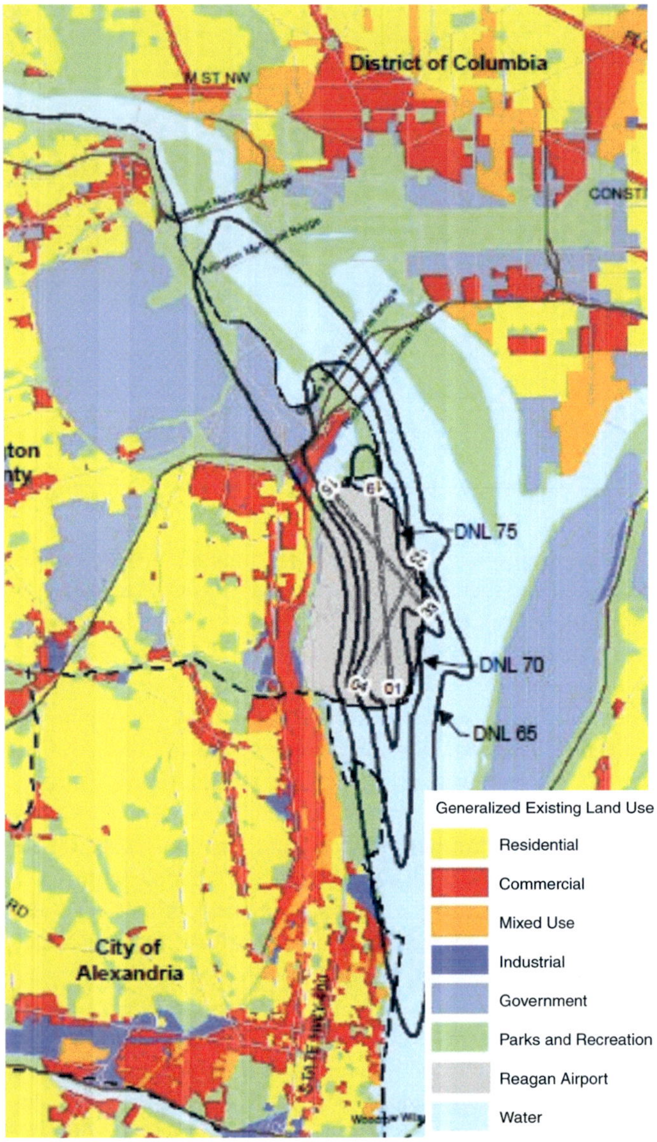

FIGURE 7-1 Contour map showing Day-Night Average Sound Levels (DNL) around Ronald Reagan National Airport in Washington, D.C. Source: Reprinted with permission of EA Engineering, Science and Technology.

rate ways. The committee also recognized that some likely variables for surface transportation noise, such as long-term noise reduction characteristics, installation costs, and maintenance costs of quiet pavements would not be available.

FHWA has developed a software tool called the Traffic Noise Model (TNM; FHWA, 2009b). The TNM is a useful tool for estimating sound pressure levels at various distances from a highway in terms of traffic mix, speeds, and other factors. Using various scenarios, noise reductions in decibels can be predicted by the model. However, no attempts have been made to monetize those benefits in terms of home values, sleep disturbance, or other measures of impact. The TNM could also provide information showing that quiet pavements could eliminate the need to construct a barrier or that a less expensive barrier would provide enough noise reduction, as measured at the property line.

The remainder of this chapter describes how environmental economic analysis techniques have been used by EPA and the FAA for purposes of cost-benefit analysis and how such analyses might be used for surface transportation noise. In addition, European efforts to conduct cost-benefit analyses on highway noise are reviewed, as is pavement research that will lead to lower noise levels and will be a vital input to any cost-benefit analysis model.

ENVIRONMENTAL ECONOMIC ANALYSIS

Microeconomics (i.e., the study of disaggregated entities and behaviors) is generally used rather than macroeconomics (i.e., the study of aggregates) to analyze environmental issues. When resources are scarce, economic analyses can help planners compare options to determine which uses of those resources will generate improvements in well-being for people who live near busy highways. Environmental economic analyses provide a rigorous, quantitative approach to support these decisions.

To compare relative values requires metrics that are comparable in terms of the outcomes each alternative produces (i.e., decibel reductions) as well as in terms of costs, both in dollars and negative effects (including eliminating potential desirable outcomes or benefits). Economists use monetary value to compare alternatives, and the conversion to monetary values of physical or social effects that are not naturally denominated in dollars is called *monetization*. The primary framework used to compare monetized positive and negative attributes of alternative policies and investment choices is called cost-benefit analysis (CBA) or benefit-cost analysis—the terms are interchangeable.

General guidelines for government entities conducting a CBA have been published by the Office of Management and Budget (OMB) in the form of circulars. For example, Circular A-94, first published in 1992 and updated annually, provides discount rates (OMB, 1992). Because OMB's audience includes all federal agencies, the guidelines are general, rather than domain specific. Thus, Circular A-94 encourages monetization but does not offer specific guidance on monetization techniques.

Individual agencies often publish their own more detailed domain-specific guidelines. For instance, EPA published *Guidelines for Preparing Economic Analyses*, which covers nuances of CBA in the environmental arena and provides detailed guidelines for monetization (EPA, 2000). In comparison to the 21-page Circular A-94, EPA's *Guidelines* includes more than 200 pages. One area in which EPA has authority to engage in CBA is noise (42 USC 65, Section 4913).

Because the reader may not be familiar with CBA as practiced by economists or with terms such as willingness to pay, willingness to accept, revealed preference, stated preference, and others, a summary of CBA is included in this report as Appendix F. In the appendix, OMB guidance on CBA is mentioned, but emphasis has been placed on EPA procedures because of the agency's experience with this subject. Where possible, suggestions have been made as to how EPA procedures would apply to CBA for noise issues.

The FAA has also been developing CBA tools for use around airports. A summary of these activities is given in the next section to provide an introduction to what FHWA might develop for CBA of noise reduction along the nation's highways.

COST-BENEFIT ANALYSIS OF AIRCRAFT NOISE

The methods being developed by FAA to perform a CBA of measures for mitigating aircraft noise illustrate useful applications of the general concepts described earlier in this chapter.[1] It is well documented that aircraft noise has a range of undesirable impacts, primarily felt by people living around airports. These include physical effects, such as annoyance (e.g., interference in speech communication and activities), sleep disturbance, impacts on school learning and academic achievement, physical and mental health effects, building rattling and other noise, and compromised work performance (WHO, 2004). These effects result in monetary impacts, such as lower property values, health costs, and personal and business economic costs. To perform CBA, aircraft noise must be related to these impacts.

[1]The FAA Office of Environment and Energy, in collaboration with Transport Canada and the National Aeronautics and Space Administration, is developing a comprehensive suite of software tools for a thorough assessment of the environmental effects and impacts of aviation noise. The main purpose is to develop a new capability to characterize and quantify interdependencies among aviation-related noise and emissions, impacts on health and welfare, and industry and consumer costs, under different policy, technology, operational, and market scenarios. The three main functional components of the tools suite are the Environmental Design Space (EDS), which is used to estimate aircraft CAEP/8 performance trade-offs for different technology assumptions and policy scenarios; the Aviation Environmental Design Tool (AEDT), which takes as input detailed fleet descriptions and flight schedules and produces estimates of noise and emissions inventories at global, regional, and local levels; and the Aviation Environmental Portfolio Management Tool (APMT), which is the framework within which policy analyses are conducted and which provides additional functional capabilities. APMT functional capabilities include an economic model of the aviation industry, with inputs of different policy and market scenarios and existing and potential new aircraft types (the latter from EDS or other sources). It then simulates the behavior of airlines, manufacturers, and consumers, producing a detailed fleet and schedule of flights for each scenario year for input to AEDT. APMT also takes the outputs from AEDT (or other similar tools) and performs comprehensive environmental impact analyses for global climate change, air quality, and community noise. These environmental impacts are quantified using a broad range of metrics (including, but not limited to, monetized estimates of human health and welfare impacts, thereby enabling both cost effectiveness and cost-benefit analyses). Additional information can be found in ICAO (2007) and on the FAA website.

The sound at a given point from one aircraft in flight is typically measured (or estimated) and then expressed in decibels in a metric called the effective perceived noise level. This metric is used by the FAA as a measure of airplane noise *emission*. This metric takes into account the nonuniform response of the human ear, tonal corrections, and other factors. Then the noise from a representative sample of flights (typically for one day) can be combined into a measure, such as the standard DNL metric, in which the sound energy from multiple events is averaged, and a 10-dB correction is made for flights that occur between 10 p.m. and 7 a.m. DNL and other average measures have been shown to correlate with community response to aircraft noise, as shown, for example, in Table 7-1.

It is important to recognize that responses to aircraft noise vary widely among people and communities, as illustrated in Figure 7-2. Note that for aircraft noise levels typical of communities within 5 miles of airports (55 to 65 dB DNL), the proportion of the population "highly annoyed" varies from 0 to 75 percent. This variability in personal and community response suggests that monetization methods based on statistical distributions, or that accept ranges of inputs, may be most relevant. Thus, the DNL metric is most useful for summary assessments but may not adequately describe the effects of noise on a specific impacted population; it is also sometimes difficult to explain the DNL concept to the public. Information on this subject can be found in a report by the National Research Council Transportation Research Board (Eagan, 2007).

Because it is difficult to assess independent impacts of noise on annoyance, sleep, health, school learning, and so on, it is typical to use one of two methods as surrogates for the total impact of noise. The first of these is the change in property value associated with aircraft noise. Many studies have statistically analyzed this relationship, typically presenting it in terms of a noise depreciation index (NDI) with units of percentage of property value loss per decibel. The results of many of these studies are shown graphically in Figure 7-3 (left). Figure 7-3 (right) shows the results of willingness-to-pay (WTP) studies based on carefully designed surveys of people who live near airports; the typical metric is euros per decibel per household per year. The "X" marks an equivalent value between the two measures (assuming an appropriate average house price and depreciation level).

Both measures of economic impact reflect the wide variability that is characteristic of personal and community responses to noise. Nevertheless, both methods (observing real estate transactions and surveying people) produce similar results in terms of overall value and a similar range of values from low to high. Thus, they provide a basis for estimating the economic impacts of aircraft noise—as a surrogate for estimating the large number of individual impacts, many of which overlap in meaning and are difficult to value (e.g., the relationship between sleep disturbance, stress, and school or work performance).

TABLE 7-1 Relationship between Day-Night Average Sound Level and Impacts

		Annoyance		
Day-Night Average Sound Level (dB)	Hearing Loss Qualitative Description	Percentage of Population Highly Annoyed	Average Community Reaction	General Community Attitude
≥ 75	May occur	37	Very severe	Noise is likely to be the most important adverse aspect of the community environment.
70	Not likely to occur	22	Severe	Noise is one of the most important adverse aspects of the community environment.
65	Will not occur	12	Significant	Noise is one of the important adverse aspects of the community environment.
60	Will not occur	7	Moderate to slight	Noise may be considered an adverse aspect of the community environment.
≤ 55	Will not occur	3	Moderate to slight	Noise is considered no more important than other environmental factors.

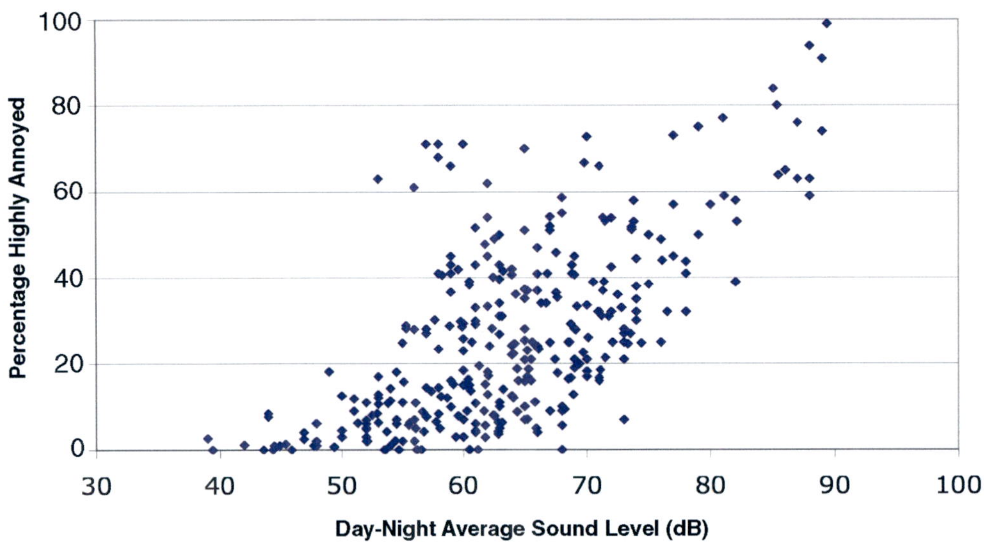

FIGURE 7-2 Relationship between percentage of population highly annoyed and DNL level, in decibels. Sources: Kish (2008) and Fidell and Silvati (2004).

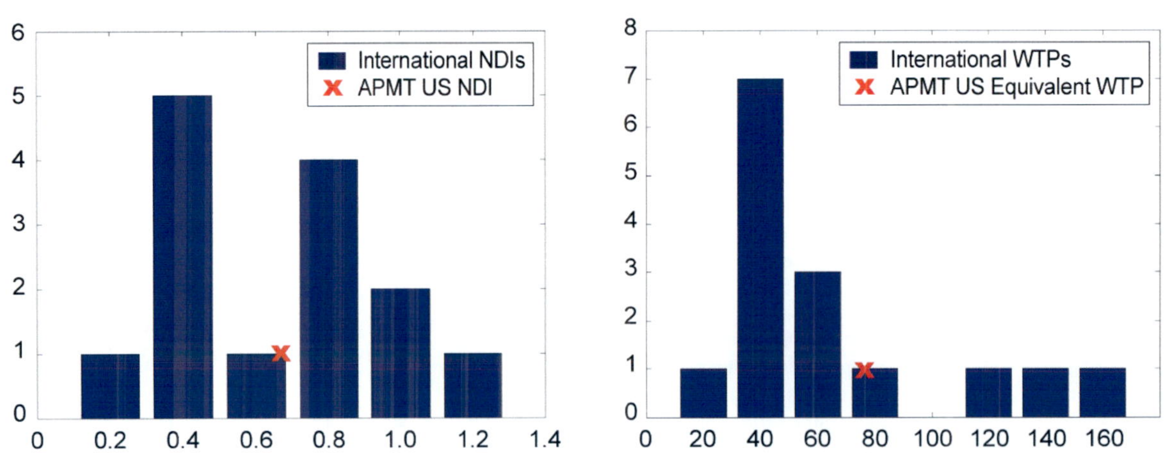

FIGURE 7-3 (left) Noise depreciation indices (NDI) (percent of property value loss per decibel); (right) willingness-to-pay (WTP) values (Euros/household/dB/year) based on a number of North American, European, Japanese, and Australian studies of aircraft noise. APMT = Aviation Environmental Portfolio Management Tool. For reference, "X" marks equivalent values (assuming an average housing price and depreciation value). Source: Kish (2008).

The FAA has recently developed methods that overlay contours of noise levels with census data describing populations and housing values (FAA, 2008; Kish, 2008). With these, statistical distributions and ranges of NDIs and WTP values are used to provide monetized estimates of the negative impacts of noise. These monetized estimates are then compared to policy implementation costs, industry costs, and costs and benefits associated with changes in other interdependent environmental impacts. These tools have recently been developed and, to date, have been used only in sample cost-benefit analyses of technology, operations, and policy options. Nonetheless, the intention is to use them for real analyses after further research and development. More information can be found at *http://web.mit.edu/aeroastro/partner/apmt/* and *http://web.mit.edu/aeroastro/partner/apmt/noiseimpact.html*.

COST-BENEFIT ANALYSIS FOR HIGHWAY NOISE

Since the 1980s, few major CBAs have been done for highway noise in the United States.[2] A meta-analysis in 1982 of 17 hedonic pricing estimates for the United States and Canada showed a range of NDIs of 0.16 to 0.63 percent, with a mean value of 0.40 percent per decibel (Nelson, 1982). New studies, using CBA techniques described in this report and economic terms, such as hedonic pricing, stated preference, and WTP, are needed to assess the costs and benefits of both sound barriers and quieter road surfaces with respect to noise abatement, especially to compare the two to ensure that funds currently provided for noise mitigation are being well spent.

The FHWA policy for highway noise abatement includes an implied CBA in determining the "reasonableness" of the abatement method (i.e., sound walls). Following the process outlined in FHWA noise policy 23 CFR 772, each state develops a cost allowance associated with any noise-impacted residence for a proposed highway project (FHWA, 2006). These cost allowances range from a low of $10,000 per residence to a high of $50,000. Some states allow increases in these values based on the severity of the predicted impact and, in some cases, the predicted noise reduction. The cost allowance for all "benefited" residences that receive a 3- to 5-dB reduction from a proposed sound wall are then totaled, and this cost is compared to the cost of the sound wall, using a process specific to individual states.

FHWA policy does not now allow quieter pavement to be considered as a noise abatement method, and therefore it is not included in the CBA. However, a National Cooperative Highway Research Program project (NCHRP 10-76) is under way to develop methodologies for including quieter pavement in CBAs. One of the problems encountered so far is that the noise reduction from quieter pavements typically degrades over time and must be rehabilitated, on some cycle, throughout the life of the highway. Barriers, on the other hand, are typically assumed to have minimal ongoing costs.

Policies on Noise Barriers

According to official FHWA policy, "the use of specific pavement types or surface textures must not be considered as a noise abatement measure" (FHWA, 2009c). Thus, wherever highway noise mitigation is required, noise barriers should be used. A summary of the number of barriers that have been constructed and their costs is given below. Benefits are achieved only relatively close to the highway and are generally measured in terms of a reduction in A-weighted sound pressure level.[3] In addition, noise barriers are not feasible in many areas—for example, to protect homes on a hillside above a busy highway.

Barriers are constructed from a variety of materials, including wood, concrete block, precast concrete, brick, and other materials. Earth berms may also be used as noise barriers. Construction of barriers is a cooperative effort between FHWA and the state in which the barrier is constructed, in determination of both the requirements and the costs. FHWA defines two types of highway projects for which barriers are considered. A Type I highway project is a planned new construction project or construction to increase the capacity of an existing highway. Federal laws require that a noise impact statement be prepared, and if noise levels exceed an established limit, noise abatement must be considered. The limits, set by the state, range from 64 to 67 dB(A) in response to the FHWA requirement that abatement be provided for levels "approaching" 67 dB(A) for the loudest hour predicted for the highway project. Given typical urban region traffic patterns, this worst level of 67 dB(A) can result in day-night levels of 69 dB(A) or more (Greene, 2002).[4] Once a sound wall is designed, the state determines if it is "reasonable" (cost-effective) and "feasible" (technically) to construct the barrier. *Feasible* in this context equates to the requirement that the barrier achieve at least a 5-dB noise reduction. If not, the barrier is not feasible, and no abatement is implemented in the project.

Barriers for Type II projects, those undertaken in response to noise complaints, are voluntary. FHWA will provide matching funds for Type II projects, although the requirements for this are often difficult to meet. As a result, construction of barriers for existing highways is rare, and cost is a major factor.

---

[2]Nelson, J. 2007. Cost-Benefit Analysis and Transportation Noise. Presentation at an NAE-sponsored workshop on cost-benefit analysis, Cambridge, Massachusetts.

[3]Typically, noise barriers are most effective within 200 feet (FHWA, 2009a).

[4]Donavan, P.D. 2009. Analysis based on Greene (2002). Private communication, September 17.

Design and Performance

Information on the technical aspects of barrier design and evaluation are available in I-INCE (1999) and FHWA (2009d). In the I-INCE document (1999), the best estimate by the working group that prepared it was that barrier insertion loss (the difference in A-weighted sound pressure level before and after installation of a barrier) typically ranges from 5 to 12 dB. FWHA (2009d) classifies the insertion loss (attenuation) as follows:

5 dB = simple
10 dB = attainable
15 dB = very difficult
20 dB = nearly impossible

The fundamental quantity in barrier design is the Fresnel number (the difference in path length from source to receiver with and without the barrier, measured in half-wavelengths of the sound). High frequencies have a high number and more attenuation; low frequencies have a lower number and are more difficult to attenuate.

Barriers are most effective when constructed near the highway or near the receiver (which tends to maximize the path-length difference); the exact range of barrier effectiveness depends a great deal on the terrain. For example, in a rising ground level the effectiveness can be small (low Fresnel number), whereas if the ground level goes down, the barrier is more effective. FHWA (2009a) estimates that barriers are most effective within 200 feet of a highway FHWA. Thus, only a few rows of homes are protected by a barrier.

Cost

The costs of barrier construction, as documented by FHWA, are summarized here (FHWA, 2007). [5] Costs vary from project to project, and methods of reporting costs are not uniform from state to state. However, available data are a good starting point.

The obvious variables are barrier height and length. According to Polcak (2003), the most reliable cost breakdown is for Type II barriers and can be divided into seven categories:

- *Preliminary.* This category includes mobilization costs, clearing and grubbing, field office setup, and other preparatory activities that must be done before construction begins.
- *Drainage.* This category includes everything related to maintaining and facilitating drainage of the barrier site, including, but not limited to, inlets, pipes, underdrain systems, ditch treatments, rip-rap, and stormwater management facilities.
- *Excavation.* This category includes grading and excavation ditches, benching, construction roads, and other access features.
- *Guardrail.* This category includes traffic control devices, signage, jersey barriers, or other protective equipment that may be used for maintenance of traffic requirements or ultimately protecting the newly installed noise barrier from vehicle impacts.
- *Utilities.* This category includes temporary or permanent relocations of overhead or underground utilities that may be affected by the noise barrier construction.
- *Barrier system.* This category includes basic physical elements of the structural barrier system, including posts, panels, and foundations. Also included are grade beams; special panels; architectural, decorative, or aesthetic finishes; or absorptive-surface treatments. There might also be special foundation requirements to accommodate subsurface conditions or retaining walls.
- *Landscaping.* This category includes site restoration when construction is complete, trees and shrubs, seeding, mulching, and so forth.

FHWA requests information every three years from the states on the number of miles of barrier constructed and the costs. Through the end of 2004, 45 states and the Commonwealth of Puerto Rico had constructed 2,205 miles of noise barriers at a cost of $3.4 billion (FHWA, 2009a). Thus, the average cost per mile is approximately $1.54 million in 2004 dollars. Table 7-2 shows the cost breakdown. The apparent discrepancy between the numbers above and below is because not all states are included in the table; data from California for 1998 to 2004 are missing.

Cost elements used to determine project costs vary greatly from state to state; some states report the total bid cost, others just use the cost of the barrier "system." Even the items included in the reported barrier system may differ. States that use the same or similar approaches may use different underlying assumptions. Thus, detailed comparisons are difficult to make.

If, in the upper left table, Minnesota were eliminated and, in the lower left table, Colorado were eliminated (to be able to compare the same nine states), the average barrier cost per square foot for nine states would be $18.29. The highest cost is for Pennsylvania ($24.88), and the lowest cost is for California ($13.04). However, recent data for California are not included, so Ohio should be considered the low-cost state ($13.51). The 10-state average is thus $1.75 million per linear mile. Table 7-3 shows data for the states in Table 7-2 using common data converted to metric units.

Costs from earlier FHWA data, published by Polcak (2003), show costs per project for many states. For example, Figure 7-4 shows construction costs in Maryland for precast concrete barriers and for all barriers. Note that the vertical

[5] The summary is for the years up to 2004.

TABLE 7-2 Noise Barrier Construction by State, through 2004

Square Feet (thousands)		Linear Miles	
California	30,644	California	482.8
Virginia	11,227	Arizona	155.1
Arizona	11,226	Virginia	127.5
New Jersey	9,440	Ohio	112.4
Ohio	8,675	New Jersey	96.9
Maryland	8,422	Colorado	92.5
Minnesota	7,187	New York	90.7
New York	7,011	Pennsylvania	87.0
Florida	6,700	Minnesota	83.7
Pennsylvania	6,415	Maryland	81.8
10-State Total	106,946		1,410.4
Actual Cost at Time of Construction ($ millions)		**Cost in 2004 Dollars ($ millions)**	
California	399.6	California	592.8
Arizona	258.7	Arizona	284.6
New Jersey	202.4	New Jersey	277.5
Maryland	200.9	Maryland	253.6
Virginia	169.6	Virginia	225.3
New York	165.9	New York	207.3
Pennsylvania	159.6	Pennsylvania	197.8
Florida	150.7	Florida	175.9
Ohio	117.2	Ohio	139.0
Colorado	80.0	Minnesota	107.7
10-State Total	1,904.5		2,461.4

scales are costs per square meter. Similar data for Virginia, including all construction materials, are shown in Figure 7-5. As the figures show, there is a great deal of variability in cost from project to project, and the data are only weakly dependent on barrier length. A major factor in the variability is that FHWA data for the states include barriers for both Type I and Type II projects, which by their nature are more likely to have different elements included in the cost figures.

TABLE 7-3 Summary of Barrier Construction and Costs, by State

State/ Total/Average	Barrier Area (m^2)	Cost per Square Meter ($)	Barrier Length (km)	Barrier Cost per Kilometer ($ thousands)
California	2,847	140.36	777.0	0760
Arizona	1,043	248.05	249.6	1140
New Jersey	877	230.78	155.9	1780
Maryland	782	256.76	131.6	1930
Virginia	1,043	162.60	205.2	1100
New York	651	254.70	146.0	1420
Florida	622	242.11		—
Ohio	806	145.42	180.9	0770
Pennsylvania	596	267.80	140.0	1410
TOTAL	9,268		1,986.3	
AVERAGE		196.87		1100

Quiet Pavement Design

FHWA policy supports research related to quiet pavements. However, predictions of highway noise, and the criteria for whether noise mitigation is allowable for federal cost sharing, are based on an average of all pavement types. Thus, even if the noise characteristics of a particular pavement type are known, they are not used in highway noise predictions. Modifications of source data to account for quieter pavements are allowable only under the stringent requirements of the FHWA Quiet Pavement Pilot Program (FHWA, 2005). In addition, because there are no acceptance tests in place to ensure that a pavement meets planned noise levels, there are no incentives for state or local agencies to build or maintain quieter pavement that would benefit the public. Other issues related to the design and implementation of low-noise road surfaces include measurement of the noise reduction at the source and its relationship to noise measurements in the community, the technology of the design of road surfaces, safety, and durability. Many of these issues were discussed at a workshop sponsored by the National Academy of Engineering (NAE) in February 2007.[6]

Long-term studies have been under way for several years on the durability of road surfaces (CDOT, 2005; Rochat, 2002), and the results of a 52-month study were recently published (Rochat and Read, 2009). In addition, there have been many studies in Europe (see section below) and other studies in the United States (Corbisier, 2005; Donavan, 2005a,b, 2006; Rasmussen et al., 2007; Reyff, 2007a,b). FHWA maintains a home page on the subject (FHWA, 2009e) as well as guidance for the development of quiet pavement programs (FHWA, 2009c). The Tire-Pavement Noise Research Consortium is funded by eight states and FHWA (TPNRC, 2009). FHWA also sponsors a workshop, "Tire-Pavement Noise 101," that has been given in many states; the essence of the course material is available in the *The Little Book of Quieter Pavements* (Rasmussen et al., 2007a,b).

Quiet pavements reduce noise by controlling the surface characteristics of the pavement. Much less documentation is available on the costs for pavement modifications by resurfacing and grinding than on the costs of noise barriers. The literature to date provides data only on pilot projects, with the emphasis on onboard noise measurements, the correlation of these data with pass-by noise, surface characteristics and their relationship to noise emission, and the durability of road surfaces (Donavan, 2006).[7] Costs for quiet pavements vary with the extent of treatment (e.g., grinding), the addi-

[6]Donavan, P.D. 2007. Reductions in Noise Emissions from Porous Highways. Presentation at an NAE workshop on Cost-Benefit Analysis of Transportation Noise Control Technology, Cambridge, Massachusetts, February 22.

[7]See also Donavan, P.D. 2007. Reductions in Noise Emissions from Porous Highways. Presentation at an NAE workshop on Cost-Benefit Analysis of Transportation Noise Control Technology, Cambridge, Massachusetts, February 22.

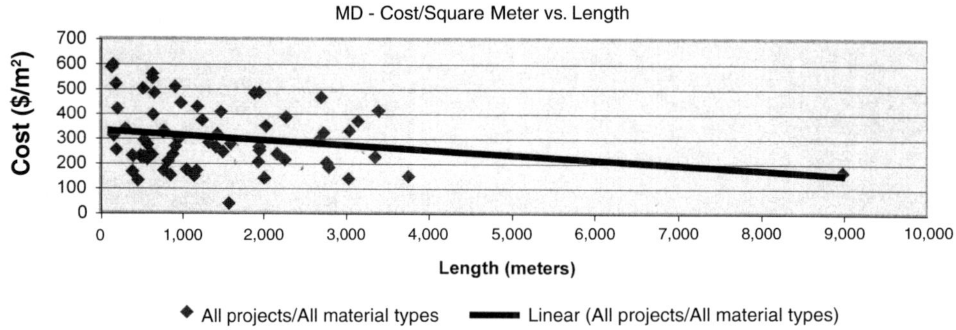

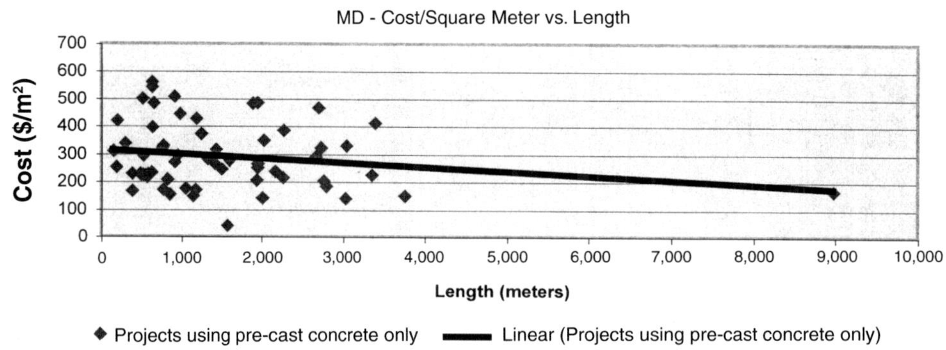

FIGURE 7-4 Cost of barriers per square meter in Maryland for all projects (upper) and for precast concrete (lower). Source: Polcak (2003).

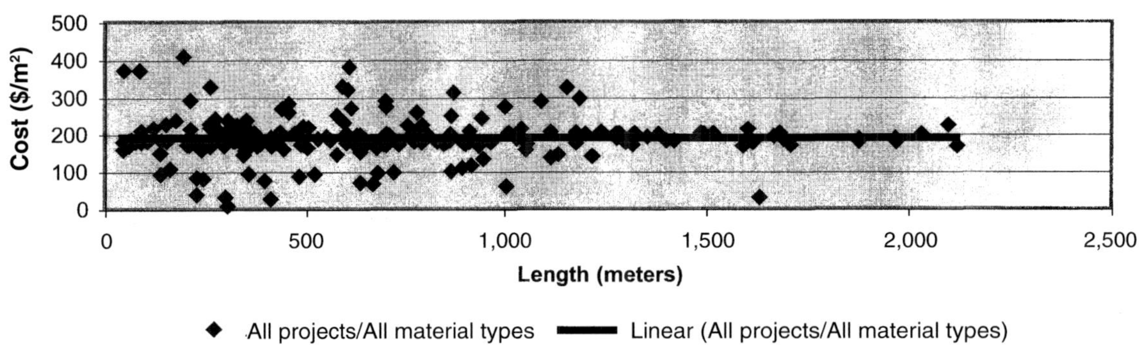

FIGURE 7-5 Cost of barriers per square meter in Virginia for all projects. Source: Polcak (2003).

tion of a thin (25-millimeter) porous layer, the removal of and complete replacement of pavement, and the construction of a new road.

Scofield provides some information on diamond grinding; at an average cost of $3.52 per square yard, the cost for 1 mile would be $61,952 per 30 feet of highway width.[8] According to Arizona guidelines from 2007, the cost of a 25-millimeter asphalt rubber asphaltic concrete friction course is $6.55 per square yard;[9] thus, the cost for 30 feet of highway width is still well below the cost per mile of a noise barrier. As discussed in the previous section, the average cost of a noise barrier in 10 states was estimated to be $1.75 million per linear mile. According to the Transportation Research Board (Alexandrova et al., 2007), asphalt rubber friction

[8]Scofield, L. 2009. E-mail communication, American Concrete Pavement Association.

[9]McDaniel, B. 2009. E-mail communication, Becky McDaniel, Purdue University.

course overlays also have a positive impact on tire wear, emissions, and air quality.

Based on data compiled by the California Department of Transportation, some analysis has been done comparing the typical costs of barriers to the costs of quieter pavement options (Donavan, 2005b). Assuming that barriers are 16 feet high, the maximum allowed in the state, and that they line both sides of a freeway, the cost was estimated at $5 million per mile. Assuming a six-lane freeway, the cost of a quieter pavement overlay, such as rubberized open-graded asphalt, was estimated at $210,000 to $270,000 per mile. For Portland cement concrete pavement surfaces, the cost of grinding the pavement to reduce tire/pavement noise was estimated at $320,000 to $600,000 per mile.

Noise barriers protect only the first few rows of houses, whereas pavement treatments, which essentially reduce noise *emissions*, can provide protection at greater distances. Complaints about highway noise may come from long distances from the highway. This was documented in a report by the Transportation Research Board in 2006 (Herman et al., 2006):

> A portion of 1-76 near Akron, Ohio, was reconstructed by the Ohio Department of Transportation with concrete pavement to replace the previous asphalt surface. During reconstruction, the concrete surface was textured with random transverse grooves. After construction, residents living in the project area as far as 800 m (2,600 ft) from the roadway perceived an undesirable increase in noise level, which they attributed to the new concrete pavement in the reconstruction project. Therefore, another project was initiated to retexture the pavement surface by diamond grinding. The transverse grooves were replaced with longitudinal grooves. Traffic noise measurements were made before and after grinding at five sites in the project area, at distances of 7.5 m (24.6 ft) and 15 m (49.2 ft) from the center of the near travel lane. The average reduction in the A-frequency-weighted broadband noise levels at 7.5 m (24.6 ft) was 3.5 dB, and the average reduction at 15 m (49.2 ft) was 3.1 dB. Spectrum analysis showed that the greatest reduction in noise occurred at frequencies above 1 kHz and that the retexturing had little to no effect on frequencies less than 200 Hz.

Unfortunately, the report does not note if complaints ended after the grinding or whether before/after noise measurements were made at long distances. A detailed CBA will be necessary to determine if the extensive use of low-noise road surfaces will have a general benefit for people who live near busy highways.

With current technology, noise reduction from tire/road interaction is not as effective as can be achieved by noise barriers. However, because larger reductions are achieved only near the barrier, relatively few people benefit from the reduction. Reduction of the tire/road interaction noise provides a smaller benefit, but it is a noise reduction at the source and therefore can benefit a larger number of people. CBA is an approach to making this kind of trade-off.

EUROPEAN COST-BENEFIT ANALYSES

Many CBAs of mitigation options for aircraft, road, and rail noise have been done in Europe. This section provides a brief summary of selected activities.

In 2001 a workshop on CBA, "A Billion Euro Question," was held in The Hague, Netherlands, in conjunction with the 2001 International Congress and Exposition on Noise Control Engineering (INTER-NOISE 01).[10] The focus of the workshop was on how much should be paid for noise control, and several presentations included descriptions of how aircraft, road, and rail noise were valued. In December 2001 the European Commission sponsored a second workshop, "State-of-the-Art in Noise Valuation," and a workshop report was published in 2002 (Vainio and Paque, 2002). The workshop participants came to the following conclusions:

- Contingent valuation and revealed preference (including the hedonic price method) were acceptable methods for valuing the benefits of noise reduction, with the caveat that these methods be followed rigorously to ensure that the results are meaningful.
- A day-evening-night sound level of 55 dB should be an interim lower cutoff point for noise valuation.
- A rough assessment of the cost per household per decibel per year for levels above 55 dB should be between 5 and 50 Euros.

On April 14, 2002, a 68-page report was delivered to the European Commission Directorate General Environment on the theoretical basis and valuation techniques for cost-benefit reviews and other studies of noise valuation for road traffic, aircraft noise, rail noise, and industrial noise (Navrud, 2002).

Strategies and Tools to Assess and Implement Noise Reducing Measures for Railway Systems (STAIRRS) was a project to review strategies for reducing noise around railways (Oertli et al., 2002). The program used to determine the costs and benefits in some railway noise emission situations was described by Lenders and Hecq (2002). The results of the study allow the calculation of costs and benefits in any geographical area of Europe. Noise barriers were shown to have a poor (high) ratio of costs and benefits.

The European Commission (EC) issued a 49-page draft report in 2006 (EC, 2006) that included information on several European Union (EU) projects related to CBA. In 2008 the consulting firm CE Delft produced a report for the EC detailing external costs for a number of items in the transportation sector—including noise. The report provides an overview of a number of studies related to noise costs and benefits and

[10] Vainio, M., G. Paque, B. Baarsma, P. Bradburn, H. Nijland, S. Rasmussen, and J. Lambert. 2001. A Billion Euro Question: How Much Should We Pay for Noise Control, and How Much Is It Worth? Presentation at Workshop on Costs and Benefits Analysis in Noise Policy. INTER-NOISE 01, The Hague, The Netherlands, August 29, 2001. Final Report, December.

summarizes the results of 11 studies of WTP. Some of these studies relate WTP to per capita income; others use a noise depreciation sensitivity index. The report leans heavily on HEATCO (*H*armonized *E*uropean *A*pproaches for *T*ransport *CO*sting and Project Assessment) studies (HEATCO, 2009). The CE Delft report recommends that, "to value the disutility due to traffic noise, it is recommended to use an annual WTP-value equal to 0.09%–0.11% of capita income per dB, which is in line with the range of WTP-values recommended to the EU in 2002 by Navrud."

CBA was the subject of two presentations by Ulf Sandberg of the Swedish National Road and Transport Research Institute at an NAE workshop in 2007.[11] Sandberg's talk focused on CBAs in Norway, Sweden, and Denmark. He said that the 1996 Green Paper published by the EU included an estimate that the total cost of transportation noise in 17 European countries was €38 billion, which amounted to 0.65 percent of gross domestic product. One of the expert groups established to follow up on the Green Paper was the Working Group on Health and Socioeconomic Aspects, which published a paper on noise valuation in 2003, reflecting the opinions of the majority of members of the group (Working Group, 2003). Although this was not an official EU document, the group recommended that, when using the day-evening-night sound level (L_{den}) metric, a value of 25 euros per decibel per household per year be used to evaluate transportation noise. Swedish studies, he said, indicate that much higher values should be used for day-evening-night sound levels of more than 60 dB.

Sandberg described a CBA he conducted in 2001 that assumed the cost of a low-noise road surface of $5 per square meter (reasonably consistent with a Danish study [Larsen and Bendtsen, 2002]), a barrier cost of $500 per meter (lower than costs in the United States), and a road length of 200 meters (Sandberg, 2001). In Sandberg's analysis the cost of a barrier for a 10-meter-wide roadway was $100,000, whereas the cost for pavement was $10,000. Note that the estimate of $5 per square meter was based on conditions in 2001 for a single-layer porous asphalt pavement. In addition, the estimate did not take into account the expected shorter acoustical lifetime of a quiet pavement. Cost estimates in 2008–2009 for more efficient double-layer porous pavements are three to four times higher, and lifetimes are shorter than for conventional pavements.

The HEATCO project, completed in 2006, included a six-country contingent valuation study by a contractor in Norway, E-CO Tech. The data are given in Euros per person per year, and for road traffic range from €37 for "little annoyed" persons to €85 for "highly annoyed" persons. In contrast, the corresponding numbers are €38 to €59 for rail traffic. The numbers varied greatly from country to country.

Sandberg also referred to a seminar on road noise abatement sponsored by the Danish Road Institute and a 2006 report that included his presentation on a study of tire noise by the Forum of European National Highway Research Laboratories (FEHRL, 2006). The goal of the project, which included a CBA, was to provide information for the EC on the effects of more stringent tire noise limits. Assuming that the benefits of reduced tire noise would be in effect sometime between 2010 and 2022, FEHRL determined that the monetary benefits of a conservative reduction of 0.9 dB(A) in tire noise would be €48 billion, and the benefits of an optimistic reduction of 2.3 dB(A) would be €123 billion (FEHRL, 2006a).

Sandberg also described SILVIA, a study name based on the Latin *Silenda via* (the road must be silent), better known as the "Sustainable Road Surfaces for Traffic Noise Control Study." Based on SILVIA, a *Guidance Manual for the Implementation of Low-Noise Road Surfaces* was produced. Task 3.3 in the manual was the monetization of costs and benefits of quiet pavements. These included the creation of low-noise pavement, including the pavement itself, maintenance, and indirect costs; no charges were necessary against changes in rolling resistance or accident costs, because quiet pavements were found to be neutral in this respect, but there may be some differences in water pollution between standard and porous road surfaces. SILVIA concluded that quiet pavements were justified from a cost-benefit point of view in areas where many people along the road were impacted by high noise levels (FEHRL, 2006b).

At the INTER-NOISE 05 meeting, Jacques Lambert summarized CBAs (Lambert, 2005). He gave an overview of the various methods of doing cost-benefit analysis in Europe and reported on 12 European studies of willingness to pay for noise reduction in several countries. He also presented noise values for six different European countries. These are shown in Table 7-4.

FINDINGS AND RECOMMENDATION

As this brief review shows, much activity in Europe has focused on the costs and benefits of noise control. Despite differences in results among these studies, and even though some were not based on the most recent dose-response data, they make a compelling case for noise reduction. The United States would benefit from similar studies on all sources of transportation noise—road, rail, and air.

EPA has expertise in CBA and the authority to study the economics of noise mitigation. The FAA has a head start on using CBA techniques in evaluating noise around airports.

The FHWA and states have expertise in measuring noise from highway traffic and determining road surface costs. The reported cost of barrier construction varies from state to state for reasons related to building costs and the methods

[11] Sandberg, U. 2007. Discussion of European Activities Related to Cost Benefit Analysis and Highway Noise, and Future Technology for Design of Quiet Tires, and European Specifications for Tire/Road Noise. Presentations at an NAE workshop on Cost-Benefit Analysis of Noise Control Technology, Cambridge, Massachusetts, February 23.

TABLE 7-4 Noise Values for Selected European Countries

Country	Valuation Technique	Recommended or Official Noise Value in Euros, 2002	Application
Germany	HP	46.7 €/dB/person affected/ year for Leq day > 55 dB	Transport project
France	HP	0.4 to 1.1%/dB for Leq day > 55 +30% for Leq day > 70 dB & Leq night > 65 dB	Road and rail project
Norway	CV HP	1,000 to 1,170 €/affected person/year (according to the mode of transport)	Road, rail, and air project, and environmental protection project
Netherlands	HP	32.2 €/dB/person affected/ year for Leq day > 55 dB	Transport project
Sweden	HP	0 € at 50 dB to 1,810 € at 85 dB (Leq 24h)/person affected/year	Road project
Switzerland	CV	500 €/person affected/year for Leq day > 55 dB and for Leq night >45 dB	Road project

states use for reporting to FHWA. Costs also vary by state with prevailing construction costs, design requirements (barrier dimensions of height and length), and the definition of a cost basis for each state. Nevertheless, there appear to be sufficient data to predict costs when the specifics of a building site are known.

Present FHWA policy limits noise mitigation around highways to the construction of barriers, so the relative merits and costs of noise reduction from the installation of quieter road surfaces, although currently being investigated, are not part of noise mitigation policy.

Recommendation 7-1: A formal cost-benefit analysis should be performed to compare the costs and benefits of using pavement technology for noise reduction with the costs and benefits of installing noise barriers. This cost-benefit analysis should be a cooperative effort of the Federal Highway Administration, U.S. Environmental Protection Agency, and the several states with technology programs in road surface design. Inputs to the analysis should include data from analyses of noise reduction efforts around airports.

REFERENCES

Alexandrova, O., K.E. Kaloush, and J.O. Allen. 2007. Impact of asphalt rubber friction course overlays on the tire wear emissions and air quality models for Phoenix, Arizona, airshed. Transportation Research Record, 2011: 98–106. Available online at *http://bit.ly/c7z88A*.

CDOT (California Department of Transportation). 2005. I-80 Davis OGAC Pavement Noise Study: Traffic Noise Levels Associated with Aging Open Grade Asphalt Concrete Overlay. Available online at *http://www.dot.ca.gov/hq/env/noise/pub/IH80_davis_ogacpvmntwtudy_7yrrpt.pdf*.

CE Delft. 2008. Handbook on Estimation of External Costs in the Transport Sector. Available online at *http://ec.europa.eu/transport/costs/handbook/doc/2008_01_15_handbook_external_cost_en.pdf*.

Corbisier, C. 2005. Roadmap to Quieter Highways. Proceedings of NOISE-CON 05, The 2005 National Conference on Noise Control Engineering, Minneapolis, MN, October 17–19. Available online at *http://www.bookmasters.com/marktplc/00726.htm*.

Donavan, P.D. 2005a. Overview of the Arizona Quiet Pavement Program. Proceedings of NOISE-CON 05, The 2005 National Conference on Noise Control Engineering, Minneapolis, MN, October 17–19. Available online at *http://www.bookmasters.com/marktplc/00726.htm*.

Donavan, P.D. 2005b. Reducing Traffic Noise with Quieter Pavements. Proceedings of NOISE-CON 05, The 2005 National Conference on Noise Control Engineering, Minneapolis, MN, October 17–19. Available online at *http://www.bookmasters.com/marktplc/00726.htm*.

Donavan, P.D. 2006. Generation of Noise by Truck and Car Tires on Various Types of Asphalt Concrete Pavements. Proceedings of INTER-NOISE 06, The 2006 International Congress and Exposition on Noise Control Engineering, Honolulu, Hawaii, December 3–6. Available online at *http://www.bookmasters.com/marktplc/00726.htm*.

Eagan, M.E. 2007. Supplemental metrics to communicate aircraft noise effects. Transportation Research Record, 2011: 175–183. Available online at *http://bit.ly/bTjRBo*.

EPA (U.S. Environmental Protection Agency). 1981. Noise in America: The Extent of the Noise Problem. EPA/ONAC Report No. 550/9-81-101. Washington, DC: EPA. Available online at *http://www.nonoise.org/epa/Roll6/roll6doc7.pdf*.

EPA. 2000. Guidelines for Preparing Economic Analyses. Available online at *http://yosemite.epa.gov/ee/epa/eed.nsf/webpages/Guidelines.html*.

EC (European Commission). 2006. Noise Classification of Road Pavements. Task 2: Cost-Effectiveness of Low Noise Pavements. Available online at *http://www.cowiprojects.com/noiseclassification/docs/noiseclass_task2report.pdf*.

FAA (Federal Aviation Administration). 2008. See, for example, *http://faa.gov/about/office_org/headquarters_offices/aep/models/history*.

FEHRL (Forum of European National Highway Research Laboratories). 2006a. Final report S12.408210 Tyre/Road Noise, Volume 1.25, Brussels. Available online at *http://ec.europa.eu/enterprise/automotive/projects/report_tyre_road_noise1.pdf*.

FEHRL. 2006b. SILVIA: Sustainable Road Surfaces for Traffic Noise Control. Available online at *http://www.trl.co.uk/silvia/silvia/pdf/silvia_guidance_manual.pdf*.

FHWA (Federal Highway Administration). 2005. Guidance on Quiet Pavement Pilot Programs and Tire/Pavement Noise Research. Available online at *http://www.fhwa.dot.gov/environment/noise/qpppeml.htm*.

FHWA. 2006. Highway Traffic Noise in the United States: Problem and Response. Available online at *http://www.fhwa.dot.gov/environment/usprbrsp.pdf*.

FHWA. 2007. Summary of Noise Barriers. Available online at *http://www.fhwa.dot.gov/environment/noise/barrier/summary.htm*.

FHWA. 2009a. Highway Traffic Noise Barriers at a Glance. Available online at *http://www.fhwa.dot.gov/environment/keepdown.htm*.

FHWA. 2009b. FHWA Traffic Noise Model. Available online at *http://www.fhwa.dot.gov/environment/noise/tnm/index.htm*. See also *http://www.fhwa.dot.gov/environment/noise/qpppmem.htm*.

FHWA. 2009c. Highway Traffic Noise Analysis and Abatement Policy and Guidance. Available online at *http://www.fhwa.dot.gov/environment/polguid.pdf*.

FHWA. 2009d. FHWA Highway Noise Barrier Design Handbook. Available online at *http://www.fhwa.dot.gov/environment/noise/design/index.htm*.

FHWA. 2009e. Tire-Pavement Noise Home Page. Available online at *http://www.fhwa.dot.gov/environment/noise/tp_noise.htm*.

FICON (Federal Interagency Committee on Noise). 1992. Federal Agency Review of Selected Airport Noise Analysis Issues. Available online at *http://www.fican.org/pdf/nai-8-92.pdf*.

Fidell, S., and L. Silvati. 2004. Parsimonious alternative to regression analysis for characterizing prevalence rates of aircraft noise annoyance. Noise Control Engineering Journal 52(2):56–68.

Greene, M. 2002. Typical Diurnal Traffic Noise Patterns for a Variety of Roadway Types. Proceedings of INTER-NOISE 02, The 2002 International Congress and Exposition on Noise Control Engineering, Dearborn, MI, August 19–21.

HEATCO (Developing *H*armonized *E*uropean *A*pproaches for *T*ransport *Co*sting and Project Assessment). 2009. Sixth Framework Programme 2002–2006. Available online at *http://heatco.ier.uni-stuttgart.de*.

Herman, L., J. Withers, and E. Pinckney. 2006. Surface Retexturing to Reduce Tire-Road Noise for Existing Concrete Pavements. Transportation Research Record 1983: 51–58. Available online at http://bit.ly/96UE5C.

ICAO (International Civil Aviation Organization). 2007. Environmental Report 2007. Montreal, Canada: ICAO. Available online at *http://www.icao.int/env/pubs/env_report_07.pdf*.

I-INCE (International Institute of Noise Control Engineering). 1999. Technical Assessment of the Effectiveness of Noise Walls. Publication 99-1. See *http://www.i-ince.org*.

Kish, C. 2008. An Estimate of the Global Impact of Commercial Aviation Noise. Master's thesis, Massachusetts Institute of Technology, Cambridge.

Lambert, J. 2005. Valuation of the Benefits of Transportation Noise Reduction. Proceedings of INTER-NOISE 2005, The 2005 International Congress and Exposition on Noise Control Engineering, Rio de Janeiro, Brazil. Available online at *http://scitation.aip.org/journals/doc/INCEDL-home/cp/*.

Larsen, L.E., and H. Bendtsen. 2002. Noise Reduction with Porous Asphalt—Costs and Perceived Effect. Lyngby, Denmark: Danish Transport Research Institute. Available online at *http://bit.ly/bTl9xO*.

Lenders, A., and W. Hecq. 2002. The Cost and Benefit Functions in the Project STAIRRS: Strategies and Tools to Assess and Implement Noise Reducing Measures for Railway Systems. Proceedings of INTER-NOISE 2002, The 2002 International Congress and Exposition on Noise Control Engineering, Dearborn, Michigan. Available online at *http://www.bookmasters.com/marktplc/00726.htm*.

Navrud, S. 2002. The State-of-the-Art on Economic Valuation of Noise. Final Report to the European Commission DG Environment. Available online at *http://ec.europa.eu/environment/noise/pdf/noise_monetisation.pdf*.

Nelson, J. 1982. Highway noise and property values: a survey of recent evidence. Journal of Transport, Economics and Policy 16(2):117–138.

Oertli, J., F. Elbers, and P. van der Stap. 2002. The STAIRRS project: A cost-benefit analysis of different measures to reduce railway noise on a European scale. Proceedings of INTER-NOISE 2002, The 2002 International Congress and Exposition on Noise Control Engineering, Dearborn, MI, August 19–21. Available online at *http://www.bookmasters.com/marktplc/00726.htm*.

OMB (Office of Management and Budget). 1992. Guidelines and Discount Rates for Benefit-Cost Analysis of Federal Programs. Circular A-94. Washington, DC: OMB. Available online at *http://www.whitehouse.gov/omb/circulars/a094/a094.pdf*.

Polcak, K.D. 2003. Highway traffic noise barriers in the U.S.—construction trends and cost analysis. Noise/News International 11(3):96–108. Available online at *http://www.noisenewsinternational.net/archives_idx.htm*.

Rasmussen, R., R. Bernhard, U. Sandberg, and E. Mun. 2007a. The Little Book of Quieter Pavements. Washington, DC: Federal Highway Administration. Available online at *http://www.tcpsc.com/LittleBookQuieterPavements.pdf*.

Rasmussen, R.O., E.P. Mun, and R.E. DeDios. 2007b. Tire-Pavement and Traffic Noise Research in the State of Colorado. Proceedings of NOISE-CON 07, The 2007 National Conference on Noise Control Engineering, Reno, NV.

Reyff, J.A. 2007a. Reduction of Traffic and Tire/Pavement Noise: 3rd Year Results of the Arizona Quiet Pavement Program—Site III. Proceedings of NOISE-CON 07, The 2007 National Conference on Noise Control Engineering, Reno, NV. Available online at *http://www.bookmasters.com/marktplc/00726.htm*.

Reyff, J.A. 2007b. REMEL Database Developed for Different PCC Pavement Surfaces. Proceedings of NOISE-CON 07, The 2007 National Conference on Noise Control Engineering, Reno, NV. Available online at *http://www.bookmasters.com/marktplc/00726.htm*.

Rochat, J.L. 2002. Long-Term, Multiple Pavement Type, Tire/Road Noise Study. Presentation at INTER-NOISE 2002, Dearborn, Michigan, August 19–21. Available online at *http://bit.ly/cESj9P*.

Rochat, J.L., and D.R. Read. 2009. Noise benefits of asphalt pavements—trends at ages up to 52 months. Noise Control Engineering Journal 52(2):84–93.

Sandberg, U. 2001. Tyre/Road Noise—Myths and Realities. Proceedings of INTER-NOISE 01, The 2001 International Congress and Exposition on Noise Control Engineering, The Hague, Netherlands. Available online at *http://scitation.aip.org/journals/doc/INCEDL-home/cp/*.

TPNRC (Tire-Pavement Noise Research Consortium). 2009. Transportation Pooled Fund Program. Available online at *http://bit.ly/9p8SSo*.

Vainio, M., and G. Paque. 2002. Highlights of the Workshop on the "State-of-the-Art in Noise Valuation." Final report. July. Available online at *http://bit.ly/cwTUIf*.

Waitz, I., R.J. Bernhard, and C.E. Hanson. 2007. Challenges and promises in mitigating transportation noise. The Bridge 37(3):25–32.

Working Group. 2003. Valuation of Noise. Position paper of the Working Group on Health and Socio-Economic Aspects. Available online at *http://bit.ly/9HIyU1*.

WHO (World Health Organization). 2004. Transport-Related Health Effects with a Particular Emphasis on Children: Noise. Geneva: WHO Transport, Health and Environment Pan-European Program.

8
The Role of Government

This chapter provides a review of federal, state, and local government responsibilities for noise emission levels. Although state and local governments do not have the authority to control noise from some sources that have been federally preempted, such as transportation (e.g., aircraft, new motor vehicles, railroads), they play an important role in preparing environmental impact assessments for new transportation projects and other federally funded construction. State and local governments are also responsible for controlling other sources of environmental noise, such as noise from industrial and commercial facilities.

NOISE-RELATED ACTIVITIES BY FEDERAL AGENCIES

Currently, research and other activities related to noise abatement and control by federal agencies are poorly coordinated. Because each agency has its own methodology for dealing with noise problems, there are few uniform descriptors, criteria, or approaches to noise control on the federal level. For example, the Federal Highway Administration, the largest modal agency in the U.S. Department of Transportation (DOT), has adopted immission criteria based on a single hour of noise exposure, in contrast to the nearly universal use of the day-night average sound level.

A few government organizations, however, meet regularly to coordinate noise research and activities. Perhaps the most notable of these is the Federal Interagency Committee on Aircraft Noise (FICAN), which was chartered in 1993 to "carry out interagency coordination on matters related to aviation noise research in the United States." FICAN meets quarterly and is chaired by one of its member agencies on a two-year rotating basis.

Another example of an active organization is the National Research Council Transportation Research Board (TRB) Committee ADC40 (Transportation-Related Noise and Vibration), which has subcommittees that focus on noise from aviation, highway, and rail transportation. ADC40, like all other TRB committees and subcommittees, meets annually in January in Washington, D.C., and holds additional meetings elsewhere during the year. TRB committees are not involved in policymaking and generally do not coordinate noise research, but they do provide a venue for the presentation and discussion of research by other organizations. Committee members are drawn from the federal government (primarily DOT), academia, private consulting organizations, and state departments of transportation. In addition, a large number of "friends" are kept informed about ADC40 activities (ADC40, 2009).

Two of the most active organizations involved in the coordination of noise issues on the federal level are the Partnership for Air Transportation Noise and Emissions Reduction (PARTNER) and the Joint Program and Development Office (JPDO). PARTNER's primary areas of focus are noise, air quality, and climate change. JPDO, which deals with the Next Generation Air Transportation System (NextGen), is mainly focused on the capacity of air transportation and only peripherally on noise.

A third organization, a DOT working group chaired by Arnold Konheim of the Office of the Secretary of Transportation (OST), meets annually and on an ad hoc basis when urgent noise issues arise. Representatives of all DOT modal agencies attend these meetings.

In the past, noise-related activities on the federal level were better coordinated. In the Noise Control Act of 1972 (NCA 72), Congress assigned the task of coordinating "the programs of all federal agencies relating to noise research and noise control" to the U.S. Environmental Protection Agency (EPA). This coordinating function is not being carried out. More details on EPA activities are given in a later section of this chapter.

Federal Interagency Committee on Aviation Noise

Background

Two programs for coordinating federal activities on noise issues—the Federal Interagency Committee on Urban Noise (FICUN) in the late 1970s and the Federal Interagency

Committee on Noise (FICON) during the 1980s and early 1990s—preceded the creation of FICAN in November 1993. FICUN developed land use noise compatibility guidelines for all modes of transportation in 1980, and FICON focused on airport noise in a 1992 report, *Federal Agency Review of Selected Airport Noise Analysis Issues* (FICON, 1992). Among FICON's recommendations was that a standing federal interagency committee be formed to assist agencies in research and development (R&D) related to aviation noise. The Federal Aviation Administration (FAA) committed itself to establishing such a committee in its report to Congress in 1993 (see *http://www.fican.org/pdf/FICAN_Charter.pdf*).

Membership

FICAN meetings are held quarterly and at the discretion of the chair. Members are appointed by their respective agencies, each of which is obligated to send a representative to all proceedings. The chair and vice chair are selected by majority vote every two years, with the understanding that the positions will rotate among the agencies. Current members are DOT, OST, FAA, U.S. Department of Defense (DOD), U.S. Army, U.S. Navy, U.S. Air Force, U.S. Department of the Interior (DOI), National Park Service, U.S. Department of Housing and Urban Development, and EPA.

Purpose

FICAN is not involved in setting policy related to aviation noise. Its role is to review and comment on technical issues and make recommendations. FICAN is a focal point for questions on R&D on aviation noise.

Operations

Members contribute to a pooled fund to administer the committee's activities. FICAN operates on a limited budget, generally about $100,000 per year, most of which goes toward contracts with outside parties to support its activities, including administrative assistance. Harris Miller Miller & Hanson Inc., a private contractor, has provided administrative assistance since 1993.

Although FICAN conducts studies, it is not a source of major research contracts. Its efforts are directed toward reviewing ongoing research on aviation noise with an eye toward avoiding duplication. As part of its coordinating role, FICAN reviews activities related to research on aviation noise by PARTNER, JPDO, and TRB committees and programs. FICAN also prepares position statements on subjects of interest suggested by member agencies and on reports it has been asked to review. Public workshops and symposia are held from time to time.

Application of the FICAN Model

FICAN was created to study technical issues, not to make policy. Committee participants are qualified in technical aspects of the field of aviation noise, but participation in FICAN is peripheral to their jobs. They are appointed by their respective agencies but are not required to report on their activities, nor do member agencies take any action as a result of FICAN's work. Committee members tend to be less than enthusiastic and are not given extra credit for their efforts.

FICAN is made up entirely of federal agencies. Although industry participation is not included in the workings of the committee, private-sector research is included in FICAN's reviews.

Partnership for Air Transportation Noise and Emissions Reduction

Background

PARTNER, which was established in 2003, is one of the FAA's eight air transportation centers of excellence, wherein colleges and universities are given grants to conduct research on aviation issues considered important to airspace planning and airport design. Centers of excellence, which were established through enabling legislation dated November 1990 as an amendment to the Federal Aviation Act of 1958, represent a strategic research partnership of government, academia, and industry. Elements of the PARTNER program include education, research, and technology transfer in the context of an academic setting. PARTNER is the only FAA center of excellence that deals with noise issues (in addition to air quality and climate change).

Membership

PARTNER is sponsored by the FAA, the National Aeronautics and Space Administration (NASA), and Transport Canada. Consequently, it can be considered an example of federal coordination of noise research, even if the research is conducted by academic institutions and not by federal agencies. Nine collaborating universities conduct PARTNER research. The Massachusetts Institute of Technology, the lead university, provides PARTNER's program director, Ian Waitz, and an administrative office. An extensive advisory board with 53 member organizations supports PARTNER, by giving advice and by directly collaborating in the research program. All federal grant funds allocated through PARTNER must be matched one to one with nonfederal cost sharing (typically from in-kind support provided by the advisory board and other organizations collaborating on research programs).

One of PARTNER's greatest strengths is the diversity and inclusiveness of the advisory board. Members include representatives of aerospace manufacturers; airlines; airports; national, state, and local governments; professional and trade associations; nongovernmental organizations; and community groups.

Purpose

The purpose of PARTNER is to provide a forum for coordinating research in the areas of aviation noise, air quality, and climate change.

Operations

Recommendations for research topics are solicited from the advisory board, sponsoring organizations, and academic institutions. The sponsors—FAA, NASA, and Transport Canada—decide which topics should be funded. Projects are reviewed at designated semiannual meetings.

Application of the PARTNER Model

PARTNER is a good example of how research can be coordinated with input from federal agencies, academia, industry, community, and other organizations. However, some federal agencies are more active in PARTNER than others, and the centers of excellence have a strong focus on academic R&D and the development of the future workforce. PARTNER might be more effective with interagency coordination of federal research endeavors in a broader range of noise-related topics.

Joint Planning and Development Office

Background

Although JPDO is not focused on noise-related activities, its organizational structure provides a potential model for a multiagency cooperative effort to establish policy. The JPDO website (*www.jpdo.gov*) describes the background of the formation of this office:

> By 2025, U.S. air traffic is predicted to increase two to three times. The traditional air traffic control system will not be able to manage this growth. The Next Generation Air Transportation System (NextGen) is the solution. NextGen is an example of active networking technology that updates itself with real-time shared information and tailors itself to the individual needs of all U.S. aircraft. NextGen's computerized air transportation network stresses adaptability by enabling aircraft to immediately adjust to ever-changing factors such as weather, traffic congestion, aircraft position via GPS, flight trajectory patterns, and security issues. By 2025, all aircraft and airports in U.S. airspace will be connected to the NextGen network and will continually share information in real time to improve efficiency, safety, and absorb the predicted increase in air transportation.

NextGen was enacted in 2003 under VISION 100—Century of Aviation Reauthorization Act (P.L. 108-176). As part of this initiative, JPDO, which is responsible for managing a public/private partnership to bring NextGen online by 2025, is the central organization that coordinates specialized efforts by DOT, DOD, the U.S. Department of Homeland Security (DHS), the U.S. Department of Commerce (DOC), the FAA, NASA, and the White House Office of Science and Technology Policy (OSTP).

A further description of JDPO's task is "to create and carry out an integrated plan for NextGen, spearhead planning, and coordinate research, demonstrations and development in conjunction with relevant programs of other departments and agencies, and with the private sector."

Membership

JPDO is governed by federal agencies, primarily DOT, but working groups include representatives of private industry. For example, the working group on aircraft is cochaired by the FAA and Boeing.

Purpose

JPDO administers the NextGen program, guided by three planning documents:

- "Concept of Operations" describes how NextGen will work as a system.
- "Enterprise Architecture" provides structural details to make NextGen work.
- "Integrated Work Plan" describes the steps in the transition from existing conditions to the new system.

Operations

A senior policy committee directs the NextGen initiative. The committee is chaired by the secretary of transportation and includes the DOT undersecretary for policy; administrator of the FAA; administrator of NASA; secretary of the U.S. Air Force, representing DOD; deputy secretary of DOC; deputy secretary of DHS; and the director of OSTP. A board made up of senior personnel from each member agency reports to the senior policy committee. JDPO has six divisions, each headed by a division director from federal agencies (JPDO, NASA, and FAA). Finally, working groups consisting of teams of representatives from federal agencies and industry work to solve problems and make recommendations.

Working groups have the following features:

- a documented mission statement, terms of reference, structure definition—all guided by "Framework of NextGen"

- an executive committee cochaired by one person from government and one from industry
- standing committees that handle long-term issues and ongoing tasks
- study teams that address short-term tasks and can draw on expertise from other working groups and study teams, as needed

Application of the JPDO Model

Noise research is not the main focus of JPDO (as it is for FICAN). However, the Environmental Working Group is charged with "thinking green" and providing environmental protection while sustaining aviation growth. According to the JPDO website, the Environmental Working Group considers four key areas: aviation noise, air quality, water quality, and fuel consumption.

JPDO is an example of a successful multidisciplinary organization that establishes policy, coordinates research, and encourages noise control.

National Institute of Occupational Safety and Health Partnerships on Occupational Noise

The following statement is taken from an internal National Institute of Occupational Safety and Health (NIOSH) document:

> Partnerships are an important element of many aspects of the NIOSH Hearing Loss Research (HLR) program, from planning to research activity to transfer of outputs. NIOSH considers an external organization to be a partner when they are involved in the inputs and activities of our program. In partnership we have a joint effort to conduct research, develop technology, define best practices, and promulgate the knowledge gained from our research. The HLR program has an active interaction with many external partners across a wide variety of organizations and collaborations. . . . Our partners come from many organizations, including other governmental agencies. . . . The HLR program has active collaborations with sister agencies like the Mine Safety and Health Administration, the OSHA, and the Environmental Protection Agency (EPA), who use or could use program outputs for regulatory actions. HLR program outputs are also used by non-regulatory sister agencies such as the DOD and the Federal Railway Administration.

Environmental Protection Agency

EPA's responsibilities (according to current law) must be reviewed as part of any analysis of the federal government's activities related to noise. The Clean Air Act established the Office of Noise Abatement and Control (ONAC) in the EPA, and then the Noise Control Act of 1972 and the Quiet Communities Act of 1978 (QCA 78) resulted in an active noise control program within EPA. This is described in the next section.

Noise Legislation

EPA's responsibilities, under the NCA 72 and QCA 78, are codified in 42 USC 65 4901-4918. In the early 1970s the program was tilted toward regulatory activity (imposed by Congress), but the "Levels Document" was an outstanding contribution both for its definition of what constitutes acceptable community noise levels and at what point noise exposure becomes hazardous (EPA, 1974). In later years (1976 to 1981) the program was oriented more toward outreach, support of state and local activities, research, and technical assistance.

EPA activities were carried out by ONAC and have been reviewed by Maling (2003). President Reagan ordered that the EPA noise program be phased out by the end of fiscal year 1982. This action effectively disestablished ONAC. Some noise regulations still remain in the Code of Federal Regulations, and EPA's responsibilities are still spelled out in the Code. ONAC still exists on paper in the Code. Despite the 1982 phase-out, there is currently some EPA activity related to noise—as described below.

Current Activities

A brief history of very early activities can be found on the EPA website (EPA, 2009a); there is also a section on frequently asked questions (EPA, 2009b). Current EPA activities related to noise include a revision of the current regulation on hearing protectors, implementation of a congressional earmark related to railroad noise, and a modest public information program.

The Shapiro Study

Even though funding for the EPA noise program was cut off in 1981, many regulatory actions are still "on the books," and the laws have not been rescinded. EPA regulations still in place can be found in the Code of Federal Regulations (*http://ecfr.gpoaccess.gov/cgi/t/text/text-idx?sid=ee1b7c388227bc1ef430888b7849386e&c=ecfr&tpl=/ecfrbrowse/Title40/40cfrv24_02.tpl*).

In 1991 a critical analysis of the program was conducted by Sidney Shapiro on behalf of the Administrative Conference of the United States (ACUS); many of the conclusions of that report are on the Internet (*http://www.law.fsu.edu/library/admin/acus/305926.html*). As of 1995, the ACUS no longer exists.

Because laws and regulations are still in place but adequate funding has not been provided for EPA to carry out its responsibilities, the situation regarding government noise-related activities is complicated, especially with regard to preemption, which limits the ability of state and local governments to set noise emission standards. Among other things, the Shapiro report recommended that the following items be reviewed:

- scientific and technical developments since 1981
- the methodology for measurement and assessment of noise
- the allocation of responsibility among federal agencies
- federal participation in voluntary standards activities

To the committee's knowledge, the full set of ACUS recommendations has never been accepted by EPA. Nevertheless, the recommendations are still valid.

Congressional Action

Several attempts have been made to reestablish ONAC at EPA. Bills were introduced into the 107th, 108th, and 109th Congresses, but Congress has taken no action.

U.S. Code

Sections in the U.S. Code on federal programs (42 USC 4903) and quiet communities, research, and public information (42 USC 65 4913) are important for the present study.[1,2] These sections are reproduced in Appendix D, parts A and B, respectively. Portions are reprinted below:

> **Coordination of noise control activities of federal and state agencies.** Section 4903 states: "(1) The Administrator shall coordinate the programs of all Federal agencies relating to noise research and noise control. Each Federal agency shall, upon request, furnish to the Administrator such information as he may reasonably require to determine the nature, scope, and results of the noise-research and noise-control programs of the agency."

> **Portions of Section 4913, 24 USC 65.** Section 4913 states: ". . . the Administrator shall, in cooperation with other Federal agencies and through the use of grants, contracts, and direct Federal actions. . ."

> **Cost-benefit analysis of noise control technologies.** ". . . (4) investigation of the economic impact of noise on property and human activities…"

> **Public information on the benefits of low-noise products and the adverse effects of excessive noise:** "…(a) develop and disseminate information and educational materials to all segments of the public on the public health and other effects of noise and the most effective means for noise control, through the use of materials for school curricula, volunteer organizations, radio and television programs, publication, and other means. . . ."

> **Assistance to state and local community noise control programs:** "(c) administer a nationwide Quiet Communities Program which shall include, but not be limited to . . . (1) grants to States, local governments, and authorized regional planning agencies for the purpose of . . ."

The Vér Proposal

In "Proposal for a Long Range Noise Control Policy Based on Cooperation Among Government, Industry and the Research Community," Vér (1991) proposed that noise control research be jointly sponsored by the government and industry and trade associations. The former would ensure that the research topics are in line with long-range government noise policies. The latter would ensure that the research projects undertaken are relevant to the needs of industry and that the results are presented in a form and at a level most useful to design engineers.

Organizational Structure for Implementing a Noise Policy

None of the examples described above has an ideal organizational structure for implementing a comprehensive noise policy in the United States. Perhaps the JDPO model comes closest. However, each model has ideas worth considering.

A new organization to determine noise policy would require enabling legislation by Congress to provide authority and funding. The legislation would also have to rescind some of the responsibilities of the EPA under the U.S. Code. Such an organization should have the following characteristics. The organization should be supervised by a government agency or a consortium of agencies, but a lead agency would have to be identified for funding purposes. Assuming that several agencies would be tasked in the legislation to participate, a "senior policy committee" would be required to perform several tasks. These include preparation of a mission statement, a roadmap for future activities, and a work plan to implement the roadmap. Given the complexity of the noise problem and the wide variety of activities affected, it is likely that the organization would have to be organized as several divisions, each with a director and several working groups within each division. Participants in a working group would include both government and industry, and the working groups would be concerned with occupational noise; community noise, including annoyance issues; health effects of noise; and criteria for noise. Many of the policy recom-

---

[1]42USC65 4903. The Public Health and Welfare, Chapter 65, Noise Control: Sec. 4903, Federal Programs. Available online at *http://frwebgate.access.gpo.gov/cgi-bin/getdoc.cgi?dbname=browse_usc&docid=Cite:+42USC4903.*

[2]42USC65 4913. The Public Health and Welfare, Chapter 65, Noise Control: Sec. 4913, Quiet communities, research, and public information. Available online at *http://frwebgate.access.gpo.gov/cgi-bin/getdoc.cgi?dbname=browse_usc&docid=Cite:+42USC4913.*

mendations in this report could be considered by such an organization. However, it will take time for such an organization to be established, and many of the recommendations, especially concerning EPA, could be implemented now given funding by Congress.

NOISE-RELATED ACTIVITIES BY STATES

In Chapters 5 and 7 of this report, the activities of state governments on controlling highway noise are described. Research and development of pavements designed to reduce tire/road interaction noise has been a cooperative effort with significant results.

Transportation systems are generally operated by state agencies (e.g., airport commissions, highway departments, commuter rail agencies), and residents and community leaders look to them to control noise from sources in their jurisdictions. However, Congress considers that "primary responsibility for control of noise rests with state and local governments [although] Federal action is essential to deal with major noise sources in commerce, control of which requires national uniformity of treatment."[3]

Thus, state transportation agencies are caught in a bind. They are powerless to control noise directly from major transportation sources, such as airplanes, new motor vehicles, and railroads, yet citizens expect them to reduce noise that adversely affects them.

For federally funded projects, including transportation, state agencies are required to prepare environmental impact assessments before construction begins. These assessments must include the noise impact of proposed projects. Many states require similar assessments for major projects even when no federal funds are involved.

Many states have laws and regulations covering a variety of situations related to environmental noise and noise abatement in general. State laws and regulations may address noise from industrial plants, commercial facilities, and construction sites. The Noise Pollution Clearinghouse lists 12 states that have general noise regulations and nine that have watercraft noise regulations (*http://www.nonoise.org/lawlib/states/states.htm*); however, these lists and a survey of state regulations need to be updated.

Maine has a comprehensive well-written statute with regulations that have been described as the most complex in the country (Doyle, 2001). Connecticut and Illinois also have carefully written regulations (Brooks, 2001, 2003). A summary of New Jersey's noise regulations and related activities has been published (Zwerling, 2005). Other states with comprehensive environmental noise regulations are California, Minnesota, and Oregon. The noise requirements in Massachusetts, Minnesota, New York, and Oregon are administered in the general environmental permitting process.

Although there is no requirement that state noise regulations be identical, it would be helpful if requirements, including measurement procedures, were compatible from state to state. The absence of well-defined standards creates uncertainty in the minds of developers of industrial facilities and can cause delays in the approval process. Thus, most developers of industrial facilities would welcome consistent, well-written standards for noise emissions.[4] EPA has a mandate to promote the development of effective state and local noise programs (42 USC 65, Section 4913) but is not currently funded to do so.

LOCAL NOISE CONTROL PROGRAMS

Control of noise at the local level presents several challenges. First, cities and towns in the United States have different needs. Second, many noise problems do not have engineering solutions. Third, local officials often do not have the information they need to find the best methods of solving local problems.

Some large cities do have the resources to study noise problems and issue appropriate regulations. New York City, for example, has a modern noise law (NYC, 2009a,b; Maling, 2007) which states that "the making, creation or maintenance of excessive and unreasonable and prohibited noises within the city affects and is a menace to public health, comfort, convenience, safety, welfare and the prosperity of the people of the city." The new law mandates that all construction be conducted in accordance with individual noise mitigation plans and prescribes ways to lessen the noise from each type of construction equipment. The code also sets standards for noise levels created by handling containers and construction material on public streets and restricts the noise levels created by air conditioners and circulation devices. Some portions of the New York law may not be applicable to other large cities, but the law is a good starting point for upgrading existing laws or creating new ones. Chicago was one of the first cities to establish comprehensive regulations for noise from industrial and commercial facilities. Boston has adopted a comprehensive construction noise regulation as part of the Central Artery/Tunnel project; this regulation is also a model for other communities (Thalheimer, 2000; 2001).

Smaller cities, suburban towns, and rural villages have different problems. They rarely have the resources to make extensive use of professional advice in drafting local noise ordinances. Many towns and villages have common problems. At one time the EPA had a program called Each Community Helps Others (ECHO) that helped communities with common problems communicate with each other. Although

[3]42 USC 65, Section 4901(a)(3). Congressional Findings and Statement of Policy re Chapter 65, The Public Health and Welfare. Available online at *http://frwebgate.access.gpo.gov/cgi-bin/getdoc.cgi?dbname=browse_usc&docid=Cite:+42USC4901.*

[4]Wood, E.W. 2005. Community noise from new industrial plants—engineering and regulatory challenges. Presentation at the NAE Technology for a Quieter America Workshop, Washington, D.C., September 13–15.

EPA produced a model noise ordinance in 1975, it is out of date today.

The American National Standards Institute's Accredited Standards Committee S12 Noise is working to produce a standard with guidelines for developing community noise ordinances or regulations to inform communities (ANSI, 2009). The sponsor of this effort, the Acoustical Society of America (ASA) Committee on Standards S12 Working Group 41, has a draft document for comments from working group members before it is submitted for voting as an ANSI standard. If it is adopted, this standard will provide a menu of options to guide local communities in establishing enforceable, practical noise ordinances. The draft standard is expected to be voted on in 2010.

Thanks to corporate sponsorship, ASA has made ANSI S12.60 available for free download; this standard pertains to acoustics in classrooms (ASA, 2009). ASA would perform a public service if it could also provide free guidance on community noise ordinances.

SUMMARY

The noise-related activities of many federal government agencies are described elsewhere in this report. With respect to the federal government, the emphasis in this chapter is on current mechanisms of federal interagency cooperation and the characteristics of a new organization that would require congressional action to create. Less emphasis has been placed on the role of state and local governments; these activities are difficult to describe briefly and, with the exception of work on highway pavement, do not generally involve technology. Under current law, there are opportunities for EPA to provide assistance to state and local governments, which could help in the coordination of noise-related activities.

FINDINGS AND RECOMMENDATIONS

Although EPA currently has a small program related to noise, the agency has the authority under the U.S. Code to do much more. It appears that if new tasks are assigned to another department in the federal government, the law will have to be changed. EPA, however, could carry out these tasks if Congress appropriated the necessary funds.

Of several models of federal cooperation related to noise activities, the existing model most suitable for a new organization in EPA is JPDO, which is involved in policy, R&D, and cooperation with industry. As noted by the Administrative Conference of the United States (see Shapiro, 1991) and reports by others, many items related to noise regulation could be addressed. Until these items are addressed, it will be difficult to make progress on noise control.

The Congressional Research Service (CRS) provides regular reports to Congress in a series titled *Noise Abatement and Control: An Overview of Federal Standards and Regulations*; in addition, three annual reports are available (CRS, 2000, 2003, 2006). CRS should be asked to prepare a new report to Congress outlining policy options and encouraging congressional action to develop a new noise policy.

Recommendation 8-1: The Environmental Protection Agency should carry out its coordinating function under 42 USC 65, Section 4903. The agencies with noise-related activities include the U.S. Department of Defense, U.S. Department of Transportation, U.S. Department of Labor, U.S. Department of Commerce, U.S. Department of Health and Human Services, U.S. Department of Housing and Urban Development, and the National Science Foundation.

Recommendation 8-2: Congress should pass legislation and provide the necessary funds to establish the Environmental Protection Agency as the lead agency in the development of a cooperative effort on noise measurement, abatement, and control involving federal agencies, state governments, industry, consulting firms, and academia. An EPA office should implement 42 USC 65, Section 4903, and the legislation should expand the authority already given by Congress to ensure that the agency can effectively manage a program to meet the following objectives:

- coordination and cooperation among existing interagency groups concerned with noise
- clear delineation of the roles of federal agencies, as well as state and local governments
- assisting American industry in lowering noise levels in the U.S. workplace and developing industrial and consumer products with noise emissions that are competitive with foreign products
- development of international standards for the measurement and labeling of noise emissions
- active U.S. participation in the harmonization of noise emission requirements worldwide
- development of metrics for environmental noise that truly represent community response to noise
- ongoing assessment of the costs and benefits of noise control
- increased research on the health effects of noise, especially nonauditory effects

REFERENCES

ADC40 (Transportation Research Board Committee on Transportation Noise and Vibration). 2009. ADC40 Membership List. Available online at *http://www.adc40.org/adc40membershiplist.pdf*.

ANSI (American National Standards Institute). 2009. Accredited Standards Committee S12, Noise. Working Group 41 (ANSI S12/WG41), Model Community Noise Ordinances. Available online at *http://www.acosoc.org/standards/S12/ASCS12.htm*.

ASA (Acoustical Society of America). 2009. American National Standard Acoustical Performance Criteria, Design Requirements, and Guidelines for Schools. Available online at *http://asastore.aip.org/*.

Brooks, B.M. 2001. The Need for a Unified Community Noise Policy. Proceedings of NOISE-CON 01, The 2001 National Conference on Noise

Control Engineering, Portland, Maine. Available online at *http://www.bookmasters.com/marktplc/00726.htm.*

Brooks, B.M. 2003. The need for a unified community noise policy. Noise Control Engineering Journal 51(3):160–161.

CRS (Congressional Research Service). 2000. Noise Abatement and Control: An Overview of Federal Standards and Regulations. CRS Report for Congress by D.M. Bearden. Updated April 7, 2000. Available online at *http://ncseonline.org/NLE/CRSreports/Risk/rsk-52.cfm?&CFID=7463199&CFTOKEN=71822941.*

CRS. 2003. Noise Abatement and Control: An Overview of Federal Standards and Regulations. CRS Report for Congress by D.M. Bearden. Updated January 13, 2003. Available online at *http://digital.library.unt.edu/govdocs/crs/permalink/meta-crs-4552:1.*

CRS. 2006. Noise Abatement and Control: An Overview of Federal Standards and Regulations. CRS Report for Congress by D.M. Bearden. January 23, 2006. Available online at *http://www.cnie.org/nle/crsreports/06Feb/RS20531.pdf.*

Doyle, T. 2001. Twelve years with the most complex noise regulations in the United States. Proceedings NOISE-CON 01, The 2001 National Conference on Noise Control Engineering, Portland, Maine. Available online at *http://www.bookmasters.com/marktplc/00726.htm.*

EPA (U.S. Environmental Protection Agency). 1974. Information on Levels of Environmental Noise Requisite to Protect Public Health and Welfare with an Adequate Margin of Safety. EPA Report 550/9-74-004. Washington, DC: Office of Noise Abatement and Control. Available online at *www.nonoise.org/library/levels74/levels74.htm.*

EPA. 1975. Model Community Noise Control Ordinance. Report EPA 550/9-76-003. Washington, DC: Office of Noise Abatement and Control. Available online at *http://www.nonoise.org/epa/roll2/roll2doc7.pdf.*

EPA. 2009a. History: Noise Control Act. Available online at *http://www.epa.gov/history/topics/nca/index.htm.*

EPA. 2009b. Frequent Questions: EPA Web Page on Noise. Available online at *http://publicaccess.custhelp.com/cgi-bin/publicaccess.cfg/php/enduser/std_adp.php?p_faqid=1765.*

FICON (Federal Interagency Committee on Noise). 1992. Federal Agency Review of Selected Airport Noise Analysis Issues. Available online at *http://www.fican.org/pdf/nai-8-92.pdf.*

Maling, G.C. 2003. An editor's view of the EPA Noise Program. Noise Control Engineering Journal 51:143–150.

Maling, G.C. 2007. Pan-American news: New York noise code goes into effect. Noise/News International (Sept.):106–109. Available online at *http://www.noisenewsinternational.net/archives_idx.htm.*

NYC (New York City). 2009a. Local Laws of the City of New York for the Year 2005. No. 113. Available online at *http://www.lowermanhattan.info/extras/pdf/law05113.pdf.*

NYC. 2009b. Air and Noise: Noise Code (Local Law 113 of 2005). Available online at *http://nyc.gov/html/dep/html/air_and_noise/index.shtml.*

Shapiro, S.A. 1991. The Dormant Noise Control Act and Options to Abate Noise Pollution. Report for the Administrative Conference of the United States. Available online at *http://www.barkingdogs.net/actamshap1991head.shtml.*

Thalheimer, E. 2001. Proactive Construction Noise Control Policies Developed for the Central Artery/Tunnel Project's C17A6 Contract. Proceedings of NOISE-CON 01, The 2001 National Conference on Noise Control Engineering, Portland, Maine, October 29–31. Available online at *http://www.bookmasters.com/marktplc/00726.htm.*

Thalheimer, E. 2000. Construction noise control program and mitigation strategy at the Central Artery/Tunnel project. Noise Control Engineering Journal 48(5):157–165.

Vér, István L. 1991. Proposal for a Long Range Noise Control Policy Based on Cooperation Among Government, Industry and the Research Community. Distinguished lecture presented at NOISE-CON 91 in Tarrytown, New York. Reprinted in Noise Control Engineering Journal 51(3):176–181. Available online at *http://www.noisenewsinternational.net/docs/ver-1991.pdf.*

Zwerling, E.M. 2005. Regulatory Scheme for Noise Enforcement in New Jersey. Proceedings of NOISE-CON 05, The 2005 National Conference on Noise Control Engineering. Minneapolis, MN, October 15–17. Available online at *http://www.bookmasters.com/marktplc/00726.htm.*

9

Education Supply and Industry Demand for Noise Control Specialists

The Massachusetts Institute of Technology (MIT) was an early leader in noise control education. Courses in acoustics and acoustical engineering were taught there in the 1950s and 1960s, and a series of summer courses were offered. "Noise Control Engineering Design," believed to be the first undergraduate course with "noise control engineering" in the title, was taught by Professor Conrad Hemond at the University of Hartford beginning in 1971. Another course at the same university, "Engineering Acoustics," has been taught since the 1960s and is still a prerequisite for the noise control engineering course, a full-semester project course. Purdue University and others began to offer undergraduate courses in noise control in the 1970s, soon after enactment of the Noise Control Act of 1972.

The first graduate course with "noise control engineering" in the title is believed to have been taught in the early 1970s by Professor Uno Ingard in the MIT Department of Aeronautics and Astronautics. Today graduate programs have been established—for example, in the mechanical engineering department at Purdue University and in the Graduate Program in Acoustics at Pennsylvania State University (Penn State), which offers master of engineering/master of science and doctoral degrees in the field of acoustics.

UNDERGRADUATE EDUCATION IN NOISE CONTROL ENGINEERING

Most existing noise control and acoustics courses are taught either at the graduate level or are noncredit short courses. The committee believes that academic institutions should find room in their curricula to offer an undergraduate course in noise control engineering that could provide a basic knowledge and understanding of noise control. The course could be offered as an elective in a bachelor's degree program or as a course for a minor (e.g., in acoustics or interdisciplinary studies). Academic institutions could also offer capstone project courses, undergraduate research courses, honors projects, technical or free electives, and so on.

Objectives for Undergraduate Courses

Learning objectives for one (or more) undergraduate course(s) in the science and practice of noise control engineering course are offered below:

Objective 1: Understand how noise is measured, using decibels and frequency weighting, how to describe sound in frequency bands, and how to apply international and national standards.

1.1 Learn how to measure sound pressure level in decibels (dB) using the sound-level meter and the A and C frequency weighting scales.

1.2 Learn to describe sound levels in frequency bands (e.g., narrow and octave/one-third octave bands).

1.3 Understand the mechanisms of human hearing and the effects of noise on people, including noise-induced hearing loss, annoyance, perceived noisiness, speech interference, enjoyment of music, etc.

1.4 Learn to apply criteria for controlling noise and vibration in communities, buildings, vehicles, and industrial machines, based on international or national standards and recommended practices.

1.5 Examine at least one case study that shows how these principles can be used in a real-world situation.

Objective 2: Understand the nature of sound fields, noise sources, and noise control paradigms.

2.1 Learn the concepts of noise source, path, and receiver and how to use them to define a real-world problem.

2.2 Learn the basic description of sound waves, including one-dimensional plane waves and spherical waves, near- and far-field characteristics, anechoic chamber free-field concepts, and diffuse field concepts in reverberant rooms.

2.3 Understand relationships between vibration and

radiated noise in terms of sound power, radiation efficiency, and surface velocity.

2.4 Understand the basic noise sources (e.g., mechanical, airflow, electro-mechanical) and relationships between operational parameters and noise.
2.5 Learn to evaluate common noise sources in buildings, communities, industry, and vehicles and participate in an exercise in setting a noise target for at least one source.
2.6 Examine a case study that shows how these principles have been applied in a real-world setting.

Objective 3: Learn how to control structureborne, airborne, and fluidborne noise paths.

3.1 Understand the parameters of a mechanical oscillator, including natural frequency and damping ratio.
3.2 Learn the concept of vibration isolation and how to specify the stiffness of a system.
3.3 Learn the concept of resonance control and how to specify viscous or structural damping.
3.4 Understand the concepts of absorption and reflection of harmonic sound waves by solid and fluid boundaries and materials and be able to relate them to the impedances of materials or duct elements (including reactive and resistive characteristics).
3.5 Understand the concept of sound transmission through a wall and the mass law.
3.6 Learn the characteristics of sound-absorptive materials and how to specify their performance.
3.7 Learn the concepts of basic muffler elements, such as expansion chambers and side branch resonators, and how to specify their performance.
3.8 Learn how to design a simple enclosure and how to control noise in various ways.
3.9 Examine a real-life problem that illustrates how these principles have been applied and propose source or path noise control solution(s).
3.10 Critically examine professional issues, such as safety, ethics, economics, product liability, and environmental concerns via case studies and group discussions.

The objectives described above should be considered minimal requirements for a one-semester course. Issues related to prerequisites and materials would depend on the program offering the course, the educational level of the student, and other specific factors. Evaluation methods (as appropriate) include homework assignments and examinations, classroom discussion, and student-conducted noise measurements on simple noise sources. Instructors are encouraged to use modern pedagogical methods (e.g., sound visualization codes, field animation software, MATLAB (or comparable codes), Internet-based tools). Experimental demonstrations (on the nature of sources and/or the effect of simple noise and vibration control devices) should be used to engage students. Guest speakers from industry, the community, and other academic departments could be brought in to illustrate the fascinating and challenging aspects of noise control engineering.

Undergraduate Course Descriptions

A short description of an undergraduate course that meets the objectives of the previous section follows. This course deals with the fundamentals of noise control and engineering, including design criteria based on human response to noise (e.g., hearing damage, annoyance, speech intelligibility, enjoyment of music). Acoustic wave propagation and transmission phenomena are covered, along with noise measurement and reduction techniques. Applications deal with machines, building design, musical instruments, and speakers. Ideal acoustical rooms (e.g., anechoic and reverberant rooms) are demonstrated. Students are expected to conduct sound measurements on a source of their choice using a handheld sound-level meter.

Another example of a course that meets the objectives is "Noise Control in Machinery" taught at Penn State. The course covers the nature of noise sources in machine elements and systems and deals with the propagation and reduction of machinery noise and the effects of noise on people.

GRADUATE EDUCATION IN NOISE CONTROL ENGINEERING

On the graduate level, institutions have offered several engineering-science-based courses, such as engineering acoustics, aero-acoustics, continuous vibrations, and digital signal processing. However, a comprehensive search of graduate programs turned up only a few courses with "noise control" in the titles. Penn State and Ohio State offer a sequence of year-long graduate courses, and the University of Nebraska at Lincoln offers one graduate-level course. Catalog descriptions are listed below:

- Penn State, Noise Control Engineering I: The first of three courses, this course provides an orientation to the program and covers fundamentals of noise control.
- Penn State, Noise Control Engineering II: This course applies fundamentals of noise control covered in Noise Control Engineering I to noise generation, propagation, measurement, and effects.
- Penn State, Noise Control Engineering III: This course covers advanced methods for analyses of noise and vibration and treatments for control of noise and vibration.
- Ohio State, Automotive Noise, Vibration, and Harshness Control I: An integrated study of acoustics, shock

and vibration, and dynamic design issues with emphasis on automotive case studies and problem-solving methodology.

- Ohio State, Automotive Noise, Vibration, and Harshness Control II: Continuation of 777 with focus on source-path-receiver identification, modal analysis, passive/active control, and machinery diagnostics.
- Ohio State, Automotive Noise, Vibration, and Harshness Control III: Continuation of 778 with focus on advanced modeling and experimental methods, structural/acoustic interactions, and flow-induced noise and vibration.
- University of Nebraska at Lincoln, Advanced Noise Control: Characterization of acoustic sources, use and measurement of sound power and intensity, sound-structure interaction, acoustic enclosures and barriers, muffling devices, vibration control, and active noise control.

The graduate-level sequence in automotive noise, vibration, and harshness control at Ohio State was developed by the university and General Motors in the mid-1990s. The three engineering practice courses are based on an innovative case study approach (similar to the approach used in business, law, and medical schools). This course sequence teaches the integration of concepts of mechanical vibrations, acoustics, digital signal processing, and machinery dynamics. Overall, the concepts of noise control are related to product design, manufacturing, materials, performance, and economic considerations.

Sample Course Descriptions

Traditionally, topics and coverage in a graduate-level course tend to depend on the research expertise of the instructor, students' backgrounds, and the needs of ongoing research programs. The characteristics of a sample course in noise and vibration control (with the emphasis on engineering practice) listed below are based on the third course at Ohio State.

- Wave equation solutions: Three-dimensional acoustic cavities and basic sources, such as monopole and dipole.
- Noise source identification: acoustic intensity using the two-microphone method, near-field holography, structural-acoustic responses using modal expansion, operating motion surveys, and laser scanning system.
- Noise and vibration sources: (1) friction sources, such as brake squeal, belt vibration, and tire noise; (2) clearance sources, such as transmission rattle, door slam, and piston slap; (3) aerodynamic sources, such as vehicle components, alternators, and antennas
- Passive and active noise and vibration control methods applicable to fluid and structural sources and paths.
- Topics for case studies and guest lecturers include the development of experimental facilities, structureborne noise paths in products, muffler system tuning, statistical energy analysis applied to interior acoustics, international design and marketing (from the noise control perspective), ethics, and professionalism.

The following course description is based on Penn State's Noise Control Engineering III:

- Sources of noise: power transmission, electric equipment, nonturbomachinery, flow-induced, and turbomachinery.
- Outdoor noise and structural acoustics: outdoor noise propagation, transportation noise, response of propagation in and radiation from structures, coupled structures.
- Measurement and analysis: single- and two-channel frequency analyses, coherence, and transfer functions.
- Noise treatments: vibration mounting systems, damping treatments, mufflers and silencers, active noise, and vibration control.
- Modeling: finite and boundary element methods, statistical energy analysis.

Faculty should consider offering noise control courses that provide a balance between theory and engineering practice without sacrificing academic rigor. Classroom education can be augmented by field trips, guest lectures, and seminars. Industry, government laboratories, and consulting firms could provide valuable help by offering their facilities for course-related experiments or miniprojects. A graduate internship program would motivate students while building a cadre of future noise control engineers.

CONTINUING EDUCATION AND SKILL DEVELOPMENT

Distance Education

Changes in products, competitive features, and regulations continue to create a need for expertise in noise control engineering. Companies and agencies often fill these needs by calling on employees who know the business well and can assume these responsibilities in addition to or in lieu of their regular jobs. Because of a paucity of formally trained noise control engineers in most companies, these emerging requirements are often assigned to engineers with training in fields that may overlap with noise control engineering (e.g., aerodynamics, crash-worthiness, physics, mechanical engineering, vibrations, or electrical engineering) or to individuals with no previous experience with noise control engineering who are judged to have outstanding skill in other areas (e.g., product design). These new "noise control practitioners" often need opportunities for professional development outside of formal educational settings. In fact, there are many venues for professional development, such as distance education through universities, short-course offerings from universities or private sources, and conferences.

Distance education offerings are often modified versions of university courses. Ohio State, Penn State, and Purdue University offer courses via video link and over the Internet.

Ohio State offers a one-year sequence of three quarter-long courses developed as a noise and vibration control engineering sequence for General Motors. The sequence is offered in the distance learning mode (via asynchronous video recordings and synchronous webex/video conferences). In addition, the sequence is offered biennially to approximately 15 to 30 students at General Motors and 15 to 25 graduate students at Ohio State.

In the past, Penn State offered a three-course sequence in the fundamentals of acoustics and noise control, but it was discontinued when the developer and instructor of the course retired. The sequence was offered asynchronously (at each student's preferred pace) through Internet and video recordings. Total enrollment was approximately 100. However, Penn State continues to offer many courses through distance education that are fundamental to noise control engineering: Fundamentals of Acoustics, Digital Signal Processing, Electroacoustic Transducers, Acoustics in Fluid Media, Acoustical Data Measurement and Analysis, Techniques for Solving Acoustic Field Problems, Sound/Structure Interaction, Flow-Induced Noise, Audio Engineering, Sound Quality, Computational Acoustics, and Nonlinear Acoustics. More than 70 students enroll in these courses each semester. Upon successful completion of 30 credits, a student is awarded a master of engineering degree in acoustics.

Purdue offers five courses in acoustics and vibrations, through the IHETS interactive video network and by videotape, to several companies that have contracted courses through the university. Courses are offered on a two- or four-year cycle, depending on their popularity. Approximately 20 students take these courses each year.

All of these distance-learning courses, which are slightly modified versions of courses offered on campus in formal noise control engineering or acoustics programs, include homework and test requirements. Students may sign up for a graduate degree program through distance education with these courses as part of a plan of study or they may take them on a nondegree status as courses of interest. In either case, the courses are rigorous and provide a strong general background in acoustics, vibrations, and noise control engineering.

Short Courses

Short courses are available from universities and private sources. Courses run from a single day to one week and are generally intensive but offer little hands-on experience and no competency tests. University offerings tend to be adapted versions of coursework for which a need has been identified.

Short courses can be divided into two groups: (1) general courses that teach fundamental topics and (2) advanced courses specific to an emerging area of interest. General courses tend to serve the same audience as distance education courses, usually individuals who do not have access to distance education or who think a short course meets their needs in terms of logistics or learning methodology. Students in advanced courses tend to be well educated in the fundamentals of noise control engineering but need to learn about emerging or advanced topics. However, students in an advanced course often have different backgrounds and different levels of understanding, which make teaching such courses a challenging undertaking. Examples of advanced short courses include topics in signal processing, active noise control techniques, and nonlinear vibrations.

Short courses offered by private sources include both general and advanced topics. Many of these courses were developed to address common recurrent or customer problems or to educate potential customers who might use the services offered by the sponsoring company. A few courses are used as marketing tools to attract business or create new opportunities for the company. For example, a one-day course on the basics of acoustic measurement might be a demonstration of a new acoustical measurement device.

A large proportion of continuing education in noise control engineering is provided by private sources. Short courses, whether offered by universities or private companies, often attract students with diverse backgrounds, cover materials as quickly as possible with maximum possible retention, and motivate participants to learn subjects that may not have been of immediate interest.

Technical Conferences

Technical conferences are widely used as educational vehicles, perhaps more in noise control engineering than in other fields. About 1,400 people attend the biennial SAE (Society of Automotive Engineers) International Noise and Vibration Conference and Exhibition, which generally has fewer than 300, mostly practical, presentations. The educational mission of the conference is described in the brochure for the 2009 event (*http://www.sae.org/events/nvc/*):

> The SAE Noise and Vibration Conference and Exhibition—the only dedicated mobility noise, vibration and harshness event in North America—will bring together nearly 1,400 leading experts and specialists from all points of the globe to learn about, present and display the latest technological innovations all under one roof. Attendees will gain a full understanding of NVH and sound quality issues related to vehicle design, engineering and testing, learn the latest trends

> and solutions during the technical paper presentations, visit innovative organizations at the exhibition, and exchange ideas with industry peers from around the world during special networking opportunities. *This is a must-attend event!*

The event includes approximately a dozen affiliated short courses, as well as workshops, demonstrations, and a large exhibition of products and materials.

Technical conferences sponsored by the Institute of Noise Control Engineering (NOISE-CON and INTER-NOISE), the National Research Council Transportation Research Board (summer meeting of the Transportation Noise Committee), and the American Institute of Aeronautics and Astronautics (Aeroacoustics Conference) place less emphasis than SAE on learning and more on technical exchanges by technical leaders. However, practitioners who want to learn something attend all of the conferences; all of them offer short courses in conjunction with the event.

Conclusion

Because the demand for noise control engineers is much greater than the supply of formally trained engineers, distance education and continuing education play a large role in developing practitioners in the field. The strongest offerings play a valuable role and should be encouraged to continue. However, many offerings compromise quality for expediency or marketability. Nevertheless, both will continue to be important for the foreseeable future. Therefore, guidelines associated with a certification process for noise control engineering in continuing education programs would help participants gauge the content and value of courses and other offerings.

SUPPLY-SIDE CHALLENGES

A major challenge to an adequate supply of well-trained noise control engineers is that educational programs are not homogeneous. For "mainstream" engineering disciplines, organizations such as ABET (*http://www.abet.org/requirements*) dictate that a majority of engineering departments at universities across the nation offer similar courses and cover the same general content. However, there are no such requirements for noise control engineering. A number of university departments offer education in noise control engineering either as degree programs, continuing or distance education, or both. However, even though they may sound alike, these departments often look very different. In comparing university departments, two characteristics vary dramatically—the size of the department and the school or discipline in which the department is housed.

Departments that have attained a critical mass of faculty members trained in noise control engineering often offer substantial courses and research opportunities. However, a large number of other departments have only one or two faculty members with training in noise control engineering. The database of the Acoustical Society of America (ASA) identifies universities that offer programs in various subdisciplines of acoustics.[1] A search for "noise and noise control" reveals that of the 39 universities identified, only eight have more than two faculty members in the area of noise control engineering. Although these individuals may be well respected in the field, it is difficult for students in those programs to receive the same level of education as they would in a larger program.

Noise control engineering programs are also housed in a variety of departments. According to the ASA database, the majority are housed in mechanical engineering or aerospace engineering departments. However, the others can be found in departments of electrical engineering, physics, civil engineering, oceanography, architecture, communication science and disorders, recording arts and sciences, speech pathology, and audiology, and other unlikely departments, such as agriculture, otolaryngology, and biomaterials. Figure 9-1 shows the percentages of faculty members associated with departments identified as offering noise control engineering programs.

Lack of Homogeneity

The lack of homogeneity reflects the multidisciplinary nature of noise control engineering, which creates some benefits but also several challenges. The benefit is in bringing people from different backgrounds into the field who can contribute valuable new perspectives. One of the major challenges is that there is no consistent "home" for the discipline on university campuses.

In the middle of the twentieth century, most noise control engineering programs were housed in physics departments; a smaller number were housed in engineering departments—primarily mechanical and electrical engineering. Today most are housed in mechanical engineering departments, although, as indicated above, many other departments are involved in noise control engineering education. This lack of a focal point can make it difficult for employers or anyone else looking for help in the area of noise control engineering to know exactly what to look for—a mechanical engineer, a physicist, an electrical engineer, or someone else.

Another challenge is that people trained in different engineering and scientific disciplines tend to look at problems from different perspectives and use different terminologies, each of which has advantages and disadvantages. The three main (complementary) perspectives are:

[1] This database is cited because it may be less biased to a given discipline than some other databases. For example, the American Society of Mechanical Engineers (ASME) may be biased toward mechanical engineers, the American Institute of Aeronautics and Astronautics (AIAA) may be biased toward aerospace engineering, and the Institute of Electrical and Electronics Engineers (IEEE) may be biased toward electrical engineers.

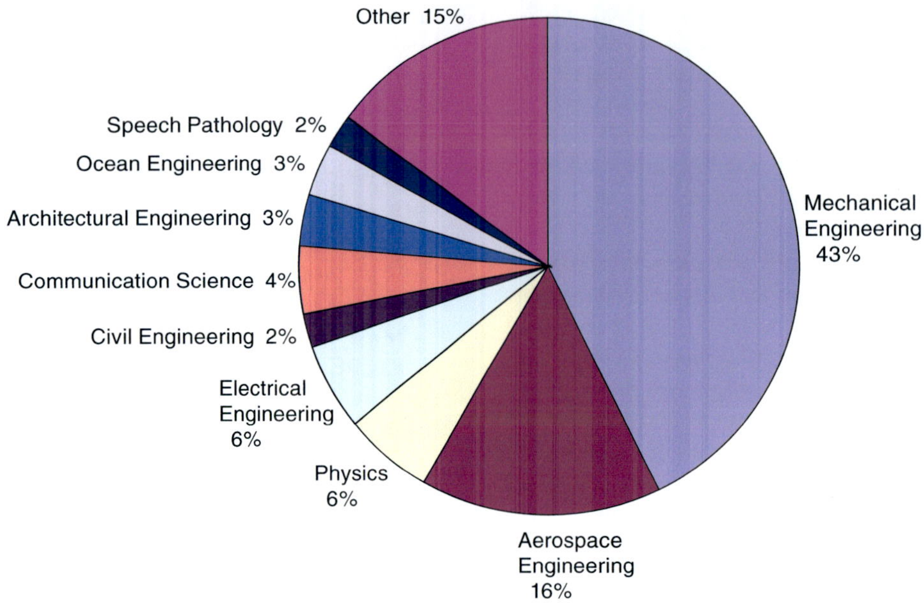

FIGURE 9-1 U.S. noise control programs in university departments. Source: Reprinted with permission from Scott D. Sommerfeldt, Brigham Young University.

- analysis in terms of acoustic or structural modes
- analysis in terms of wave propagation
- analysis in terms of sound levels and noise metrics

Many noise control applications are typically analyzed in terms of acoustic or structural modes. This approach is useful for analyzing finite structures or enclosed sound fields, and considering a problem in terms of the modal response of a system can yield considerable insight. This approach could address problems in the sound field in rooms, automobiles, aircraft, and equipment enclosures, or the vibration response of equipment, transformers, housings, and so forth. The modal approach can also be useful for analyzing sound/structure interactions (i.e., when there is significant coupling of a vibrating structure and the fluid into which the structure radiates).

However, a modal approach is not effective for addressing acoustical problems in large areas, such as community noise. For these and other applications a wave propagation approach is typically used. Wave propagation analysis is applicable to community noise problems involving source radiation and sound propagation, reflection/transmission problems, acoustical properties of porous materials, sound propagation in heating and ventilation systems, and mufflers.

Over the years, noise control terminology using frequency averaging and other attempts to account for human perceptions of sound have been developed to support an engineering approach to noise control. A substantial portion of noise control engineering involves the use of metrics based on sound levels (expressed in decibels).

Each application area in noise control engineering tends to have several metrics that are particularly useful for that application. This raises two potential difficulties. First, some people who work in noise control engineering were never taught the concept of sound levels and may be uncomfortable using metrics. Second, many who have been exposed to basic metrics like sound pressure level are unfamiliar, and hence uncomfortable, with metrics used in other application areas. In either case, additional training in metrics (both definitions and how they are used to characterize noise) would be beneficial.

These three complementary, intertwined descriptions of noise control analysis can create an obvious problem. People with different educational backgrounds may have been exposed to only one or perhaps two of these approaches (e.g., modes and sound levels) but not the third approach (metrics). Even if they have been exposed to all three, the quality of education can vary dramatically. The problem is more pronounced at institutions that do not have large programs where there are not enough resources for students to be exposed to the full range of approaches. This problem is also common for individuals not trained in acoustics and noise control who were assigned to work on noise control applications by their employers. In most cases they are familiar with only one approach and must somehow make up that deficiency to be effective. Thus, they must have access to educational opportunities, and they must take advantage of them.

The multidisciplinary nature of noise control engineering also contributes to challenges for industry employers, who may expect new employees to understand sound propagation, acoustic and structural absorption, reflection/transmission phenomena, instrumentation and measurement techniques, basic data and signal processing, computational techniques, and some basic psychoacoustics.

To address the challenges just described, it would be helpful for noise control engineering educators to establish a standard core curriculum for noise control engineering programs. Specialized coursework would also be desirable, but a standard curriculum would at least ensure that fundamental core concepts are understood by all students trained in the field. However, given the variety of departments that house noise control engineering programs, a standard curriculum is not likely to be adopted, unless an external body, such as ABET, pushes to implement a standard core curriculum.

DEMAND FROM INDUSTRY

Low-noise levels are becoming increasingly important for many products. In the cabin of vehicles, for example, low noise is strongly associated with high quality, as evidenced by the number of automobile commercials for high-end products that emphasize noise reduction features designed into the vehicle. The low-noise characteristics of appliances are also emphasized in sales literature and advertisements. We increasingly hear about public resistance to airport expansion or construction based on environmental noise. Similar resistance has been raised to expansion or construction of highways. Occupational environments are also concerned about workers' exposure to noise. Considering the growing interest in noise control in all of these areas, the demand for noise control engineers in all fields will also grow.

Automobile Companies

Automakers compete on the basis of cost, perceived quality, safety, and fuel mileage. Noise reduction has been closely associated with high quality, as has been apparent in television and print advertising over the past decade. Automobile companies have even become interested in tuning the noise of vehicles for so-called sound quality. Traditionally, the top 10 warranty issues include troublesome noises and vibrations, such as squeaks and rattles. In addition, reducing noise overall usually meant adding weight to the vehicle, which reduced fuel mileage.

Thus, noise control engineering is a significant aspect of all parts of automobile design, from the conceptual phase when targets are set and basic architecture is decided to the finishing touches. In fact, engineers are needed at all levels and in all operations of companies, ranging from noise control specialists capable of setting targets and diagnosing problems to noise-aware designers capable of incorporating noise control strategies into routine design decisions.

Currently, many automotive companies have only enough noise control engineers to staff central noise control laboratories. The staff operates in a reactive mode to fix problems after most decisions have been made. Component suppliers often do not have any staff or laboratories for noise control design and therefore depend heavily on consultants. Thus, unnecessary noise problems arise, especially for fans, motors, transmissions, and pumps.

Aircraft Companies

Aircraft companies are concerned with reducing both interior and exterior aviation noise. Because weight is an important factor in the design of airplanes, the interior of an airplane is highly susceptible to both airborne and structure-borne engine noise and wind-rush noise caused by airflow over the fuselage. Thus, noise controls must be lightweight and highly efficient.

Noise control engineering for aircraft must begin at the conceptual design phase and continue through the development of the detailed design and prototype. Airframe companies such as Boeing hire aggressively and have noise control engineering personnel throughout the company. Boeing also invests intensively in continuing education to ensure that engineering designers in general are sensitive to noise reduction methodologies.

Exterior aircraft noise has received considerable attention. Policy has dictated a 10 dB per decade reduction in aircraft noise and mandated the retirement of a major portion of the fleet. Cost estimates for achieving this goal are as high as $5 billion. The noise control engineers who carried this effort forward were employed by aircraft engine manufacturers and the National Aeronautics and Space Administration.

As aircraft engines have become quieter, attention has turned to reducing noise from the airframe itself. Thus. today airframe manufacturers are more involved in exterior noise control than they were in the past. Most of the noise control engineers involved in current studies have advanced degrees in acoustics with expertise in aerodynamics.

Noise Reduction in Other Areas

Purchasers of appliances also associate quiet with quality. Companies that plan to market their products internationally where buyers live in densely populated settings must provide quiet appliances to meet market regulations. Many appliance manufacturers have built small noise control laboratories, but they do not have critical mass to retain noise control engineers in a market in which demand greatly exceeds the supply.

In defense applications, noise control is not as uniformly important as it is in the commercial sector—with a few notable exceptions. During the cold war, acoustical detection of submarines and the suppression of the acoustical signature of submarines were high-priority technologies. During those years, very large numbers of noise control engineers, consultants, and contractors were employed in defense agencies. Noise reduction is still a significant aspect of stealth weapons systems for air, ground, and naval applications, and noise control engineering is widely used in all branches of defense. The U.S. Department of Defense (DOD) has shown significant interest in personnel who can assist with hearing protection and other occupational safety issues. Some of the largest veterans' benefits payouts are for hearing loss suffered by DOD personnel.

Consulting Companies

Based on membership in the National Council of Acoustical Consultants, it is estimated that approximately 2,000 noise control and acoustical consultants are practicing in the United States, and the number continues to grow. Consulting companies occupy a unique position between the public and either government or commercial operations. Effective noise control engineering consultants must not only understand the fundamentals of the field, but must also have the skills to understand policy and interact with a variety of clients, including real estate developers, construction companies, hospitals, municipal governments, and others. The situation was aptly described in a private communication to the committee by Senior Vice President Nicholas Miller of Harris Miller Miller & Hanson, Inc., a leading consulting company in transportation noise:

> Since there seems to be very little education at the undergraduate level in noise control or acoustics, we do not expect to find people with undergraduate degrees with knowledge of acoustics. On the other hand, by the time they have done a serious MS or PhD in acoustics, they are likely overqualified for the positions we need. We often hire people with little or no knowledge of noise and acoustics and train them internally for the skills we need. Widespread undergraduate exposure to the basics of noise and acoustics would help us identify and retain good staff.

DOES DEMAND EXCEED SUPPLY?

The answer to the question of whether demand exceeds supply is based on responses collected during a workshop held in Reno in October 2007 as part of the NOISE-CON 2007 Conference and on other sources. This issue was also a subject of discussion at a National Academy of Engineering noise control research workshop in June 2008. Both workshops and an informal survey of engineers working in the field indicate that there is a strong demand for graduates in noise control engineering.

The same results were found in a poll of key academics in U.S. institutions who say they regularly receive phone calls and e-mail asking about graduate students with skills in noise control engineering. The number of practicing engineers in continuing education classes who have backgrounds significantly outside noise control engineering also indicates an undersupply of well-qualified graduates. University departments with educational opportunities related to noise control engineering generally report that inquiries about graduates qualified in noise control engineering and related disciplines exceed the local supply. Another indication of an imbalance between supply and demand comes from educators, who report that salary offers for new graduates with backgrounds in noise control are higher than for other engineers and that these students often receive multiple offers.

The consensus at both workshops was that graduate programs in noise control engineering should be expanded and that funding for current educational programs should be increased to ensure a steady supply of young professionals entering the field. In addition, undergraduate studies in noise control engineering should be expanded to ensure the availability of workers who can perform engineering tasks, such as making measurements and design calculations at the basic engineering levels. This would not only answer a need of employers but would also free practicing engineers (generally trained in other disciplines) who have difficulty with some approaches to noise control that are counterintuitive.

For American industries to produce quieter, more competitive products for domestic and global markets, noise emission and associated issues (such as costs, environmental considerations, and system design issues) must be added to the list of product and equipment requirements. This will mean that design and manufacturing engineers must understand some elements of noise control engineering and closely related engineering disciplines. There is also a need for qualified personnel in government for policy development and enforcement.

FINDINGS AND RECOMMENDATIONS

Undergraduate education in noise engineering varies greatly from institution to institution in terms of the department in which it is housed and the courses offered. Funding for noise control engineering programs at universities is problematic, and support for graduate students to assist in research (or teaching) and to develop a new cadre of professionals is inadequate. The geographic distribution of leading programs is also a concern. The largest programs tend to be where funds for sponsored research are available rather than where industry demand for specialists is highest.

Recent reports highlighting the state of engineering education in the United States, such as *The Engineer of 2020: Visions of Engineering in the New Century* (NAE, 2004), which offers "future scenarios of the possible world conditions for the 2020 engineer," recommend changes in engineering curricula and pedagogical methods. The report recommends that practical and interdisciplinary issues that impact society and industry, such as ethics, safety, and environment, should be integrated into the undergraduate engineering curriculum. A recent report by the Carnegie Foundation for the Advancement of Teaching (Sheppard et al., 2009) finds that "American engineering education is too theoretical and not hands-on enough. . . . A widespread emphasis on theory over practice . . . discourages many potential students while leaving graduates with too little exposure to real-world problems and ethical dilemmas." The study committee of the present report believes that the promotion of noise control engineering in academia is consistent with the recommendations and visions of both reports, that it would fit the ABET criteria for engineering programs, and that it would serve the needs of related programs, such as physics, architecture, biological sciences, and speech and hearing.

The multidisciplinary nature of noise control engineering poses challenges for engineering practice and for lifelong learning. Typically, employees attempting to solve complex noise control problems must have a rigorous knowledge of noise measurement and signal processing techniques, propagation of noise though air and structures (including acoustic absorption, insulation, damping, and vibration isolation), computational techniques, and psychoacoustics. They may also need additional expertise in specific areas of noise control engineering (e.g., aero-acoustic problems are very different from problems raised by noise from machine elements). Neither undergraduate nor graduate programs are comprehensive, and the need to understand new issues and technologies over time creates a strong demand for continuing education.

Elements of noise control engineering degree programs should be formally taught in an intra- or interdisciplinary way by faculty in academic units (in engineering, physical sciences, and architecture). Major professional societies (such as AIAA, ASME, American Society of Heating, Refrigerating, and Air-Conditioning Engineers, Institute of Noise Control Engineering of the USA, SAE). and other stakeholders should organize symposia (or special sessions in regular conferences), where leading academic and industry leaders can propose and refine curricula and suggest improvements in teaching methods and delivery mechanisms. Collaboration among academic, research, and industry leaders will be necessary for the development of interesting case studies or practice modules that could then be disseminated to teachers of undergraduate courses.

Funding is particularly important for research on environmental noise that encourages interdisciplinary collaboration between acousticians, engineers, social scientists, psychologists, sociologists, and health scientists to develop improved metrics for evaluating the impact of noise, including annoyance, speech and communications interference, cognitive impairment, sleep disturbance, and health effects.

A comparison of research activity on environmental noise in Europe, Japan, and the United States clearly reveals that the level of activity in Europe and Japan far exceeds the level in the United States. Substantial funding for research in Europe and Japan has enabled very large scale and many smaller scale studies. An indirect effect of this funding is the number of graduate students in environmental noise being educated in Europe and Japan, which has resulted in widespread understanding of acoustics and environmental problems and helped inform decisions and encouraged the adoption of noise mitigation efforts and appropriate metrics.

Recommendation 9-1: Academic institutions should offer an undergraduate course in noise control engineering, broaden the scope of the engineering curriculum, and increase the pool of engineering graduates who are equipped to design for low-noise emissions. The course could be offered as an elective in a bachelor's degree program or as part of a minor (e.g., in acoustics or interdisciplinary studies).

Recommendation 9-2: Graduate-level noise control courses should provide a balance between theory and engineering practice without sacrificing academic rigor. The committee strongly encourages the establishment of graduate internships in industry and government agencies and thesis research programs to motivate students and to build a cadre of future noise control engineers.

Recommendation 9-3: Federal agencies, private companies, and foundations with a stake in noise control should provide financial support for graduate students who assist in research on, and the teaching of, noise control engineering. This support is crucial for the development of noise control professionals and noise control educators.

REFERENCES

NAE (National Academy of Engineering). 2004. The Engineer of 2020: Visions of Engineering in the New Century. Washington, DC: National Academies Press.

Sheppard, S.D., W.M. Sullivan, A. Colby, K.Macatangay, and L.S. Shulman. 2009. Educating Engineers: Designing for the Future of the Field. The Carnegie Foundation for the Advancement of Teaching. San Francisco, CA: Jossey-Bass.

10

Public Information on Noise Control

If enough people actively work to quiet a source of noise—either acting as individuals or as organized groups complaining to the noise producer—the noise will be quieted.
—Bugliarello et al., 1976
The Impact of Noise Pollution

There is no doubt that the public can be an effective force in promoting a quieter America. In the late 1950s, people noticed the difference between the noise generated by jet and propeller-driven airplanes, even though a standard sound-level meter indicated that the A-frequency weighted noise levels were the same for both. The public outcry forced authorities in New York to pressure both airplane manufacturers to reduce the noise of jet airplanes and the scientific community to find a better way of measuring human reaction to noise (Beranek, 2008).

As a result, the perceived noise level was developed, and in 1969 the Federal Aviation Administration issued regulations limiting noise emissions from airplanes. International regulations soon followed, and cooperation between manufacturers and the federal government has led to airplanes that are much quieter today than when they were first introduced.

Nevertheless, because of the enormous increase in air traffic, problems with noise around major airports continue, as does the dialogue between authorities and the public. In the area around O'Hare Airport near Chicago, for example, regular meetings are held to discuss measures to address airport noise issues (ONCC, 2009).

In the early 1970s, before noise walls became commonplace along American highways, two acoustical consultants visited Baltimore, Maryland, and recommended the construction of a noise barrier for a controversial highway construction project in anticipation of complaints from a nearby community.[1] At the time the interaction between the road surface and tires, which is now known to be the major source of highway traffic noise, was not well understood. Thus, instead of considering the reduction of noise emissions at the source, noise barriers became the solution of choice for abating traffic noise.

A survey for the appliance industry in 1999 showed that 84 percent of respondents considered "ultra-quiet" operation of dishwashers a desirable feature (KBDN, 1999). Today, many quiet dishwashers are on the market, and, although there is no uniform system for labeling noise emissions from appliances in the United States, in some cases a noise emission label is placed on products.

Despite these and other examples of responses to public concerns about noise, success stories are exceptions rather than the rule, and there is plenty of room for improvement. In a line-by-line compilation by the U.S. Census Bureau (2005), 11,757,000 households in 2001 reported that street noise and/or traffic noise was "bothersome." Of those, 4,457,000 said that noise was so bothersome they wanted to move. In another line of the report, noise was reported as a "problem" by 2,652,000 households. Other studies have shown that there is widespread dissatisfaction with noise levels and the lack of speech privacy in many offices (e.g., Center for the Built Environment, 2009; Jensen et al., 2005).

Noise from lawn care equipment is frequently the subject of citizen complaints. One approach to addressing these complaints is for citizens to pressure local authorities to enact noise control ordinances or to use another legal procedure. However, this approach immediately puts citizens groups in conflict with manufacturers or trade associations. An alternative is for the public to convince manufacturers that engineering controls for reducing noise are feasible and that there is a market for quiet outdoor equipment. In this way, public pressure could be a powerful force in driving innovation and noise reduction for consumer products.

Another source of widespread public complaints is noise from motorcycles. Although federal regulations to control noise emissions and muffler designs are in place, they are widely ignored. On September 11, 2009, the Portland (Maine) *Press Herald* reported that citizens had failed to persuade the city council to control motorcycle noise by

[1] Miller, L. 2009. Acceptance speech on receiving the INCE/USA Outstanding Educator Award. Presented at INTER-NOISE 09, The 2009 International Congress on Noise Control Engineering, Ottawa, Canada, August 26.

insisting on the installation of mufflers approved by the U.S. Environmental Protection Agency (EPA). Would the outcome be different if citizens groups had better public information? Time will tell, because this issue is sure to come up again.

WORKING TOWARD AN INFORMED PUBLIC

As the examples above show, there are many obstacles to achieving lower noise levels. Groups in favor of noisy devices for financial and other reasons will rise in opposition to noise reductions, and governments will listen to persuasive arguments on both sides of an issue and try to balance the needs of opposing groups. Manufacturers have shown that they will respond, sometimes slowly, once they are convinced there is a market for quieter products. At times, citizens become convinced that nothing can be done about noise, and they move on to other issues.

The study committee that prepared this report believes that a well-informed public has a better chance of success than a public that lodges complaints based only on subjective reactions to noise. To support that argument, the next sections review what has been done in the past and describe the current situation. The purpose here is not to list all of the stakeholders but to give a brief snapshot of some past and present activities and to suggest actions that could be taken in the future to improve public access to authoritative, accurate, and timely information that can support and inform a strong public presence in future efforts to reduce noise.

Past Efforts

In 1970, Theodore Berland, a well-known writer of popular science at the time, wrote *The Fight for Quiet*, an influential book in which he presented information on the health effects of noise, how noise is generated, and what the public can do about it. Much of the information was based on interviews with prominent scientists and engineers with expertise in noise. Berland presented data on noise levels in a wide variety of common situations. *The Fight for Quiet* is believed to have greatly influenced public policy, especially the passage of the Noise Control Act of 1972 and a decade of EPA involvement in noise issues. Robert Alex Baron, a former theater manager and head of New York Citizens for a Quieter City, wrote *The Tyranny of Noise* in 1970, a book intended to inform the public about noise issues, including many issues that had been raised by Berland.

Another influential book, *The Impact of Noise Pollution*, by George Bugliarello et al. (1976), focused on technical issues but included a discussion of the dissemination of information on noise through public service announcements by the Ad Council, an organization that produces highly effective public service announcements on a wide range of subjects (*http://www.adcouncil.org/*). At that time, however, the EPA program had taken center stage regarding noise issues, and the idea of a campaign by the Ad Council was never pursued. With the authority given to EPA by Congress under the Noise Control Act of 1972 and later the Quiet Communities Act of 1978, EPA had an active public information program. One element of the program was called ECHO (Each Community Helps Others), which gave communities with limited resources an opportunity to share ideas on what works and what does not with respect to noise. As detailed below, EPA still has that authority, although its program was curtailed by Congress in 1981.

Later, the American Speech-Language-Hearing Association convened a group of experts to study noise issues and publish the results. The report, *Combating Noise in the '90s*, was published by the association (ASHA, 1991). Working Group VII of the team that produced the report was charged with developing a strategy for educating the public and disseminating information. Target groups included preschool children, school-age children and youth, college and professional students, adult citizens and consumers, practitioners in influential professions, and specific groups at risk—in short, most of the population. Key messages would address quality-of-life issues, health effects, noise hazards to hearing, and the prevention of noise-induced hearing loss. Unfortunately, none of these outreach or educational programs was pursued, perhaps because there was little follow-up in making the recommendations known to the public or because no organization stepped in to lead efforts to implement the recommendations.

A report with a similar title, *Fighting Noise in the 1990s*, was produced in Europe by the Organisation for Economic Co-operation and Development (OECD, 1991). In this report the authors observed that "the experience of several countries (Australia, Japan, The Netherlands, and Switzerland) suggests that it is better to organize ongoing campaigns of limited scope, giving regular backing to advances in noise abatement (e.g., the introduction of new regulations or a new policy), rather than major, short-lived national campaigns unrelated to progress achieved and with no lasting effect."

Current Efforts

Although EPA currently has broad authority from Congress to develop and disseminate information on noise to the public, the agency's current program might be described as "extremely modest." However, a few others have taken up the task. Some examples are given below.

A children's book, *Listen to the Raindrops*, is being distributed by the New York City Department of Environmental Protection to children in the public schools; the book is accompanied by a teacher's guide to noise pollution (Bronzaft, 2008). The Acoustical Society of America has a publicly available guide on the acoustics of classrooms and has developed an American National Standard on Classroom Acoustics (ASA, 2009). The "Dangerous Decibels" campaign (*http://www.dangerousdecibels.org/*) is a collaborative effort by the Oregon Museum of Science and Industry and the Oregon Hearing Research Center to educate children about the dangers of hazardous noise and ultimately to reduce the prevalence of noise-induced hearing loss. The Noise Pollution Clearing House (*http://www.nonoise.org/*) maintains a repository of reports on noise and provides online information about noise activities in several states; it also maintains a short list of citizens groups concerned with noise (*http://nonoise.org/quietnet.htm*). In addition, many other organizations publish online information on noise, including some government agencies, such as EPA, the several modal agencies of the U.S. Department of Transportation, and the Centers for Disease Control and Prevention. Many professional organizations maintain websites (e.g., Institute of Noise Control Engineering, Acoustical Society of America), and many general information sites are available, such as Noise Free America, Citizens Against Noise of Hawaii, and the Alaska Quiet Rights Coalition.

The Center for Hearing and Communication (*http://www.chchearing.org/*) sponsors International Noise Awareness Day to educate the public on the dangers of excessive noise. Several resources are available for download, such as a Noise Center, which contains useful facts on noise.

Recent efforts have been made in the technical community to determine how its public information outreach can be improved. The situation in the United States was discussed in a workshop held in Dearborn, Michigan, in 2008 to investigate messages that should be communicated, the role of engineering societies, and current EPA activities (Bronzaft, 2008). A 2007 workshop in Istanbul (Moss, 2008) covered European campaigns (some of them successful) to raise public awareness. That workshop included anecdotal information about attempts by citizens to convince authorities that noise should be reduced.

Noise Action Week (*http://www.environmental-protection.org.uk/noiseactionweek*) is an annual initiative coordinated by Environmental Protection UK to raise awareness of problems caused by neighborhood noise and the solutions available to address them. This initiative provides an opportunity for local authorities, housing providers, mediation services, and all those involved in neighborhood noise management to publicize information about services available and promote practical solutions.

Online Information

The large number of references in this report to online sources attests to the importance of the Internet as a source of information on noise. Most government agencies that have missions connected to noise have websites on which they regularly post noise-related information, as do professional societies, trade associations, and citizens groups. An Internet search for "noise" using a major search engine returned about 117 million results. A search for "noise pollution" returned about 325,000 results, and "noise abatement" returned 81,000. The term "noise control engineering" returned many fewer results, about 12,300. Thus, an enormous amount of information on noise is available on the Internet. The problem for the public is how to judge the relevance and reliability of this information.

In 1945, Vannevar Bush, a professor and dean of engineering at the Massachusetts Institute of Technology and director of the U.S. Office of Scientific Research and Development from 1941 to 1945, published "As We May Think," an essay in *Atlantic Monthly*, in which he expressed concern about how engineers and scientists would find their way through the mass of technical information generated during World War II (Bush, 1945). The exponential increase in the amount of information available today has complicated this problem by orders of magnitude, but Bush's concept of "trails" can still be helpful. He conceived of a machine, which he called the "memex," which modern readers will recognize as a desktop computer and monitor with extensive local storage and a high-speed connection to the Internet. A knowledgeable user, Bush speculated, would be able to sort through masses of information and create a "trail" that could be turned over to others with an interest in the subject.

In today's terms, a "trail" would be a carefully annotated description of a subject together with hyperlinks to information resources. Even though "trails" through the mass of information on noise control do not exist, a group of persons with knowledge of the subject and a bias toward providing accurate, relevant information to the public could create a document that would inform and support the development of persuasive arguments for noise reduction.

Dissemination of Information

Although articles on noise occasionally appear in the mainstream media, they usually focus on a specific problem considered to be "news" at the moment. Currently, no concerted, coordinated efforts are being made to disseminate basic, authoritative information in an effective way. Experience has shown (e.g., OECD, 1991) that an effective noise information campaign will require a variety of messages for specific target audiences and a continuous stream of messages that highlight advances in noise reduction.

EPA has a website and a modest program related to public information (*http://www.epa.gov/air/noise.html*). However, several of the links on this site lead to sources of information that are badly outdated. EPA needs much more support and the cooperation of other agencies and organizations to provide accurate, authoritative, timely information to the public.

An alliance of stakeholders would be a major step toward the creation of a comprehensive plan to develop and disseminate public information. One of the major stakeholders in this alliance should be the engineering community, which has the capability of developing methods and technologies for reducing noise at the source. Specific interests of professional and other societies include air and surface transportation noise, noise from air-conditioning systems, noise control in buildings, aeroacoustics, flow noise, and others. Specialists from a number of engineering disciplines could help craft messages intelligible to the public about successful efforts to reduce noise. People from other disciplines could contribute information on the effects of noise on hearing and other health effects. The public should also be informed about current activities of government agencies to reduce noise, and communities should help each other by making information available about successful efforts to reduce or control noise.

SUMMARY FINDINGS AND RECOMMENDATIONS

In the past the public has been a driving force behind the establishment of noise control programs in the United States. Armed with accurate, up-to-date information, public action and opinion could again become an effective force. Unfortunately, information available to the public is currently scattered among numerous federal agencies and numerous sites on the Internet, and most of the books written to inform the public are relatively old and are based on outdated information.

EPA is the only federal agency with the authority to support public action and the capability of addressing all aspects of the noise problem. The U.S. Code requires that EPA "develop and disseminate information and educational materials to all segments of the public on the public health and other effects of noise and the most effective means for noise control, through the use of materials for school curricula, volunteer organizations, radio and television programs, publication, and other means." At this time, however, EPA does not have the internal resources to create a large public information program, and it is likely that much of the effort will have to be done through contractors.

The labeling of product noise emission levels should be a critical aspect of a program designed to benefit the public and enable people to make informed purchasing decisions. Although EPA has labeling authority, it might be more practical for professional organizations, trade associations, and standards organizations to develop labeling methodology for specific products because of the wide variety of products and noise measurement methods.

Professional organizations should take the lead in the development and dissemination of information about noise to the public. Engineering societies (e.g., Institute of Noise Control Engineering, American Society of Mechanical Engineers, SAE International) can deliver the message that, given demand by the public, engineering solutions to noise problems can be found. Other societies can deliver messages related to the effects of noise on hearing and the effects of noise on health.

Recommendation 10-1: The Environmental Protection Agency should take the following actions under the authority of 42 USC 65, Section 4913, to improve public information and education on the effects of noise and the most effective means of controlling noise:

- Conduct a survey of all activities by federal agencies related to noise, and publish URLs that provide information of interest to the public.
- Develop a categorized list of stakeholders with interests in noise (e.g., professional societies, scientific societies, citizens groups).
- Help organize a coalition of current stakeholders with the goal of improving the availability of information on noise to the public.
- Develop educational materials to inform the public of the health effects of noise, especially noise-induced hearing loss and cardiovascular effects.
- Develop information to help the public understand the benefits of using personal hearing protection devices.
- Provide information on the selection and use of hearing protection devices, making intelligent decisions about frequenting high noise exposure events, the importance of reducing noise exposures by buying quieter products, and being vigilant and active in public policy decision making about community noise zoning issues.

Recommendation 10-2: Engineering professional societies such as the American Institute of Aeronautics and Astronautics, the American Society of Mechanical Engineering, American Society of Heating, Refrigerating and Air-Conditioning Engineers, Society of Automotive Engineers, and Institute of Noise Control Engineering of the USA should develop engineering information on noise control to help the public understand techniques for reducing noise emissions.

REFERENCES

ASA (Acoustical Society of America). 2009. American National Standard Acoustical Performance Criteria, Design Requirements, and Guidelines for Schools. Melville, NY: ASA. See also booklet on classroom acoustics. Available online at *http://asa.aip.org/classroom/booklet.html*.

ASHA (American Speech-Language-Hearing Association). 1991. Combating Noise in the '90s: A National Strategy for the United States. Available online at *http://bit.ly/cL3ilk*.

Baron, R.A. 1970. The Tyranny of Noise. New York: St. Martin's Press.

Beranek, L.L. 2008. Riding the Waves: A Life in Sound, Science, and Industry. Cambridge: Massachusetts Institute of Technology Press.

Berland, T. 1970. The Fight for Quiet. Englewood Cliffs, NJ: Prentice-Hall.

Bronzaft, A.L. 2008. How Do We Stimulate Collective Action to Motivate the Public to Demand Quiet? Presented at a workshop in conjunction with NOISE-CON 08, The 2008 National Conference on Noise Control Engineering, Dearborn, Michigan.

Bugliarello, G., A. Alexandre, J. Barnes, and C. Wakstein. 1976. The Impact of Noise Pollution. New York: Pergamon Press.

Bush, V. 1945. As We May Think. Available online at *http://www.theatlantic.com/doc/194507/bush*.

Center for the Built Environment. 2009. Acoustical analysis in office environments using POE surveys. Available online at *http://www.cbe.berkeley.edu/research/acoustic_poe.htm*.

Jensen, K.L., E. Arebs, and L. Zagreus, L. 2005. Acoustical Quality in Office Workstations, as Assessed by Occupant Surveys. Available online at *http://bit.ly/d0PPnG*.

KBDN (Kitchen and Bath Design News). 1999. New Survey Pinpoints Dishwasher Usage Trends. Available online at *http://bit.ly/crxvMW*.

Moss, J. 2008. Public pressure—an effective force. Based on a workshop held in Istanbul, Turkey, as part of INTER-NOISE 07, The 2007 International Congress on Noise Control Engineering, August 29, 2007. Part I, Noise/News International, 16(1):18–29. Part II, Noise/News International 16(2):10–25. Available online at *http://www.noise/news international.net/archives_idx.htm*.

OECD (Organisation for Economic Co-operation and Development). 1991. Fighting Noise in the 1990s. Paris: OECD.

ONCC (O'Hare Noise Compatibility Commission). 2009. Available online at *http://www.oharenoise.org/*.

U.S. Census Bureau. 2005. Housing Summary for the United States, Section 20, Construction and Housing, Table No. 944. Available online at *http://bit.ly/aRHWFF*.

11

Summary Findings and Recommendations

This summary has been prepared to give readers a concise view of the contents of the report and the recommendations. Because of the large number of issues covered in this report and the likelihood that some readers may be interested in only one or a few chapters, findings and recommendations are also included in Chapters 3 through 10. The contents of this summary are aligned with the Executive Summary, an even shorter overview; the committee recommends that the Executive Summary be read before the findings and recommendations summarized below.

IMPROVE ENVIRONMENTAL NOISE METRICS

The committee studied the applications and limitations of existing environmental or community noise metrics. The current most widely accepted metric for characterizing the impact of environmental noise, the day-night average sound level[1] (DNL) measured in decibels (dB) has both strengths and weaknesses. Yet the committee agreed that DNL remains a very useful measure for understanding, communicating, and responding to potential noise impacts on communities. Extensive research has shown that a DNL of 65 dB yields a significantly higher fraction of affected populations that are "highly annoyed" (12 to 19 percent) than a DNL of 55 dB (only 3 to 8 percent). Therefore, the committee concludes that there is sufficient evidence to justify reducing the current U.S. federal agency limit on DNL from 65 to 55 dB.

Recommendation 3-1: The federal government (e.g., agencies of the U.S. Department of Transportation with responsibilities related to noise and the U.S. Department of Housing and Urban Development) should adopt as a goal the 1974 recommendation of the Environmental Protection Agency (EPA, 1974) to limit the day-night average sound level (DNL) to 55 decibels (dB) to protect the public health and welfare. Currently, DNL (DENL in Europe), the accepted metric for characterizing the impact of community noise, shows that a large proportion of the population is highly annoyed at a DNL of 65 dB or higher.

Recent advances in the collection, storage, and analysis of noise data have led to a reexamination of the metrics developed in the 1970s or earlier and the development of new community noise metrics that more accurately reflect human responses to noise.

Recommendation 3-2: Relevant agencies of the federal government (e.g., agencies of the U.S. Department of Transportation with responsibilities related to noise, the Environmental Protection Agency, and the U.S. Department of Housing and Urban Development) should fund the development of environmental noise metrics that are more transparent and more reflective of the impact of noise on an affected population than DNL. This will require improved tools for predicting community sound pressure time histories and the development of metrics that accurately reflect the sounds people hear. A more holistic model of annoyance is also needed that incorporates situational variables that can be used to generate predictions for overall response, as well as responses of vulnerable populations (e.g., elderly people, sick people, children, and noise-sensitive individuals). International cooperation in this effort will facilitate the development of national and international standards for calculating metrics and should include open-source code to facilitate broad implementation of the metrics. Certain measures should be taken to facilitate this development:

1. The international noise control engineering community should develop an open, collaborative data-sharing environment in which researchers can deposit and access data from community noise surveys (e.g., data from surveys of acoustic, environmental, community,

[1]The day-night average sound level (DNL) is the average sound level for a 24-hour day, after addition of 10 decibels to levels from midnight to 0700 hours and from 2200 hours (10 p.m.) to midnight.

and transportation systems to support comparisons of metrics and predictions by models).

2. Policy agencies should conduct extensive surveys around at least six U.S. airports to generate high-quality data to populate the database. These surveys should serve as models of good survey practices, including data recording and archiving to ensure that they are useful for future studies.

STRENGTHEN THE REGULATORY FRAMEWORK FOR HAZARDOUS NOISE

This report provides information on both occupational and nonoccupational noise that can damage hearing and assesses the technologies and regulatory framework that address hazardous noise in the workplace. Current U.S. Department of Labor limits on occupational noise exposure are higher than those recommended by EPA, the National Institute of Occupational Safety and Health (NIOSH), and hearing conservation professionals worldwide, as well as current limits written into national and international standards.

Recommendation 4-1: To comply with the recommendation of the National Institute for Occupational Safety and Health, the policy of several other government agencies, and widespread national and international scientific opinion, the U.S. Department of Labor should adopt the 85-dB(A)/3-dB limit for exposure to hazardous noise. This would replace the current 90-dB(A)/5-dB requirement.

With respect to impulsive noise (a single burst or a series of bursts closely spaced or isolated) and its associated auditory hazards, the committee concludes that current damage risk criteria in the United States and internationally are inadequate and need further study.

Recommendation 4-2: The National Institute for Occupational Safety and Health should be the lead agency and should be tasked by its parent agencies (U.S. Department of Health and Human Services/Centers for Disease Control and Prevention) to develop new damage risk criteria with assistance from the military services that have experience with high-amplitude impulsive noise.

PROMOTE THE USE OF ENGINEERING CONTROLS TO REDUCE HAZARDOUS NOISE

The original 1971 Occupational Safety and Health Administration noise regulation for general industry, 29 CFR 1910.95, accorded "engineering controls" (i.e., reducing the noise exposure of workers by reducing the noise of the machinery or equipment that generates the noise) primacy in reducing hazardous noise exposure in the workplace. Reviewing research and experience since the 1971 regulation, the committee concludes that engineering controls, "buy quiet" programs (programs that require or provide incentives for companies and government entities to purchase quieter equipment), or other means that reduce hazardous workplace noise provide significant long-term advantages over the use of individual hearing protection devices (HPDs) in the workplace.

The committee concludes that engineering controls of noise in the workplace should be the primary method of protecting workers from hazardous noise exposure. Accordingly, the committee recommends the following actions by U.S. government agencies, engineering and trade societies, and other stakeholders to promote the development and use of engineering controls.

Recommendation 4-3: The U.S. Department of Labor should revoke the Occupational Safety and Health Administration (OSHA) "100-dB Directive" of 1983, which effectively raised the action point for engineering control of noise from 90 to 100 dB by allowing the substitution of hearing protectors for noise control up to 100 dB and thereby devastated the market for quiet machinery and equipment. At the same time, OSHA should reconfirm that engineering controls should be the primary means of controlling noise in the workplace.

Recommendation 4-4: The National Institute for Occupational Safety and Health and the U.S. Department of Labor should develop and distribute widely an electronic database of noise control problems, solutions, and materials—taking into account the many handbooks and articles devoted to industrial noise control.

Recommendation 4-5: Engineering societies and trade organizations should develop guidelines for defining the relationship between noise emission specifications in terms of sound power level and/or *emission* sound pressure level and noise *immission* levels in industrial situations. They should provide a primer for buyers and sellers of machinery and equipment that includes: descriptions of how noise propagates in rooms; how to determine noise from a large number of machines; standards available to manufacturers and others for measuring noise emissions; and case histories of noise levels measured in *in situ* environments.

Recommendation 4-6: Government agencies should be instructed by a presidential directive or in congressional report language to show leadership in promoting "buy quiet" activities by developing and implementing programs for the purchase of low-noise products, as required by 42 USC 65, Section 4914. American industry should adopt "buy quiet" programs that require noise emission specifications on all new equipment and "declared values" in purchase specifications.

DEVELOP AND DEPLOY TECHNOLOGIES FOR NOISE CONTROL

The committee assessed new technologies in materials and systems for controlling noise from a large variety of sources. There are enormous disparities among programs, facilities, and resources for addressing noises of different types. For example, although engineering tools may be available for reducing aircraft noise and highway noise, the former has been deemed a national priority, while the latter has received less attention. Resources allocated for noise reduction are not always commensurate with noise exposures and impacts.

Aircraft noise control technology is much more advanced than technologies for addressing other noise sources, and the funds expended to reduce the noise of airplanes themselves as well as mitigation measures around airports is far greater than for other noise sources. Road traffic noise has been controlled mostly by constructing noise barriers, but work is being done on promising technologies for reducing noise generated by tire/road interaction. Technologies are available for reducing noise from rail-guided vehicles, and these will become more important as the nation develops light rail systems and high-speed trains. Technologies for the built environment will also become more important as building construction is driven by Leadership in Energy and Environmental Design (LEED) certification and "green" principles. Active controls of sound and vibration have been under development for many years, but few products on the market have incorporated them, and many barriers must still be overcome.

Many tools for designing and developing quieter products have become available in the past few decades, driven largely by increases in computational power and reductions in computational costs. Even so, access to new tools is as uneven as the allocation of resources; corporate budgets for capital equipment are generally tight, and there is competition between departments for available funds. Furthermore, organizations that are doing only routine testing of products according to national and international standards find expensive new tools hard to justify. Thus, even though noise mechanisms in aircraft, automobiles, rapid transit and trains, consumer products, and industrial machinery are fundamentally similar, the availability and application of tools for addressing them are not. The committee recommends that ways be found to give industry and academia access to these tools for the benefit of manufacturers, workers, and the public.

Reducing Aircraft Noise

Recommendation 5-1: The National Aeronautics and Space Administration (NASA) should continue to fund collaborative projects by engine, airframe, and aircraft systems manufacturers. Drawing on expert knowledge in research organizations and academic institutions, research should focus on the complex interrelationships between engine and airframe and the importance of reducing each constituent noise source to reduce the overall noise signature of aircraft. These projects should develop improved prediction tools, for example, for advanced propulsion designs; acoustic scattering and propagation models, including weather and terrain models; models of the effects of interactions between engine installation and airframe configuration; and benchmark measurements necessary for the development and validation of these advanced tools.

Recommendation 5-2: The Federal Aviation Administration should continue to fund the development of novel operational and air traffic management procedures to minimize noise and should work with NASA and industry to make intelligent trade-offs between competing noise mitigation and chemical pollution goals.

Reducing Road Traffic Noise

Recommendation 5-3: Current activities of the Federal Highway Administration and several states to investigate noise reduction through new pavement design should be continued and expanded to speed up development and application of new technologies. Studies on the durability of pavement surfaces are essential, because durability has a direct effect on the life-cycle costs of applying quiet pavement technology, which has the potential to reduce noise where barriers are not feasible—for example, where homes are located on a hillside overlooking a busy highway.

Reducing Rail Noise

Recommendation 5-4: Planning tools available from modal agencies of the U.S. Department of Transportation, such as the Federal Railroad Administration and the Federal Transit Administration, should be used in planning new rail transportation systems, and supplemental metrics should be developed and used to estimate the effects of noise on people. The public would benefit if warning horns were made more directional; research and development related to warning horn directivity should be undertaken to better understand the effects on safety and benefits to the public.

Reducing Noise in Buildings

Recommendation 5-5: The acoustics and noise control communities should actively promote the inclusion of noise criteria in requirements for Leadership in Energy and Environmental Design (LEED) certification of buildings, not only to improve the noise environment but also to ensure that the acoustical environment is not degraded. Design standards (e.g., building codes) must be improved to ensure that good acoustical practices are followed in the construction of buildings.

Recommendation 5-6: The National Institutes of Health and/or the Facilities Guidelines Institute should fund the development of improved materials for hospital environments, where traditionally used materials may harbor and promote the growth of bacteria and other harmful biological agents.

Advancing Active Noise Control

Recommendation 5-7: Research agencies should fund university research on active noise control to address situations where the use of traditional noise-control materials is problematic or where they are not suitable for attenuating noise in the appropriate frequency range. Investigations into hybrid active-passive and adaptive-passive noise control systems and the development of low-cost microphones and loudspeakers that can be used in hostile environments should also be funded.

DEVELOP PRODUCT NOISE EMISSION STANDARDS AND REGULATIONS

The need for noise emission standards is recognized worldwide, especially in the European Union (EU). This need has made the International Organization for Standardization, and to some extent the International Electrotechnical Commission, leaders in the standards community. ISO standards committees have superseded many American-based standards committees and organizations that U.S. manufacturers have relied on in the past. America's voice on the ISO standards committees is weakened by the lack of U.S. manufacturers' leadership in ISO working groups. America has only a single vote, the same as every member country in the EU.

The EU has been a leader in the development of noise regulations based on these standards. These regulations are more extensive than those that exist in the United States, and consequently European manufacturers have gained a competitive advantage over their U.S. counterparts in meeting demand for low-noise machinery and other products worldwide. It is important to note that, although more stringent noise requirements can sometimes be a burden for manufacturers, they can also encourage innovation. A manufacturer's desire to design a low-noise machine for sale in world markets is a positive force that could lead to the introduction of quiet products into American markets and provide an incentive for manufacturers and purchasers to cooperate in "buy quiet" programs.

At the time of purchase, consumers rank noise as one of the top five characteristics when comparing product performance. Other concerns are energy efficiency, cost, reliability, and serviceability. Noise levels for U.S. products are often buried in product literature and reported using different noise metrics, making it difficult for consumers to compare noise levels at the time of purchase. Thus, consumers are unable to make informed decisions on the noise emission of a product. This problem could be corrected if product noise levels were prominently displayed and manufacturers adopted a system of self-enforcement.

American manufacturers have the ingenuity to design quiet products. However, manufacturers and trade associations, as well as the voluntary standards community, have been unable to agree on a uniform standard for measuring and labeling product noise.

Recommendation 6-1: The Environmental Protection Agency should encourage and fund the development of a uniform system of labeling product noise. The system should be self-enforced by manufacturers but should have strict rules and penalties if products are deliberately mislabeled. The rules should specify standard methodologies for measuring product noise. Uncertainties in noise emission values should be acknowledged. Product noise labels should be prominently displayed so that consumers can make informed purchasing decisions. In a world with proliferating eco-labels and different requirements, international cooperation to develop one label recognized worldwide would be of great benefit to American manufacturers and consumers everywhere.

Recommendation 6-2: Government, trade associations, and industry should fund the participation of U.S. technical experts on standards bodies that develop international standards for determining product noise emissions.

Recommendation 6-3: The National Institute of Standards and Technology should take the lead in providing assistance to American manufacturers with noise regulation compliance by establishing a database of information on U.S. and international product noise emission standards and requirements.

Recommendation 6-4: To establish their credibility, organizations that determine noise emission data according to a certain standard as part of a voluntary labeling program should be accredited to test products. Managers at the National Institute of Standards and Technology and its National Voluntary Laboratory Accreditation Program should promote their accreditation program, especially in industrial laboratories.

USE COST-BENEFIT ANALYSIS AS A TOOL FOR NOISE MITIGATION

The committee considered cost-benefit analysis for different noise mitigation options in a broad context and in the specific context of reducing noise generated by interactions between vehicle tires and road surfaces. At highway speeds this tire/road interaction noise dominates noise emissions from vehicles, and efforts are being made to design road surfaces and tires that minimize this noise. The efforts of the Federal Aviation Administration to develop a cost-benefit approach to analyze noise around airports could help in the development of a similar project to analyze options for reducing highway noise.

Recommendation 7-1: A formal cost-benefit analysis should be performed to compare the costs and benefits of using pavement technology for noise reduction with the costs and benefits of installing noise barriers. This cost-benefit analysis should be a cooperative effort of the Federal Highway Administration, U.S. Environmental Protection Agency, and the several states with technology programs in road surface design. Inputs to the analysis should include data from analyses of noise reduction efforts around airports.

STRENGTHEN THE ROLE OF GOVERNMENT

In some areas—notably aircraft noise, occupational noise, and highway noise that can be reduced by barriers—government regulation has been instrumental in reducing noise. But this report shows that improvements can be made in other ways as well. For example, authority for cost-benefit analyses, interagency projects, and the dissemination of public information on noise was given to the EPA by Congress. Because of a lack of funding, however, EPA has been unable to carry out these activities. The study committee recommends changes that will make it easier for the federal government to improve the nation's noise climate and with it the lives of American citizens.

Recommendation 8-1: The Environmental Protection Agency should carry out its coordinating function under 42 USC 65, Section 4903. The agencies with noise-related activities include the U.S. Department of Defense, U.S. Department of Transportation, U.S. Department of Labor, U.S. Department of Commerce, U.S. Department of Health and Human Services, U.S. Department of Housing and Urban Development, and the National Science Foundation.

Recommendation 8-2: Congress should pass legislation and provide the necessary funds to establish the Environmental Protection Agency as the lead agency in the development of a cooperative effort on noise measurement, abatement, and control involving federal agencies, state governments, industry, consulting firms, and academia. An EPA office should implement 42 USC 65, Section 4903, and the legislation should expand the authority already given by Congress to ensure that the agency can effectively manage a program to meet the following objectives:

- coordination and cooperation among existing interagency groups concerned with noise
- clear delineation of the roles of federal agencies, as well as state and local governments
- assisting American industry in lowering noise levels in the U.S. workplace and developing industrial and consumer products with noise emissions that are competitive with foreign products
- development of international standards for the measurement and labeling of noise emissions
- active U.S. participation in the harmonization of noise emission requirements worldwide
- development of metrics for environmental noise that truly represent community response to noise
- ongoing assessment of the costs and benefits of noise control
- increased research on the health effects of noise, especially nonauditory effects

EDUCATE MORE NOISE CONTROL ENGINEERS

The committee reviewed the state of noise control engineering education in the United States and concludes that the nation must educate more specialists in the field and provide basic knowledge of the principles of noise control engineering to individuals trained as specialists in other engineering disciplines. Undergraduate education in noise engineering varies greatly from institution to institution, both in terms of the department in which it is housed and in the courses offered. Funding for noise control engineering programs at universities is problematic, and support for graduate students to assist in research (or teaching) and to develop a new cadre of professionals is inadequate.

The multidisciplinary nature of noise control engineering poses challenges for engineering practice and for lifelong learning. Elements of noise control engineering degree programs should be formally taught by faculty in academic units or departments (in engineering, physical sciences, and architecture) in an intra- or interdisciplinary way. Major professional societies (such as American Institute of Aeronautics and Astronautics, American Society of Mechanical Engineering, American Society of Heating, Refrigerating and Air-Conditioning Engineers, Institute of Noise Control Engineering of the USA, Society of Automotive Engineers) and other stakeholders should organize symposia (or special sessions in regular conferences) where leading academic and industry leaders can propose and refine curricula and suggest improvements in teaching methods and delivery mechanisms. Collaboration among academic, research, and industry leaders will be necessary for the development of interesting case studies or practice modules that could then be disseminated to teachers of undergraduate courses.

Funding is particularly important for research on environmental noise, which encourages interdisciplinary collaboration between acousticians, engineers, social scientists, psychologists, sociologists, and medical scientists to develop new metrics for evaluating the impact of noise, including annoyance, speech and communications interference, cognitive impairment, sleep disturbance, and health effects.

Recommendation 9-1: Academic institutions should offer an undergraduate course in noise control engineering, broaden the scope of the engineering curriculum, and increase the pool of engineering graduates who are equipped to design for low-noise emissions. The course could be offered as an elective in a bachelor's degree program or as part of a minor (e.g., in acoustics or interdisciplinary studies).

Recommendation 9-2: Graduate-level noise control courses should provide a balance between theory and engineering practice without sacrificing academic rigor. The committee strongly encourages the establishment of graduate internships in industry and government agencies and thesis research programs to motivate students and to build a cadre of future noise control engineers.

Recommendation 9-3: Federal agencies, private companies, and foundations with a stake in noise control should provide financial support for graduate students who assist in research on, and the teaching of, noise control engineering. This support is crucial for the development of noise control professionals and noise control educators.

IMPROVE PUBLIC INFORMATION ON THE EFFECTS OF NOISE AND NOISE CONTROL

The U.S. Code (42 U.S.C. Section 4913) requires that EPA "develop and disseminate information and educational materials to all segments of the public on the public health and other effects of noise and the most effective means for noise control, through the use of materials for school curricula, volunteer organizations, radio and television programs, publication, and other means." At this time, however, EPA does not have the internal resources to create a large public information program, and it is likely that much of the effort will have to be done through contractors.

The labeling of product noise emission levels should be a critical aspect of a program designed to benefit the public and enable people to make informed purchasing decisions. Although EPA has labeling authority, it is more practical for professional organizations, trade associations, and standards organizations to develop labeling methodology for specific products because of the wide variety of products and noise measurement methods.

Recommendation 10-1: The Environmental Protection Agency should take the following actions under the authority of 42 USC 65, Section 4913, to improve public information and education on the effects of noise and the most effective means of controlling noise:

- Conduct a survey of all activities by federal agencies related to noise, and publish URLs that provide information of interest to the public.
- Develop a categorized list of stakeholders with interests in noise (e.g., professional societies, scientific societies, citizens groups).
- Help organize a coalition of current stakeholders with the goal of improving the availability of information on noise to the public.
- Develop educational materials to inform the public of the health effects of noise, especially noise-induced hearing loss and cardiovascular effects.
- Develop information to help the public understand the benefits of using personal hearing protection devices.
- Provide information on the selection and use of hearing protection devices, making intelligent decisions about frequenting high noise exposure events, the importance of reducing noise exposures by buying quieter products, and being vigilant and active in public policy decision making about community noise zoning issues.

Recommendation 10-2: Engineering professional societies such as the American Institute of Aeronautics and Astronautics, the American Society of Mechanical Engineering, American Society of Heating, Refrigerating and Air-Conditioning Engineers, Society of Automotive Engineers, and Institute of Noise Control Engineering of the USA should develop engineering information on noise control to help the public understand techniques for reducing noise emissions.

REFERENCE

EPA (U.S. Environmental Protection Agency). 1974. Information on Levels of Environmental Noise Requisite to Protect Public Health and Welfare with an Adequate Margin of Safety. Document 550/9-74-004. Available online at *http://www.nonoise.org/library/levels74/levels74.htm*.

Appendixes

Appendix A

Basic Concepts in Acoustics and Noise

This appendix is not intended to give a complete description of all of the quantities used in acoustics and noise control—information that is available in a wide variety of textbooks and handbooks (e.g., Rossing, 2007; Vér and Beranek, 2006; Crocker, 2007).

A few key concepts are described in this appendix:

- *immission* and *emission*
- quantities used in noise control
- frequency weighting
- levels and the decibel

IMMISSION VERSUS EMISSION

When most people mention noise levels, they are speaking of *immission*—the sound they hear. The sound may come from a specific source or from a number of sources at the same time. There is little distinction between the two. That is why the sound pressure level in decibels is used as the descriptor. It is, however, necessary to make a clear distinction between sound emitted by a source (i.e., noise *emission*) and the sound heard by an observer (i.e., noise *immission*). The former is relatively independent of the environment in which the noise source is located (outdoors, in a room, etc.). There are standard methods of determining the noise emission of stationary sources as well as of moving sources such as cars, trucks, and airplanes.

Noise *immission* may come from several sources and is always dependent on the environment in which the sources are located. The position of a source in a room, the size of the room, and the amount of sound absorption in the room all influence noise *immission*. Outdoors, *immission* levels can be influenced by the nature of the terrain, sound absorption by the ground, and wind and temperature gradients—among other effects.

Quantities Used in Noise Control

Sound pressure is the small variation above and below atmospheric pressure created by the passage of a sound wave; this is what most people think of as noise. Pressure sensed by a microphone on a sound-level meter is generally converted to a mean square pressure or pressure level by the measuring instrument. The level indicated by the sound-level meter fluctuates depending on the averaging time of the measuring system. More details are given in the section below.

In some cases, the sound pressure can be used as a metric for the noise *emission* of a source. The sound pressure may be converted into a metric that more closely relates to human response—such as the *effective perceived noise level* used to specify the noise emissions of airplanes. Or it may be the *maximum sound pressure level* during a vehicle pass-by under controlled measurement conditions. A more common descriptor of noise emission for stationary sources is the *sound power level*, a measure of the total sound energy emitted by a source. *Sound intensity*, the power per unit area, can be determined—usually by a measurement of pressure gradient—by instrumentation systems and is now used to determine noise emission by tire/road interaction. The method is called the onboard sound intensity method.

Frequency Weighting

The *sound pressure* as measured by a microphone varies in time and can also be described in terms of the frequency of the sound. The ear has different sensitivities to sounds of different frequencies, and a frequency weighting is often applied to the signal to make it more representative of the sound perceived by a listener. The most common weighting is *A-weighting*, which was originally derived in the 1930s by determining the loudness of sounds. The A-weighting curve is described in most textbooks and handbooks on acoustics and noise control, and frequency weighting in general—including weighting curves—is described in the online encyclopedia *Wikipedia* (*http://en.wikipedia.org/wiki/Frequency_weighting*). Octave and one-third octave frequency bands (*http://www.diracdelta.co.uk/science/source/o/c/octave/source.html*) are also used for a more complete description of the frequency spectrum of noise (see examples in this report).

The Decibel

The *decibel*, unfortunately for public comprehension, is used in a variety of ways in noise control and other branches of engineering. That it involves a logarithm makes math-averse individuals uncomfortable. The decibel was originally used in the Bell telephone system to describe the attenuation of a mile of "standard cable." It is also commonly used to describe the gain of an amplifier and the power delivered to an electrical load. The online encyclopedia *Wikipedia* is a good source of information for a basic understanding of the concept (*http://en.wikipedia.org/wiki/Decibel*).

The decibel is firmly entrenched in the language of noise, as in "how many decibels of noise is that?" Noise "thermometers" are frequently published showing the decibel level of noise for various sources. Examples are given in Chapter 1. These levels are almost always measures of noise *immission*.

Fundamentally, the decibel is a unit of level and is defined as $10 \log Q/Q_{ref}$, where Q is a quantity related to energy and Q_{ref} is a reference quantity. It is the fact that both Q and Q_{ref} can be different quantities (squared pressure, power, intensity, etc.) that makes general use of the decibel even more confusing to the public. The mean square pressure is the quantity most commonly used to describe noise, and its corresponding reference quantity is (20 micropascals)^2 or, in terms of Newton per meter squared, $(2 \cdot 10^{-5}\ N/m^2)^2$. Given the range of mean square pressures commonly encountered when dealing with noise, the sound pressure level generally ranges from about 0 to 140 dB. The corresponding pressures are only a tiny fraction of atmospheric pressure. Although the A-frequency weighting described above applies to the signal and not to the unit (dB), the A-weighted sound pressure level is often expressed as dB(A) or dBA.

Even with one definition of Q as the mean square pressure, different averaging times lead to different decibel values—which causes further complication. For example, in the evaluation of hazardous noise in the workplace, an 8-hour average is commonly used. For environmental noise outdoors, a day-night average sound level is computed by using A-frequency weighting and averaging the mean square pressure over 24 hours with an increase in the amplification of the measuring system of 10 dB during the nighttime hours.[1] This quantity is the day-night average sound level, L_{dn} (DNL). To add further complication, it is common European practice to use a 5-dB amplification in the measuring system during the evening hours and a 10-dB gain during the nighttime hours. The result is the day-evening-night level, L_{den}.

Another important quantity is sound exposure and the corresponding *sound exposure level* in decibels. This measure is useful for assessing the noise produced by single events such as an airplane flyover or vehicle pass-by. Here, the quantity Q is the time integral of the squared pressure over the time interval of the event. The reference quantities are 20 micropascals as the reference pressure and 1 second as the reference time.

The decibel is also used in noise control for *sound intensity* and *sound power*, which are common descriptors of noise *emission*. For sound intensity level, the quantity is sound intensity and the reference quantity is $10^{-12}\ W/m^2$. For sound power level, the quantity is sound power and the reference quantity is $10^{-12}\ W$. In the information technology industry, the sound power level is commonly expressed in bels, B (10 dB = 1 B) to avoid confusion between sound pressure level and sound power level. This has not been widely adopted, however. For example, European requirements on outdoor equipment are based on the sound power level in decibels.

Sound Level

Throughout this report, the terms *sound pressure level*, *sound intensity level*, and *sound power level* are used to clarify which level is being discussed. The term *sound level* is sometimes used when sound pressure is implied—such as in day-night average sound level. It is also used in connection with instruments—such as sound-level meter—and when the quantity being discussed could be either pressure, intensity, or power.

REFERENCES

Crocker, M.J., ed. 2007. Handbook of Noise and Vibration Control. Hoboken, N.J.: John Wiley and Sons.

Rossing, T., ed. 2007. Springer Handbook on Acoustics. New York: Springer Science+Business Media LLC.

Vér, I.L., and L.L. Beranek, eds. 2006. Noise and Vibration Control Engineering. New York: John Wiley and Sons. See, for example, Beranek, L.L., Chapter 1, Basic Acoustical Quantities: Levels and Decibels.

[1]Two different uses of the decibel in one sentence!

Appendix B

International Activities Relative to Quiet Areas

The *soundscape*, as defined in *Wikipedia*, is a sound (or combination of sounds) that forms or arises from an immersive environment (*http://en.wikipedia.org/wiki/Soundscape*). Some dimensions of the soundscape can be quantified, and others cannot. In Chapters 2 and 3, U.S. activities related to noise in quiet areas is described. This appendix describes two international efforts to describe preferences and tranquility in quiet areas.

COUNTRYSIDE PREFERENCES IN HONG KONG

A recent study was done in Hong Kong of human preferences in countryside soundscapes. Based on questionnaires, interviews, and recordings taken during interviews (Lam et al., 2008), there was a clear preference for countryside sound sources; natural sounds were preferable to man-made sounds. The order of preference was found to be:

- running water
- bird
- wave
- waterfall
- wind
- insect
- other animals
- human
- road traffic

Aircraft noise is not listed, perhaps because the Hong Kong airport is on Lantau Island, not Hong Kong Island.

The sound recordings were also analyzed according to A-weighted levels and sound quality metrics, but no strong correlation between preference and acoustical quantities was found. This does not mean that acoustical quantities are unimportant; it may mean that the appropriate metric for these quantities has not been found. The authors conclude:

> In summary, the study of countryside soundscapes in Hong Kong shows that the sound pressure level and other acoustical and sound quality parameters are not good indicators of soundscape preference. The presence or absence of natural and man-made sounds is a more important determinant of human preference for countryside soundscapes.

TRANQUILITY IN ENGLAND

In the United Kingdom, the Campaign to Protect Rural England (CPRE) has done extensive work related to tranquility (CPRE, 1995). Tranquility—partly the landscape, partly the soundscape, and partly human experience—is a difficult concept to express in numerical terms. Nevertheless, the CPRE has developed a method based on the results of questionnaires and the identification of factors that contribute to tranquility. Although the algorithm used to determine the numerical value is not given on the Internet site, an attempt was made to assign a tranquility value for every 500 X 500 meter area of England. Maps are given on the Internet site, and sounds may be downloaded.

Based on surveys, the Internet site defines the 10 top factors that contribute and do not contribute to tranquility:

What tranquility is:

1. Seeing a natural landscape
2. Hearing a bird sing
3. Having peace and quiet
4. Seeing natural-looking woodland
5. Seeing the stars at night
6. Seeing streams
7. Seeing the sea
8. Hearing natural sounds
9. Hearing wildlife
10. Hearing running water

What tranquility is not:

1. Hearing constant noise from cars, lorries, and/or motorbikes
2. Seeing lots of people

3. Seeing urban development
4. Seeing overhead light pollution
5. Hearing lots of people
6. Seeing low-flying aircraft
7. Hearing low-flying aircraft
8. Seeing power lines
9. Seeing towns and cities
10. Seeing roads

CONCLUSIONS

A physical description of the soundscape is one input to the assessment of the human experience, even though it may be described as an overall good experience, as a preference, or as a tranquil environment. One has to distinguish clearly between man-made sounds and natural sounds in determining their acoustical impact in rural and naturally quiet areas. Amplitude and duration are also important. For example, a bubbling brook and waves crashing into the seacoast may, on average, be equally preferable, even though the amplitude of the latter is much greater than the amplitude of the former.

REFERENCES

CPRE, 1995. Tranquility. Campaign to Protect Rural England. Available online at *http://www.cpre.org.uk/campaigns/landscape/tranquillity*.

Lam, K-C., K-C. Chau, L.M. Marafa, and L. Brown. 2008. Human Preference for Countryside Soundscapes. Presentation at INTER-NOISE 2008, The 2008 International Congress and Exposition on Noise Control Engineering, Shanghai, China, October 26–29. Available online at *http://scitation.aip.org/journals/doc/INCEDL-home/cp/*.

Appendix C

Additional Information on Standards Activities

Accredited Standards Committees

The Acoustical Society of America (ASA) administers four accredited standards committees under contract with the American National Standards Institute (ANSI). See *http://www.acosoc.org/standards*. The four committees deal with noise and its effects:

- Accredited Standards Committee S12 Noise,
- Accredited Standards Committee S1 Acoustics,
- Accredited Standards Committee S2 Mechanical Vibration and Shock, and
- Accredited Standards Committee S3 Bioacoustics.

These committees are referred to as accredited standards committees (ASCS) because they are accredited by ANSI and function according to operating procedures approved by ANSI. To maintain their accredited status, the operation of the committees is subject to periodic audits by ANSI. These committees are composed of organizational members who vote on proposed consensus standards; criteria and dues for membership are established by each committee, and balance of interested parties is required. As of November 2008, S12 had 48 organizational members, S1 had 17 organizational members, S2 had 38 organizational members, and S3 had 27 organizational members. Each ASA ASC also has individual experts who assist each committee by providing technical expertise.

ASC S12 Scope: Standards, specifications, and terminology in the field of acoustical noise pertaining to methods of measurement, evaluation, and control, including biological safety, tolerance and comfort, and physical acoustics as related to environmental and occupational noise.

ASC S1 Scope: Standards, specifications, methods of measurement and test, and terminology in the field of physical acoustics, including architectural acoustics, electroacoustics, sonics and ultrasonics, and underwater sound, but excluding those aspects which pertain to biological safety, tolerance, and comfort.

ASC S2 Scope: Standards, specification, methods of measurement and test, and terminology in the field of mechanical vibration and shock, and condition monitoring and diagnostics of machines, including the effects of exposure to mechanical vibration and shock on humans, including those aspects which pertain to biological safety and tolerance and comfort.

ASC S3 Scope: Standards, specifications, methods of measurement and test, and terminology in the fields of psychological and physiological acoustics, including aspects of general acoustics, which pertain to biological safety and tolerance and comfort.

ASA pays a substantial annual fee to ANSI to be allowed to administer these committees. The ASA does this as a public service in furtherance of its mission to "increase and diffuse the knowledge of acoustics and promote its practical application."

Other organizations have ANSI-approved accredited standards committees that deal with noise; some of their standards are listed below. Many of these standards rely on basic acoustic or noise standards developed by ASA ASC S1, S12, S2, and S3.

Association of Home Appliance Manufacturers (AHAM)
Air Movement and Control Association International, Inc. (AMCA)
Air-Conditioning and Refrigeration Institute (ARI)
Alliance for Telecommunication Industry Solutions (ATIS)
American Society of Heating, Refrigeration and Air-Conditioning Engineers (ASHRAE)

American Society for Testing and Materials (ASTM)
Society of Automotive Engineers (SAE)
Institute of Electrical and Electronics Engineers (IEEE)
Underwriters Laboratory (UL)

Other organizations promulgate standards or recommended practices that are not accredited by ANSI and are not American National Standards. Such organizations include the Institute of Noise Control Engineering, the Audio Engineering Society (AES), and others, including trade associations.

Role of U.S. Technical Advisory Groups

The ASA also administers nine U.S. Technical Advisory Groups (U.S. TAGs). These U.S TAGs review international documents—primarily draft standards—and develop the U.S. position on them. They also provide the pool of candidates for appointment to international working groups and volunteers to coordinate and prepare draft U.S. positions (*http://www.acosoc.org/standards/*). The TAGs are also ANSI-accredited. The nine TAGs are:

- U.S. TAG for ISO/TC 43, Acoustics
- U.S. TAG for ISO/TC 43/SC 1, Noise
- U.S. TAG for ISO/TC 108, Mechanical vibration, shock, and condition monitoring
- U.S. TAG for ISO/TC 108/SC 2, Measurement and evaluation of mechanical vibration and shock as applied to machines, vehicles, and structures
- U.S. TAG for ISO/TC 108/SC 3, Use and calibration of vibration and shock measuring instruments
- U.S. TAG for ISO/TC 108/SC 4, Human exposure to mechanical vibration and shock
- U.S. TAG for ISO/TC 108/SC 5, Condition monitoring and diagnostics of machines
- U.S. TAG for ISO/TC 108/SC 6, Vibration- and shock-generating systems
- U.S. TAG for IEC/TC 29, Electroacoustics

As with the S committees, ASA pays a substantial annual fee to ANSI to be allowed to administer the U.S. TAGs in these subject areas. The TAGs are also accredited by ANSI and must follow their accredited operating procedures.

ISO Technical Committees with an Interest in Noise or Sound

The following committees of the International Organization for Standardization (*http://www.iso.org*) have an interest in noise or sound:

TC 4	Roller bearings
TC 21/SC 3	Equipment for fire protection and fire fighting/fire detection and alarm systems
TC 22/SC 22	Motorcycles
TC 23/ SC 2	Tractors and machinery for agricultural forestry/common tests
TC 23/SC 3	Tractors and machinery for agricultural forestry/safety and comfort
TC 23/SC 17	Tractors and machinery for agricultural forestry/manually portable forest machinery
TC 36	Cinematography
TC 39/SC 6	Machine tools/noise of machine tools
TC 43/SC 1	Acoustics/noise
TC 43/SC 2	Acoustics/building acoustics
TC 60	Gears
TC 70	Internal combustion engines
TC 72/SC 8	Textile machinery and accessories/safety requirements for textile machinery
TC 86	Refrigeration and air conditioning
TC 86/SC 3	Testing and rating of factory-made refrigeration systems (excluding systems covered by Subcommittees 5, 6, and 7)
TC 86/SC 5	Refrigeration and air conditioning/testing and rating of household refrigeration appliances
TC 86/SC 6	Factory-made air-cooled air-conditioning and air-to-air heat pump units
TC 108/SC 2	Measurement and evaluation of mechanical vibration and shock as applied to machines, vehicles, and structures
TC 115	Pumps
TC 117	Industrial fans
TC 118/SC 3	Compressors and pneumatic tools, machines, and equipment/pneumatic tools and machines
TC 118/SC 6	Compressors and pneumatic tools, machines, and equipment/air compressors and compressed air systems
TC 127/SC 2	Earth-moving machinery/safety, ergonomics, and general requirements
TC 131/SC 8	Fluid power systems/product testing
TC 160/SC 2	Glass in buildings/use considerations
TC 188	Small craft

ISO TCs that Develop Standards Related to Noise

TC 2	Rotating machinery
TC 5	Steam turbines (in standby)
TC 59	Performance of household and similar appliances
TC 65	Industrial process measurement, control, and automation
TC 88	Wind turbines

IEC-Developed Standards Related to Noise

Examples of standards related to noise developed by the International Electrotechnical Commission (*http://www.iec.ch*) are:

IEC 60034-9, 60034-9am1, Noise limits—rotating electrical machines (developed by TC 2)

IEC 61063 Acoustics—Measurement of airborne noise emitted by steam turbines and driven machinery (Based on ISO 3766, but under good conditions ISO 3744 may be used; developed by TC 5)

IEC 60704 Household and similar electrical appliances—Test code for the determination of airborne acoustical noise (based on ISO 3743-1, 3743-2, and ISO 3744; developed by IEC TC 59, Performance of household and similar appliances, as well as various subcommittees)

Parts of Standard

-1	General requirements
-2-1	Vacuum cleaners
-2-2	Forced draught convection heaters
-2-3	Dishwashers
-2-3am1	Dishwashers
-2-4	Washing machines and spin extractors
-2-5	Thermal storage room heaters
-2-6	Tumble dryers
-2-7	Fans
-2-8	Electric shavers
-2-9	Electric care appliances
-2-10	Ovens, grills, etc.
-2-11	Electrically operated food preparation devices
-2-13	Range hoods
-2-2-13am1	Range hoods
-3	Procedures for determining and verifying declared noise emission values

IEC 60534-8-1/-8-2/-8-3, Industrial processes—control valves. Measurement of noise by control valve noise–aerodynamic/hydrodynamic/flow through (developed by TC 65 and various subcommittees)

IEC 61400-11 Wind turbine generator systems, Part 11. Acoustic noise measurement techniques (developed by TC 88: Wind Turbines)

Appendix D

Relevant Portions of the U.S. Code

PART A. SECTION 4903 OF 42USC65

From the U.S. Code Online via GPO Access
[*www.gpoaccess.gov/uscode*]
[Laws in effect as of January 3, 2006]
[**CITE**: **42USC4903**]

TITLE 42—THE PUBLIC HEALTH AND WELFARE

CHAPTER 65—NOISE CONTROL

Sec. 4903. Federal programs

(a) Furtherance of Congressional policy

The Congress authorizes and directs that Federal agencies shall, to the fullest extent consistent with their authority under Federal laws administered by them, carry out the programs within their control in such a manner as to further the policy declared in section 4901(b) of this title.

(b) Presidential authority to exempt activities or facilities from compliance requirements

Each department, agency, or instrumentality of the executive, legislative, and judicial branches of the Federal Government—

(1) having jurisdiction over any property or facility, or

(2) engaged in any activity resulting, or which may result, in the emission of noise,

shall comply with Federal, State, interstate, and local requirements respecting control and abatement of environmental noise to the same extent that any person is subject to such requirements. The President may exempt any single activity or facility, including noise emission sources or classes thereof, of any department, agency, or instrumentality in the executive branch from compliance with any such requirement if he determines it to be in the paramount interest of the United States to do so; except that no exemption, other than for those products referred to in section 4902(3)(B) of this title, may be granted from the requirements of sections 4905, 4916, and 4917 of this title. No such exemption shall be granted due to lack of appropriation unless the President shall have specifically requested such appropriation as a part of the budgetary process and the Congress shall have failed to make available such requested appropriation. Any exemption shall be for a period not in excess of one year, but additional exemptions may be granted for periods of not to exceed one year upon the President's making a new determination. The President shall report each January to the Congress all exemptions from the requirements of this section granted during the preceding calendar year, together with his reason for granting such exemption.

(c) Coordination of programs of Federal agencies; standards and regulations; status reports

(1) The Administrator shall coordinate the programs of all Federal agencies relating to noise research and noise control. Each Federal agency shall, upon request, furnish to the Administrator such information as he may reasonably require to determine the nature, scope, and results of the noise-research and noise-control programs of the agency.

(2) Each Federal agency shall consult with the Administrator in prescribing standards or regulations respecting noise. If at any time the Administrator has reason to believe that a standard or regulation, or any proposed standard or regulation, of any Federal agency respecting noise does not protect the public health and welfare to the extent he believes to be required and feasible, he may request such agency to review and report to him on the advisability of revising such standard or regulation to provide such protection. Any such request may be published in the Federal Register and shall be accompanied by a detailed statement of the information on

which it is based. Such agency shall complete the requested review and report to the Administrator within such time as the Administrator specifies in the request, but such time specified may not be less than ninety days from the date the request was made. The report shall be published in the Federal Register and shall be accompanied by a detailed statement of the findings and conclusions of the agency respecting the revision of its standard or regulation. With respect to the Federal Aviation Administration, section 44715 of title 49 shall apply in lieu of this paragraph.

(3) On the basis of regular consultation with appropriate Federal agencies, the Administrator shall compile and publish, from time to time, a report on the status and progress of Federal activities relating to noise research and noise control. This report shall describe the noise-control programs of each Federal agency and assess the contributions of those programs to the Federal Government's overall efforts to control noise.

(Pub. L. 92-574, Sec. 4, Oct. 27, 1972, 86 Stat. 1235.)

Codification

In subsec. (c)(2), "section 44715 of title 49" substituted for "section 611 of the Federal Aviation Act of 1958 (as amended by section 7 of this Act)" on authority of Pub. L. 103-272, Sec. 6(b), July 5, 1994, 108 Stat. 1378, the first section of which enacted subtitles II, III, and V to X of Title 49, Transportation.

Termination of Reporting Requirements

For termination, effective May 15, 2000, of provisions in subsec. (b) of this section relating to annual report to Congress, see section 3003 of Pub. L. 104-66, as amended, set out as a note under section 1113 of Title 31, Money and Finance, and item 7 on page 20 of House Document No. 103-7.

PART B. SECTION 4913 OF 42USC65

From the U.S. Code Online via GPO Access
[www.gpoaccess.gov/uscode/]
[Laws in effect as of January 3, 2006]
[**CITE**: **42USC4913**]

TITLE 42—THE PUBLIC HEALTH AND WELFARE

CHAPTER 65—NOISE CONTROL

Sec. 4913. Quiet communities, research, and public information

To promote the development of effective State and local noise control programs, to provide an adequate Federal noise control research program designed to meet the objectives of this chapter, and to otherwise carry out the policy of this chapter, the Administrator shall, in cooperation with other Federal agencies and through the use of grants, contracts, and direct Federal actions—

(a) develop and disseminate information and educational materials to all segments of the public on the public health and other effects of noise and the most effective means for noise control, through the use of materials for school curricula, volunteer organizations, radio and television programs, publication, and other means;

(b) conduct or finance research directly or with any public or private organization or any person on the effects, measurement, and control of noise, including but not limited to—

(1) investigation of the psychological and physiological effects of noise on humans and the effects of noise on domestic animals, wildlife, and property, and the determination of dose/response relationships suitable for use in decisionmaking, with special emphasis on the nonauditory effects of noise;

(2) investigation, development, and demonstration of noise control technology for products subject to possible regulation under sections 4905 and 4907 of this title and section 44715 of title 49;

(3) investigation, development, and demonstration of monitoring equipment and other technology especially suited for use by State and local noise control programs;

(4) investigation of the economic impact of noise on property and human activities; and

(5) investigation and demonstration of the use of economic incentives (including emission charges) in the control of noise;

(c) administer a nationwide Quiet Communities Program which shall include, but not be limited to—

(1) grants to States, local governments, and authorized regional planning agencies for the purpose of—

(A) identifying and determining the nature and extent of the noise problem within the subject jurisdiction;

(B) planning, developing, and establishing a noise control capacity in such jurisdiction, including purchasing initial equipment;

(C) developing abatement plans for areas around major transportation facilities (including airports, highways, and rail yards) and other major stationary sources of noise, and, where appropriate, for the facility or source itself; and,

(D) evaluating techniques for controlling

noise (including institutional arrangements) and demonstrating the best available techniques in such jurisdiction;

(2) purchase of monitoring and other equipment for loan to State and local noise control programs to meet special needs or assist in the beginning implementation of a noise control program or project;

(3) development and implementation of a quality assurance program for equipment and monitoring procedures of State and local noise control programs to help communities assure that their data collection activities are accurate;

(4) conduct of studies and demonstrations to determine the resource and personnel needs of States and local governments required for the establishment and implementation of effective noise abatement and control programs; and

(5) development of education and training materials and programs, including national and regional workshops, to support State and local noise abatement and control programs;

except that no actions, plans or programs hereunder shall be inconsistent with existing Federal authority under this chapter to regulate sources of noise in interstate commerce;

(d) develop and implement a national noise environmental assessment program to identify trends in noise exposure and response, ambient levels, and compliance data and to determine otherwise the effectiveness of noise abatement actions through the collection of physical, social, and human response data;

(e) establish regional technical assistance centers which use the capabilities of university and private organizations to assist State and local noise control programs;

(f) provide technical assistance to State and local governments to facilitate their development and enforcement of noise control, including direct onsite assistance of agency or other personnel with technical expertise, and preparation of model State or local legislation for noise control; and

(g) provide for the maximum use in programs assisted under this section of senior citizens and persons eligible for participation in programs under the Older Americans Act [42 U.S.C. 3001 et seq.].

(Pub. L. 92-574, Sec. 14, Oct. 27, 1972, 86 Stat. 1244; Pub. L. 95-609, Sec. 2, Nov. 8, 1978, 92 Stat. 3079.)

References in Text

The Older Americans Act, referred to in subsec. (g), probably means the Older Americans Act of 1965, Pub. L. 89-73, July 14, 1965, 79 Stat. 218, as amended, which is classified generally to chapter 35 (Sec. 3001 et seq.) of this title. For complete classification of this Act to the Code, see Short Title note set out under section 3001 of this title and Tables.

Codification

In subsec. (b)(2), "section 44715 of title 49" substituted for reference to section 7 of this Act, meaning section 7 of Pub. L. 92-574, which generally amended section 611 of the Federal Aviation Act of 1958 (49 App. U.S.C. 1431), on authority of Pub. L. 103-272, Sec. 6(b), July 5, 1994, 108 Stat. 1378, the first section of which enacted subtitles II, III, and V to X of Title 49, Transportation.

Amendments

1978—Pub. L. 95-609 completely revised and restructured existing provisions, inserting provisions relating to authorized use of grants and direct action, investigation of economic impact of noise, administration of Quiet Communities Program, development of noise assessment program, establishment of regional centers, technical assistance to State and local governments, and use by senior citizens of these programs.

Appendix E

Modern Instrumentation for Environmental Noise Measurement

The science of environmental noise measurement has progressed rapidly in the past decade as computer technology has come online to provide rapid data acquisition and analysis in small portable packages. The end result has been a revolution in the type and complexity of measurements and calculations that can be made in analyzing environmental noise. The bulk of this discussion will be focused on the capabilities of modern measurement systems. Both sound-level meters and monitoring systems will be discussed in some detail. Finally, a summary of the findings from this analysis will be presented as to the capabilities and limitations of current measurements. Whether there are restrictions on the types of metrics that could be utilized in defining and limiting environmental noise will also be discussed.

The equations used to calculate the various metrics are not discussed here. The American National Standards Institute has a series of standards, the S12.9 series, listed as references in this appendix. Part 4 is particularly relevant to the mathematical definition of metrics for community noise. These standards are developed by ANSI Committee S12—Noise and are available through the Acoustical Society of America (ASA, 2010).

Sound-level meters and related filter characteristics have been standardized by the American National Standards Institute and are also available through the Acoustical Society of America (*http://asastore.aip.org/shop.do?cID=7*). International standards on the same subjects are developed by the International Electrotechnical Commission Technical Committee 29—Electroacoustics (IEC, 2010).

SOUND-LEVEL METERS

The Brüel & Kjær Type 2270 sound-level meter is a modern instrument and will be used as the typical example for this discussion. Other manufacturers make instruments with similar capabilities. This is an integrating sound-level meter with the ability to compute sound energy summations. This is the standard sort of capability found in high-end sound-level meters. There is a large amount of computing power using microprocessors built into the unit. This allows for sophisticated analysis, data communication, and programming. Figures E-1 and E-2 show examples of the screen display and use of this meter.

The types or measurements possible with this meter are listed below:

- for display and storage: L_{dn}, L_{den}, L_{day}, $L_{evening}$, and L_{night}
- selectable day, evening and night periods and penalties

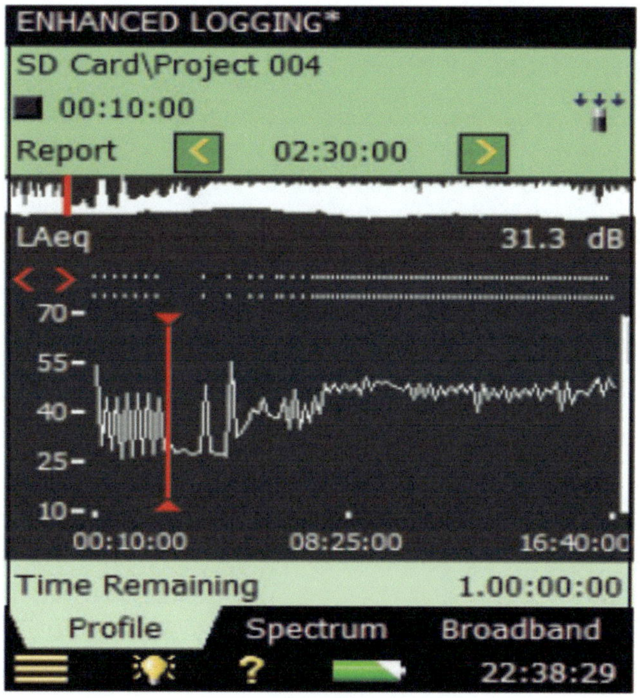

FIGURE E-1 Screen display of discrete frequency analysis for Type 2270 monitor. Copyright © Brüel & Kjær.

FIGURE E-2 Type 2270 meter in use. Copyright © Brüel & Kjær.

- report period: from 1 minute to 24 hours with 1-minute resolution
- all broadband data and statistics stored at each reporting interval
- all spectrum data stored at each reporting interval
- spectral statistics stored at each reporting interval
- logging time: from 1 second to 31 days with 1 second resolution or continuous
- data are saved in separate projects for every 24 hours of logging

NOISE MONITORING SYSTEMS

The sound-level meter market can be divided into three levels. The high end is the integrating analyzer with a tremendous amount of computing power. Some manufacturers use a laptop computer attached to the instrument for this category. At the lowest level, these instruments are meters that can merely report a sound level.

The Brüel & Kjær Type 3639 Noise Monitoring Terminal will be used as the example for this discussion. There are several competing products by other manufacturers with very similar capabilities. The Type 3639 is designed for use in all climate environments, as well as industrial, urban, and rural conditions. It can be left unattended as part of an environmental noise monitoring system for permanent, mobile, or semipermanent monitoring. The Noise Monitoring Terminal can be controlled by a remote PC. This unit is shown in Figure E-3.

Typical capabilities for these types of monitoring systems are summarized here:

- logging of broadband and 1/3-octave parameters every second or half-second

FIGURE E-3 Type 3639 monitoring station. Copyright © Brüel & Kjær.

- postprocessing that can create periodic statistical reports down to 1 minute, including L_N data
- GPS support
- sound recording
- weather data monitoring
- camera support
- remote operation via LAN, public telephone lines, mobile phone, or General Packet Radio Service (GPRS)

The definition of the measurements that can be done with these types of monitoring stations is quite lengthy. A short summary is presented here of typical calculations.

Broadband values:

- X = frequency weightings A and C, or A and Linear, or C and Linear (two weightings simultaneously)
- Y = time weightings Fast, Slow and Impulse (all simultaneously)
- L_{Xeq}, L_{Xpeak}, L_{Xim}, L_{XYinst}, $L_{XYmax/SPL}$, L_{XYmin}

Spectrum values:

- equivalent continuous level (L_{eq}) and I-weighted value also selectable (L_{AIeq})
- 1/3-octave frequency range: 12.5 to 20 kHz

Events:

- settings: hourly, user-defined on the following parameters:
- detection: separate event start/stop triggers
- event start trigger: L_{eq} or SPL with minimum threshold exceedence duration
- event stop trigger: L_{eq} or SPL with minimum threshold exceedence duration

Table E-1 is a summary of the hardware options offered by the Brüel & Kjær system to illustrate the variety available. In addition, there is a wide array of software options as shown in Table E-2. One will note that all the energy summation options and most of the other metrics noted previously are included in the available software.

MONITORING DEMAND DRIVERS—URBAN NOISE

The drivers of the demand for monitoring systems have been studied extensively by manufacturers, including Brüel & Kjær (Denmark), Lochard (Australia, mostly focused on monitoring airport noise), 01dB-Metravib (France, mostly focused on monitoring urban noise), and Norsonic/Topsonic (Germany and Norway, a software partner of Norsonic on large systems).

A primary influence is the need to meet the monitoring requirement dictated by legislation and standards. Implementation of European Union environmental noise directive 2002/49/EC and similar statutes in other parts of the world has also been a major driver for the acquisition and use of monitoring systems. Cities, counties, countries, and industries are obliged to follow the national and local legislation and the standard that defines measurement and estimation of noise in the environment. In many instances the demand is driven by a desire to have a positive relationship with the public. There is also increased attention to quality of life. Public pressure on noisy transportation systems (roads and rail), industries (metal, chemical, mining, and construction), and communities to manage and inform on environmental issues has been a driver for the use of monitoring systems in the urban environment.

MONITORING DEMAND DRIVERS—AIRPORT NOISE

For airports, the major driver in the use of monitoring systems is to optimize profit or capacity. By carefully monitoring noise, airports can increase movements and hence profits by increasing the environmental capacity. It allows them to optimize the capacity utilization. An airport can also postpone or even avoid the need for new infrastructure such as runways, taxiways, or terminals by maximizing use of the available land and runways.

It is also necessary to manage relationships with regulatory bodies. This includes the use of monitoring equipment with regard to legislation, standards such as ISO 1996 Environmental Noise Assessment, Part 1 (definitions) published

TABLE E-1 Hardware Options for Brüel & Kjær Monitoring Systems

Application	Key Features	Products and Their Key Features #)
Airport Noise Monitoring	1) L_{Aeq} Lmax	4198 Outdoor Microphone Unit 6)
Urban Noise Monitoring	2) Statistics LN	4184 Weatherproof Microphone Unit 6)
Plant Noise Monitoring	3) 1/3 Octave spectra	3631 Portable Noise Monitoring Terminal 1) 2) 16) 17)
	4) Event trigger	3637 Portable Noise Monitoring Terminal 1) 2) 3) 4) 5) 6) 7) 9) 16) 17)
	5) Sound/Video recording	3597 Permanent Noise Monitoring Terminal 1) 2) 3) 4) 5) 6) 7) 9) 16)
	6) Automatic Calibration (CIC)	7802 Noise Monitoring Software 4) 8) 10) 14) 15) 16)
	7) Automatic location (GPS)	7840 Noise Monitoring Software 8) 10) 14) 15) 16)
	8) Communication with NMT	7832 Reporting Module 11)
	9) EPNL	7833 Complaints Module 12)
	10) Database management	7804 Flight Tracking Option 13) 14)
	11) Reporting	7834 INM Link 15)
	12) Complaints handling	
	13) Correlation with flights	
	14) GIS Interface	
	15) Prediction	
	16) Weather information	

TABLE E-2 Software Options for Brüel & Kjær Monitoring Systems

Application	Key Features	Products and Their Key Features #)
Modelling outdoor noise	Noise level mapping Environmental impact assessments Scenario comparison Fulfil EU IPPC 1996/61/EEC	7810 Predictor 7812 Lima
Large-scale noise mapping	Fulfil EU END 2002/49/EC (for example, interim methods) Large-scale data handling Interface with external databases and software	7812 Lima
Measuring sound powers of noise sources Validating calculations	1) Sound intensity method 2) Sound pressure method 3) L_{Aeq} 4) L_{DEN} 5) GPS position	2260 Investigator 1) 2) 3) 5) 2260 Observer 2) 3) 5) 2250 Hand-held Analyzer 2) 3) 7816 Acoustic Determinator 2) 3) 3637 Noise Monitoring Terminal 3) 5) 3631 Noise Monitoring Terminal 3) 3597 Noise Monitoring Terminal 3) 5) 7802/40 Noise Monitoring Software 3) 4) 5)
Airport noise maps	Noise level mapping Import of actual flight information	7834 INM Link
Modelling aircraft noise	Noise level mapping Footprints and time histories	7812 Lima

2003, Part 2 (assessment techniques) DIS 2005, and ISO 20906 Aircraft Noise Monitoring (major revision of ISO 3891-1978). Finally, there is the need to manage relationships with adjacent communities. One way to accomplish this is to monitor and be able to provide noise levels to refute complaints and to demonstrate action to monitor and control noise levels.

MONITORING MARKET SEGMENTATION

One way to understand how this market is segmented is to look at the interests of customers:

1. Airport noise
2. Urban noise
3. City noise
4. Road noise
5. Railway noise
6. Industry—internal (facilities) and external (products)
7. Construction sites
8. Recreational areas

In each customer segment the buyer can be either the final customer, a consultant, or a system integrator. Another way to break down the marketplace is to look at solution segments:

1. short-term monitoring
2. long-term monitoring
3. permanent monitoring

Each segment represents a need for different types of software and hardware. In the case of short-term monitoring, a sound-level meter may be sufficient. For permanent monitoring a self-contained monitoring unit is required, and on-board analysis capabilities are probably desirable.

SUMMARY

A large number of metrics are currently being used, ranging from A-weighted sound levels to day-evening-night average sound pressure levels with various corrections. There are still some issues when it comes to low-frequency noise, impulsive sounds, and certain sources—special cases may require unique metrics. Undoubtedly, new and more complex metrics will be developed.

Sophisticated modern sound-level meters and monitoring devices have the capability to record and report any metric that can be programmed. The level of sophistication currently available is sufficient to perform measurements and calculations required by all current metrics and some of the metrics used in product sound quality evaluation. These sound quality metrics may become more widespread in the future for the evaluation of community noise. The use of modern computer technology has effectively eliminated any limitations on measurement equipment in terms of the metrics that can be used.

Data management is now much easier. Embedding large amounts of memory in instrumentation is relatively inexpensive and wireless connection capability also means that a large amount of data can now be collected and stored automatically for future processing, which greatly facilitates testing of proposed new metrics. It also means that the noise measurement component of large community surveys can be approached in a very different way from the way it was done in the past when data collection, memory, and storage capabilities were very limited.

In the competitive marketplace for sound-level meters and monitoring systems, the same sort of capability is available from several vendors. Prices and performance will continue to improve.

REFERENCES

ASA. 2010. American National Standards Institute, ASC S1—Physical Acoustics. ASC S1 standards are available online from the Acoustical Society of America at *http://asastore.aip.org/shop.do?cID=7*. The following is a list of the standards available in the S12.9 series:

- ANSI S12.9-1988 (R 2003) American National Standard Quantities and Procedures for Description and Measurement of Environmental Sound, Part 1
- ANSI/ASA S12.9-1992/Part 2 (R2008) American National Standard Quantities and Procedures for Description and Measurement of Environmental Sound, Part 2: Measurement of Long-Term, Wide Area Sound
- ANSI/ASA S12.9-1993/Part 3 (R2008)—American National Standard Quantities and Procedures for Description and Measurement of Environmental Sound, Part 3: Short-Term Measurements with an Observer
- ANSI S12.9-2005/Part 4 American National Standard Quantities and Procedures for Description and Measurement of Environmental Sound—Part 4: Noise Assessment and Prediction of Long-Term Community Response.
- ANSI/ASA S12.9-2007/Part 5 American National Standard Quantities and Procedures for Description and Measurement of Environmental Sound—Part 5: Sound-Level Descriptors for Determination of Compatible Land Use
- ANSI/ASA S12.9-2008/Part 6 American National Standard Quantities and Procedures for Description and Measurement of Environmental Sound—Part 6: Methods for Estimation of Awakenings Associated with Outdoor Noise Events Heard in Homes

IEC. 2010. Technical Committee 29—Electroacoustics, International Electrotechnical Commission, Geneva, Switzerland. Available online at *http://www.iec.ch/cgi-bin/procgi.pl/www/iecwww.p?wwwlang=e&wwwprog=TCpubs.p&progdb=db1&committee=TC&css_color=purple&number=29*.

Appendix F

Guidance for Environmental Economics

Environmental economics is the application of microeconomic tools to environmental problems. These include an analytical framework for describing or modeling the economic conditions out of which environmental problems arise. These conditions are called externalities (i.e., when parties involved in a market transaction impose costs on others that are *external* to [not involved in] the transaction). For example, a factory in a neighborhood produces and sells goods but also fouls the air for nearby residents. This externality implies an external cost that is not borne by the factory or its customers but is shifted to the residents who incur health and welfare costs associated with breathing polluted air.

The optimal solution, from an economic perspective, must take into account both the burden of pollution on the community and the value of the output generated by the factory. At first glance this may seem unfair to the neighbors. However, consider the case of road noise pollution. The obvious way to ensure silence near large roads is to ban all traffic except for bicycles and pedestrians. Similar statements could be about safety; an expedient way to stop all loss of life in automobile accidents is to ban driving. Because most activities that have some undesirable consequences also have some value to society, either extreme is in some way detrimental to society. Thus, the solution must include trade-off(s).

One purpose of economic analysis is to tabulate trade-offs in an explicit way by putting monetary values on the harm done to the community and the environment, as well as the cost to the factory, the government, or others associated with mitigating the harm. If the cost to society of a specific plan for mitigating harm is greater than the harm itself (e.g., banning driving), that form of mitigation is not justifiable from an economic perspective. Monetization of the harm incurred from externalities is necessary to make this explicit.

Critics of environmental economics may argue that it puts the environment at a disadvantage because the costs of mitigation can often be easily expressed in monetary terms while the harm cannot be expressed that way (e.g., the value of the loss of a species). Economists might respond that attempting to place a value on the environment can often be more helpful than not. Moreover, environmental economic analysis is just one of several inputs to decision making. Other inputs, such as equity and political considerations, will also influence decisions.

Cost-benefit analysis (CBA) is a formal framework for comparing harm with the cost of mitigating harm. In CBA, explicit and implicit costs associated with pursuing a course of action are all included in making a decision. Costs can be in dollars or opportunity cost (i.e., the value of the next best use of the time and resources involved). Benefits for the purposes of CBA are all of the positive gains for society associated with a course of action, whether they are naturally expressed in monetary terms or not. This can become confusing because eliminating an external cost (e.g., the burden of pollution) is considered a benefit. To avoid this confusion, costs are considered everything that is given up associated with a policy or an investment; any benefits are all positive consequences for society. In making decisions about policies or investments that are not motivated by an environmental purpose, such as the addition of a runway at an airport, environmental consequences are typically considered costs.

The scope of a CBA is society at large. Ideally, geographic or categorical boundaries are only those determined by the scope of the expected impact of a decision. This means that government, individuals, and businesses, as well as natural or environmental amenities valued by society, should all be included in the analysis. Similarly, the timescale for the analysis should, as much as possible, encompass the full length of time over which costs and benefits occur. In general, benefits of environmental policy or investment take the form of harm avoided but may also provide other indirect positive outcomes for society. The U.S. Environmental Protection Agency notes that these benefits should be handled on an "effect by effect" basis (EPA, 2000, p. 59). Costs typically include private compliance costs (for regulation), government investment costs, government regulatory costs, social welfare losses (impacts that result in higher prices), and transitional costs associated with regulation (which may include job losses and other consequences) (EPA, 2000, p. 16).

Other types of economic analysis commonly used to inform policy and public investment decisions include distributional analysis and cost-effectiveness analysis. Cost-effectiveness analysis tabulates costs associated with different methods of accomplishing a specific goal. This may be an appropriate tool if all methods being considered result in very similar outcomes. Otherwise, "cost-effectiveness analysis does not necessarily reveal what level of control is reasonable, nor can it be used to directly compare situations with different benefit streams" (EPA, 2000, p. 178). Distributional analysis, which might also be called economic impact analysis, differs from CBA in that it focuses primarily on costs and on different segments of society, rather than on society as a whole. Equity assessment is a variant that focuses on impacts on vulnerable segments of society (EPA, 2000, p. 20).

Explicit direct costs are usually easy to capture monetarily, because they are included in the budget for the investment and its maintenance or the ongoing costs of enforcement and monitoring of a regulation. Broad social costs are typically easier to monetize than benefits because they may include an increase in production costs for firms, prices for consumers, or other factors that can be readily monetized.

Measuring the benefits of environmental policy or mitigation investments requires first understanding the direct physical impacts of the policy or investment, whether measured in tons of effluent, decibels, wildlife population, or any other direct environmental metric. For many types of environmental impact, the metric may require further analysis to translate it into relevant consequences, such as increased incidence of cancer or asthma as an impact for airborne emissions. In the case of noise, the initial physical impact may be on a geographic area; but it becomes relevant to a CBA when population exposure is involved.

Once the impact has been described, it can be monetized. EPA guidelines state: "To the extent feasible, and warranted by their contribution to the results, as many of the effects of a policy as possible should be monetized. This enhances the value of the conclusions to policy makers weighing the many, often disparate consequences of different policy options and alternatives" (EPA, 2000, p. 176). Thus, the rationale for monetizing environmental impacts is to put them in terms that can be compared to the cost of policies to improve environmental quality or the benefits of actions that cause environmental harm.

Projects and actions often have different environmental effects (changes in noise level, air quality, climate, water quality); thus, another reason for monetizing these changes is so they can be compared with one another. Comparing noise annoyance and sleep awakenings with the incidence of asthma or cardiopulmonary disease and the long-term harm of climate change can be difficult. However, ultimately these comparisons must be made, and making an attempt to quantify these effects in a single comparative measure (typically monetary), while carefully accounting for uncertainty in the estimates, can be a valuable aid in decision making.

The applicable economic concepts of value for environmental benefits are (1) willingness to pay (WTP) for environmental improvements and (2) willingness to accept (WTA) compensation to endure degraded environmental quality (EPA, 2000, p. 60). WTP and WTA are not necessarily equal. Both are based on how society feels about particular environmental amenities, whether in the form of nuisance, health, aesthetics, existence (of species or natural feature), or legacy value for future generations. In CBA, environmental features are not valued in themselves; they derive their value from how highly society values them.

An estimate of WTP for environmental improvements varies by the nature of the impact associated with the improvement. An environmental amenity has "use value" when the environmental feature interfaces with relevant members of society—the interfaces may be direct or indirect, as well as market or nonmarket (meaning a transaction takes place or does not) interfaces (EPA, 2000, p. 70). Nonuse value includes "existence value," when society derives value from knowing an environmental amenity exists, and "legacy value," when society values knowing that an environmental amenity will be available to future generations (EPA, 2000, p. 71). Typically, because use values are associated with direct interactions with the environment, they are easier to monetize.

Monetization methods can be divided into three categories: (1) market methods, (2) revealed preference, and (3) stated preference (EPA, 2000, p. 72). When a good is traded in a market, the market prices, as well as supply and demand curves, are used to value it, consistent with microeconomic principles. Even when a good or environmental amenity is not directly traded, there may be data on actual market transactions that can be used to infer WTP for it. A relevant example is lower housing values in areas with high noise compared to values in areas with lower noise.

Among the many types of revealed preference techniques, hedonic analysis is the most relevant for noise (EPA, 2000, pp. 73–83). Hedonic analysis attempts to statistically decompose the market price of a good into the segments of that price associated with features or characteristics of the good using regression analysis (EPA, 2000, p. 77). For instance, if two cars are identical except for color, but the market price of a red car is $1,000 more, society has WTP of $1,000 for red. In environmental noise studies, the relevant market data that drive the analysis are real estate transactions. Controlling for other property characteristics, the difference in price or rent between a quiet property and a noisy one reveals the value a community places on quiet.

Stated preference methods range from survey techniques to constructed market techniques that attempt to incorporate perceived economic gain and loss to make survey results more plausible. For example, one might survey residents in different noise environments to find out how much they would value a reduction in noise compared to a more easily valued benefit (like a reduction in property taxes).

More than one monetization technique is used in a particular CBA because any survey may generate a collection of distinct use and non-use benefits. The benefits to a community associated with a reduction in road noise are typically use-oriented, nonmarket benefits (due to the direct experience of noise or quiet). Because market transactions can be used to infer the WTP for quiet versus noise, the road noise example is a good candidate for revealed preference methods.

Two other valuation methods may be considered relevant to the road noise problem. The first is known as the averting behavior method (EPA, 2000, p. 70), the cost people are willing to incur to defend against a particular environmental problem—for instance, wearing a filter mask when walking outdoors in a city with especially polluted air. In the case of road noise, this might be voluntary installation of sound insulation by a homeowner. The drawback of this method is that unless there are continuous, incremental opportunities, people will not be able to spend up to a level that expresses their true WTP.

The second alternative to hedonic analysis would be using either a hypothetical or government cost to purchase and install sound insulating material for homeowners (e.g., the Federal Aviation Administration Residential Sound Insulation Program). Because in either case the material would not be voluntarily paid for by the homeowner (if it were, it would be averting behavior), the price is even further disconnected from the actual WTP than the averting behavior technique because there is no evidence that the homeowner values quiet as highly as the cost of installing sound insulating materials.

Benefits transfer is the technique of applying benefits valuation estimates from past studies to new analyses (EPA, 2000, p. 85). However, given potential variations among communities, this technique should be used with care—although it should not be ruled out. Sometimes, budget constraints or the lack of relevant local data may make carrying out hedonic studies unrealistic. In that situation a benefits transfer analysis that takes into account uncertainty is preferable to no benefits study at all.

Both the EPA (EPA, 2000, p. 27) and the Office of Management and Budget (OMB, 1992, p. 10) guidelines cite the importance of being explicit and detailed in describing the sources and nature of uncertainties in an analysis, whether the uncertainties are about outcomes in the future or estimates of variables or parameters used in the analysis. Both EPA and OMB also suggest sensitivity analysis. EPA describes the use of probabilistic approaches such as a Monte Carlo analysis, as well as looking for "switch points" for important inputs to a CBA. For instance, if a CBA study were carried out using benefits transfer and a positive gain to society was found, a switch point would describe how much lower the community's values would have to be relative to the community in the original study of benefits to reverse the conclusions of the study.

Beyond calculating benefits and costs on an effect-by-effect basis, and monetizing, there are a host of tabulation issues that must be handled sensitively, some of which may also be sources of uncertainty. To the extent that benefits and costs are incurred in streams across time, a discount rate must be applied to reduce future dollars into current dollars (because a dollar is worth more now than later). Inflation also has to be handled consistently to ensure that all final results in the analysis are present in the same constant-year dollars.

Although the trade-off of dollar values in the future for dollar values today over a time period within the current generation may be a small source of uncertainty, crossing generations introduces even more uncertainty (EPA, 2000, p. 48), partly because of the nature of the net present value calculation to carry out the discounting. The farther into the future an outcome is, the less it is worth. Mathematically, the differences can be dramatic, making it appear that society places almost zero value on consequences for our grandchildren's grandchildren. EPA suggests doing a sensitivity analysis on the discount rate itself, including a presentation of a case with a zero discount rate, to address this problem. This particular guideline is more relevant for climate change studies than for noise, which has immediate effects. However, it does provide some perspective on uncertainty potentially relevant to highway studies. In the face of the enormous uncertainty about how to analytically trade off our own well-being against future generations, variations in a community's value of noise and variances within the statistical estimates may not seem so daunting.

REFERENCES

EPA. 2000. Guidelines for Preparing Economic Analyses. Available online at *http://yosemite.epa.gov/ee/epa/eed.nsf/webpages/Guidelines.html*.

OMB (Office of Management and Budget). 1992. Guidelines and Discount Rates for Benefit-Cost Analysis of Federal Programs. Circular A-94. Washington, DC: OMB. Available online at *http://www.whitehouse.gov/omb/circulars/a094/a094.pdf*.

Appendix G

Regulations and Voluntary Use of Hearing Protection Devices

Both industrial workers and military personnel, the largest users of hearing protection devices (HPDs), are governed by regulations. In fact, *regulation* is a major reason the use of HPDs has proliferated. This appendix reviews the history of these regulations, focusing on the laws that affect the majority group (i.e., U.S. workers).

OSHA GENERAL INDUSTRY AND CONSTRUCTION REGULATIONS CIRCA 1971

OSHA—General Industry

The Occupational Safety and Health Administration (OSHA) Noise Standard for General Industry (29 CFR 1910.95(a)) specifies: "Protection against the effects of noise exposure shall be required when the sound levels exceed those shown in Table G-16 when measured on the A-scale of a standard sound level meter at slow response." Table G-16 specifies a 90-dB(A) time weighted average (TWA) "criterion level" for an 8-hour exposure, including a 5-dB exchange rate between increased noise exposures and allowable exposure durations per day. So, for example, 95-dB(A) TWA is allowed for 4 hours, 100-dB(A) TWA is allowed for 2 hours, and so on, with a not-to-exceed 140-dB peak sound pressure level for impulsive or impact noise.

29 CFR 1910.95(b)(1) further states: "When employees are subjected to sound exceeding those listed in Table G-16, feasible administrative or engineering controls shall be utilized. If such controls fail to reduce sound levels within the levels of Table G-16, personal protective equipment shall be provided and used to reduce sound levels within the levels of the table." Thus, this *first* OSHA regulation, via the words "shall be utilized," required that feasible administrative or engineering controls take priority over hearing protection. However, a significant weakness was that the word "feasible" was not defined specifically in terms of technical, economical, or other criteria. This left room for industries to claim infeasibility.

Nevertheless, only if engineering or administrative controls fail to reduce noise to within the limits of Table G-16 are hearing protectors to be relied on under OSHA (1971b). Thus, in the earliest OSHA general industry regulations, HPDs were regulated as an augmentation to, and not a replacement for, administrative or engineering noise controls. In practice, however, HPDs are relied on in many industrial plants as the first line of defense against noise hazards to workers' hearing, which violates the letter of the OSHA law.

OSHA—Construction

The law for construction work, 29 CFR 1926.52, cites Table D-2 (a duplicate of Table G-16) for exposure limits; it also includes the same statement about administrative and engineering controls having priority over HPDs (OSHA, 1971a). However, the construction regulation in 29 CFR 1926.101 has additional stipulations: (a) "Whenever it is not feasible to reduce the noise levels or duration of exposures to those specified in Table D-2, Permissible Noise Exposures, in 1926.52, ear protective devices shall be provided and used." (b) "Ear protective devices inserted in the ear shall be fitted or determined individually by competent persons. (c) Plain cotton is not an acceptable protective device" (OSHA, 1971b). Subparts (b) and (c) may be a slight improvement over the general industry standard. However, overall, the construction standard became much weaker because it was never updated, as the general industry standard was.

OSHA GENERAL INDUSTRY—HEARING CONSERVATION AMENDMENT CIRCA 1983

The Hearing Conservation Amendment significantly improved the original OSHA noise standard by specifying that a multifaceted hearing conservation program is required when daily TWA noise exposures exceed 85 dB(A) (equivalent to a 50 percent noise dose; OSHA, 1983). The priority of engineering and administrative noise controls remained; in addition to other facets of a hearing conservation program (including noise monitoring, employee notification, audiometric testing, worker training, access to information and training materials, and exposure and audiometric recordkeeping), the amendment specified the use of HPDs in more detail.

Perhaps the most significant addition, at 29 CFR 1910.95(i), was (1) that "employers shall make hearing protectors available to all employees exposed to an 8-hour time-weighted average of 85 decibels or greater at no cost to the employees. Hearing protectors shall be replaced as necessary" and (2) that "employers shall ensure that hearing protectors are worn: (i) by an employee who is required by paragraph 1910.95(b)(1) of this section to wear personal protective equipment [i.e., mandatory HPD use at exposures equal to or greater than 90 dB(A) TWA] and (ii) by any employee who is exposed to an 8-hour TWA of 85 decibels or greater, and who" has not had a baseline audiogram, or who has experienced a standard threshold shift (as defined by OSHA). In addition, employers were required, under paragraph 3, to provide "a variety of suitable hearing protectors" for the employee to select from; under paragraph 4 to provide training in the use and care of all hearing protectors; and under paragraph 5 to ensure proper initial fitting and supervision in the correct use of all hearing protectors. Finally, in part (j) the amendment specified computational procedures for evaluating HPDs for adequacy of protection in specific noise exposures, with the requirement that the protected exposure levels be brought to less than or equal to 90-dB(A) TWA, or to less than or equal to 85-dB(A) TWA if the worker has experienced a standard threshold shift.[1]

The OSHA Hearing Conservation Amendment greatly impacted the requirements for hearing protection, and the numbers of HPDs supplied in occupational settings dramatically increased as a result. Although engineering or administrative controls were still required for TWA exposures above 90 dB(A), the amendment provided, at no cost to workers, a selection of HPDs to everyone exposed to 85-dB(A) TWA or above. The 5-dB(A) difference between the 90-dB(A) OSHA "criterion" level imposed as a result of OSHA (1971) and the 85-dB(A) OSHA "action" level imposed as a result of the 1983 OSHA Hearing Conservation Amendment defined an exposure window wherein thousands of workers who had not been protected by law were now to be supplied with a selection of suitable HPDs.

In this sense the new 85-dB(A) TWA action level was a major step forward in protecting workers against the hazards of noise exposures; however, the OSHA Hearing Conservation Amendment should not be understood as an indication that HPDs are preferable to engineering noise controls, which do not require human intervention to protect workers' hearing and prevent noise-induced hearing loss.

[1]The reader is referred to OSHA (1983) and Casali (2006) for more details on computing HPD adequacy.

DATA AND LABELING REGULATIONS

Labeled Versus In-Field Attenuation Performance

The labeling of HPDs has been the subject of debate for more than two decades, much of it about the differences between on-package EPA-required attenuation data and the actual protection provided for users in the field (Berger and Casali, 1997; Casali and Robinson, 2003). To comply with OSHA (1983) and other applications, the adequacy of an HPD for a given noise exposure is determined by subtracting, in a prescribed way, the attenuation data required by the U.S. Environmental Protection Agency (EPA) from the TWA noise exposure for the affected worker (see OSHA, 1983, Appendix B: Methods for Estimating the Adequacy of Hearing Protector Attenuation).

Attenuation data are obtained from psychophysical real-ear-attenuation-at-threshold tests at nine 1/3-octave bands with centers of 125 to 8,000 Hz performed on human listeners; the signed, arithmetic difference between thresholds with the HPD and without it constitutes the attenuation at a given frequency. Both the spectral attenuation statistics (means and standard deviations) and the broadband single-number noise reduction rating (NRR), which is computed therefrom, are provided, and either of them can be used to estimate HPD adequacy for a given exposure, per OSHA (1983) Appendix B.

Labeled ratings are the primary means by which end users compare different HPDs and determine if they will provide adequate protection and OSHA compliance in a given noise environment. Therefore, the accuracy and validity of label ratings are very important.

Current EPA-Required Labeling and Cited Test Standards

The labeling of hearing protectors is controlled by EPA via federal law per 40 CFR Part 211, Subpart B, which was promulgated in September 1979 and remains in effect as of this writing. This section of the law applies to "any device or material, capable of being worn on the head or in the ear canal, that is sold wholly or in part on the basis of its ability to reduce unwanted sound that enters the user's ears" (40 CFR Part 211, Subpart B). Unfortunately, the law references an outdated, superseded ANSI standard (1974) for obtaining the real-ear attenuation of threshold data on which the EPA label, which includes an NRR, is based.

The data on HPD packaging are obtained under optimal laboratory conditions with properly fitted protectors worn by trained, well-practiced human subjects. However, numerous research studies (e.g., Berger et al., 1998; Berger and Casali, 1997; Park and Casali, 1991) have shown that the "experimenter-fit" protocol and other aspects of the EPA-required test procedure do not represent the conditions under which HPDs are selected, fitted, and used in the workplace. Therefore, the attenuation data used in the octave band or NRR formulas are highly inflated and cannot be assumed to represent the protection achieved in the field.

Figure G-1 shows the results of a review of research studies in which manufacturers' on-package NRRs (in the background) were compared against NRRs computed from actual subjects using HPDs in field settings (in the foreground). Clearly, there are large differences between laboratory and field estimates, especially for earplugs (Berger, 2003). HPD consumers must take this difference into account when selecting protectors.

"Proposed" Revisions for Labeling and Cited Test Standards

ANSI Working Group S12/WG11 developed a new testing standard, ANSI S12.6-1997(R2002), which includes both a "Method A" provision for experimenter-supervised fitting of an HPD and a "Method B" provision for self-fitting of the HPD and with test subjects who have not been trained. The new standard has much improved experimental controls and human factors protocol over the current standard. Nevertheless, even though the Method B (subject-fit) testing protocol has been experimentally demonstrated to yield attenuation data that are more representative of those achievable under workplace conditions (Berger et al., 1998), as of this writing it appears that Method A (experimenter-supervised fit) is likely to be adopted by EPA for a revised regulation.

EPA has given notice (see EPA Docket OAR-2003-0024) in public workshops and presentations of a plan to revise the 1979 labeling regulation, in conjunction with the requirement to replace ANSI S3.19-1974 with the current ANSI

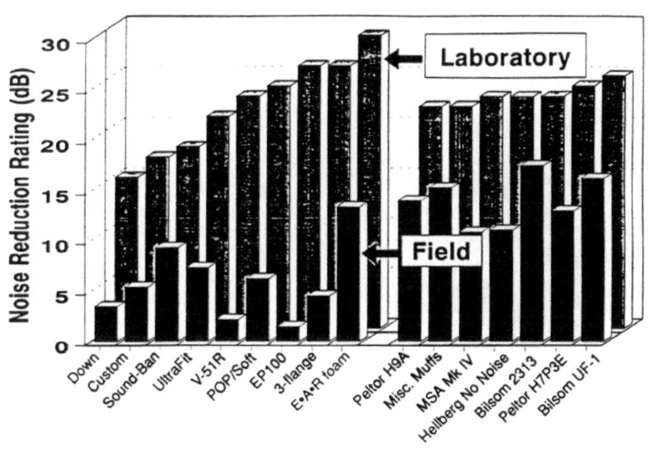

FIGURE G-1 Comparison of hearing protection device NRRs by device type: manufacturers' laboratory data versus real-world "field" data. Adapted with permission from Berger (2003), Fig. 10.18, p. 421.

standard to obtain passive attenuation data along with a new means of broadband rating. This rating is likely to be called the single-number rating, even though it will probably provide a range of values. This differs from NRR, which provided a single number.

The proposed regulation is also likely to include elements of another testing standard, ANSI S12.42 (ANSI, 2004), to enable physical, microphone-based testing in real ears and acoustical test fixtures; this will enable comprehensive testing of active noise cancellation, as well as certain other HPD types that are currently not amenable to the 1979 EPA regulation for labeling and thus cannot currently be marketed as hearing protectors. Elements of ANSI S12.68 (ANSI, 2007) are also likely to be added to prescribed methods of estimating protected exposure levels under HPDs. At the time of this writing, none of the details of the proposed EPA revised labeling regulation had been finalized. For updates the reader should go to *www.regulations.gov* (docket number: EPA-HQ-OAR-2003-0024).

REFERENCES

ANSI (American National Standards Institute). 1974. Method for the Measurement of Real-Ear Protection of Hearing Protectors and Physical Attenuation of Earmuffs. ANSI S3.19-1974. New York: ANSI.

ANSI. 2004. Microphone-in-Real-Ear and Acoustic Test Fixture Methods for the Measurement of Insertion Loss of Circumaural Hearing Protection Devices. ANSI S12.42-1995(R2004). New York: ANSI.

ANSI. 2007. Methods of Estimating Effective A-Weighted Sound Pressure Levels When Hearing Protectors Are Worn. ANSI S12.68-2007. New York: ANSI.

Berger, E.H. 2003. Hearing protection devices. Pp. 379–454 in The Noise Manual, Revised 5th Ed., edited by E.H. Berger, L.H. Royster, J.D. Royster, D.P. Driscoll, and M. Layne. Fairfax, VA: American Industrial Hygiene Association.

Berger, E.H., and J.G. Casali. 1997. Hearing protection devices. Pp. 967–981 in Encyclopedia of Acoustics, edited by M. Crocker. New York: John Wiley.

Berger, E.H., J.R. Franks, A. Behar, J.G. Casali, C. Dixon-Ernst, R.W. Kieper, C.J. Merry, B.T. Mozo, C.W. Nixon, D. Ohlin, J.D. Royster, and L.H. Royster. 1998. Development of a new standard laboratory protocol for estimating the field attenuation of hearing protection devices, Part III: The validity of using subject-fit data. Journal of the Acoustical Society of America 103(2):665–672.

Casali, J.G., and G.S. Robinson. 2003. Augmented Hearing Protection Devices: Active Noise Reduction, Level-Dependent, Sound Transmission, Uniform Attenuation, and Adjustable Devices—Technology Overview and Performance Testing Issues. EPA Docket OAR-2003-0024. Washington, DC: U.S. Environmental Protection Agency.

Casali, J.G. 2006. Sound and noise. Pp. 612–642 in Handbook of Human Factors, 3rd Ed., edited by G. Salvendy. New York: John Wiley.

OSHA (Occupational Safety and Health Administration). 1971a. Occupational Noise Exposure (Construction Industry). 29 CFR 1926.52. Federal Register.

OSHA. 1971b. Hearing Protection (Construction Industry). 29 CFR 1926.101. Federal Register.

OSHA. 1983. Occupational Noise Exposure; Hearing Conservation Amendment; Final Rule. 29 CFR 1910.95. Federal Register.

Park, M.Y., and J.G. Casali. 1991. A controlled investigation of in-field attenuation performance of selected insert, earmuff, and canal cap hearing protectors. Human Factors 33(6):693–714.

Appendix H

Acronyms and Abbreviations

%HA	percentage (of persons) highly annoyed
AC	asphalt concrete
ACARE	Advisory Council for Aeronautical Research in Europe
ACTIVE	International Symposium on Active Control of Sound and Vibration
ACUS	Administrative Conference of the United States
ADC40	Transportation Research Board Committee on Transportation-Related Noise and Vibration
ADOT	Arizona Department of Transportation
A|D|S	Aerospace|Defence|Security (see SBAC)
AHAAH	*Auditory Hazard Assessment Algorithm for Humans*
AHAM	American Home Appliance Manufacturers
AIAA	American Institute of Aeronautics and Astronautics
AIP	American Institute of Physics
AMCA	Air Moving and Conditioning Association
AMT	Association for Manufacturing Technology
ANASE	*Attitudes to Noise from Aviation Sources in England*
ANR	active noise reduction
ANSI	American National Standards Institute
ANTLE	advanced near-term low emissions
AREMA	American Railway Engineering and Maintenance-of-Way Association
ARI	Air-Conditioning and Refrigeration Institute
ASC	Accredited Standards Committee
ASEL	A-weighted sound exposure level
ASHA	American Speech-Language-Hearing Association
ASHE	American Society for Healthcare Engineering
ASHRAE	American Society of Heating, Refrigerating and Air-Conditioning Engineers
ASME	American Society of Mechanical Engineers
AST	advanced subsonic transport
ASTM	American Society for Testing and Materials
ATIS	Alliance for Telecommunications Industry Solutions
AUD	Australian dollar
BBN	Bolt, Beranek and Newman
CAD	computer-aided design
CAETS	International Council of Engineering and Technical Societies
CALM	Community Noise Research Strategy Plan (European Union)
CBA	cost-benefit analysis
CDoT	California Department of Transportation
CE	*Conformité Européenne*
CEI	Central European Initiative
CEN	European Committee for Standardization
CENELEC	European Committee for Electrotechnical Standardization
CENYC	Council of the Environment of New York City
CFD	computational fluid dynamics
CFR	Code of Federal Regulations
CPRE	Campaign to Protect Rural England
CPSC	Consumer Product Safety Commission
CRRs	commuter railroads
CRS	Congressional Research Service

CU	Consumers Union
dB	decibel
DEFRA	Department for Environment, Food and Rural Affairs
DENL	day-evening-night average sound level
DEP	Department of Environmental Protection (New York City)
DHHS	U.S. Department of Health and Human Services
DIA	Denver International Airport
DNL	day-night-average sound level
DNW	Dutch Anechoic Wind Tunnel
DOD	U.S. Department of Defense
DOL	U.S. Department of Labor
DOT	U.S. Department of Transportation
DRA	duct resonator array
EA	European Co-operation for Accreditation
EC	European Commission
ECMA	formerly the European Computer Manufacturers Association, now only ECMA
END	Environmental Noise Directive
EPA	Environmental Protection Agency
EPNdB	effective perceived noise level decibels
EPNL	effective perceived noise level
ER	exchange rate
EU	European Union
FAA	Federal Aviation Administration
FEHRL	Forum of European National Highway Research Laboratories
FGI	Facility Guidelines Institute
FHWA	Federal Highway Administration
FICAN	Federal Interagency Committee on Aircraft Noise
FICON	Federal Interagency Committee on Noise
FICUN	Federal Interagency Committee on Urban Noise
FRA	Federal Railroad Administration
FTA	Federal Transit Administration
FTAG	Federal Transportation Advisory Group
GAO	Government Accountability Office
GDP	gross domestic product
GPSG	German Equipment and Product Safety Act
GSA	General Services Administration
GSIG	Global Standards and Information Group
HARMONOISE	Project to predict environmental noise levels caused by road and railway traffic (European Union)
HEATCO	Harmonized European Approaches for Transport Costing and Product Assessment Studies (consortium)
HIPAA	Health Insurance Portability and Accountability Act
HPD	hearing protection device
HSR	high-speed passenger railroads
HWB	hybrid wing body
ICAO	International Civil Aviation Association
ICBEN	International Commission on the Biological Effects of Noise
IEC	International Electrotechnical Commission
IEEE	Institute of Electrical and Electronics Engineers
IIC	Impact-Insulation Class
I-INCE	International Institute of Noise Control Engineering
ILAC	International Laboratory Accreditation Cooperation
IMAGINE	Project to produce noise maps (European Union)
INCE/USA	Institute of Noise Control Engineering of the U.S.A.
INTER-NOISE	International Congress on Noise Control Engineering
ISEA	International Safety Equipment Association
IOM	Institute of Medicine
ISO	International Organization for Standardization
ISVR	Institute of Sound and Vibration Research (United Kingdom)
IT	information technology
ITD	integrated technology demonstrators
JPDO	Joint Planning and Development Office
KBDN	*Kitchen and Bath Design News*
LEED	Leadership in Energy and Environmental Design
LEMA	Laboratory for Electromagnetics and Acoustics
LHH	League for the Hard of Hearing (now Center for Hearing and Communication)
LRT	light rail transit
MIT	Massachusetts Institute of Technology
MSHA	Mine Safety and Health Administration

NAE	National Academy of Engineering
NASA	National Aeronautics and Space Administration
NAW	Noise Action Week
NCAC	National Council of Acoustical Consultants
NCEJ	*Noise Control Engineering Journal*
NCHRP	National Cooperative Highway Research Program
NCSI	National Center for Standards Certification Information
NDI	noise depreciation index
NDSI	noise depreciation sensitivity index
NEF	noise exposure forecast
NIH	National Institutes of Health
NIHL	noise-induced hearing loss
NIOSH	National Institute for Occupational Safety and Health
NIPTS	noise-induced permanent threshold shift
NIST	National Institute of Standards and Technology
NNI	noise and number index
NNI	*Noise/News International*
NOISE-CON	National Conference on Noise Control Engineering
NPS	National Park Service
NRR	noise reduction rating
NVLAP	National Voluntary Laboratory Accreditation Program
OECD	Organisation for Economic Co-operation and Development
OGAC	open-grade asphalt concrete
OMB	Office of Management and Budget
ONAC	Office of Noise Abatement and Control
ONCC	O'Hare Noise Compatibility Commission
OSHA	Occupational Safety and Health Administration
OSHRC	Occupational Safety and Health Review Commission
OSTP	Office of Science and Technology Policy
PARTNER	Partnership for Air Transportation Noise and Emissions Reduction
PCC	Portland cement concrete
PEL	permitted exposure level
PIRG	Public Information Research Group
PnDB	unit of perceived noise level
PNL	perceived noise level
PPE	personal protective equipment
QPPP	Quiet Pavement Pilot Program
QTD	Quiet Technology Demonstrator Program
R&D	research and development
RRT	rapid rail transit
SAE	Society of Automotive Engineers (SAE International)
SAX-40	Silent Aircraft Initiative
SBAC	Society of British Aerospace Companies (through a merger, now A|D|S)
SEL	sound exposure level
SILENCE	Aircraft Noise Reduction Project (European Union)
SILVIA	based on *silenda via* (the road must be silent)
SMA	stone matrix asphalt
SME	Society of Mechanical Engineers
STAIRRS	Strategies and Tools to Assess and Implement Noise-Reducing Measures for Railway Systems
STC	sound transmission class
TC43/SC1	(ISO) Technical Committee 43 Subcommittee 1 (Noise)
TGV	train à grande vitesse (France)
TNM	Traffic Noise Model
TPNRC	Tire-Pavement Noise Research Consortium
TRB	Transportation Research Board
TTS	temporary threshold shift
TÜV	Technischer Überwachungs-Verein (Germany)
TWA	time-weighted average
TWINS	track-wheel interaction noise system
UHBR	ultra high bypass ratio
UL	Underwriters Laboratories
UNECE	United Nations Economic Commission for Europe
USAF	U.S. Air Force
USC	U.S. Code
USCB	U.S. Census Bureau
USD	U.S. dollar
USTR	U.S. Trade Representative
VCA	Vehicle Certification Agency (United Kingdom)
VITAL	EnVIronmenTALly
WG1	Working Group on Noise Indicators
WG2	Working Group on Socio-Economic Aspects of Noise
WHO	World Health Organization
WTA	willingness to accept
WTP	willingness to pay

Appendix I

Glossary of Selected Terms[1]

***Acoustical holography**—An inspection method using the phase interference between sound waves from an object and a reference signal to obtain an image of reflections in the test object.

***Action level**—The cumulative work-shift noise dose at which a hearing conservation program is mandated by the Occupational Safety and Health Administration (OSHA). An 8-hour time-weighted average of 85 dB measured with A-weighting and slow response or the equivalent, a dose of 50 percent. *See* hearing conservation program.

***Active control**—Reducing sound and secondary sources of excitation to cancel, or at least reduce, the response of a system to prime noise sources; also to suppress self-excitation oscillations of an unstable system.

†**Ambient noise**—All-encompassing sound at a given place, usually a composite of sounds from many sources near and far.

***Annoyance**—A person's internal response to a noise. Annoyance is quantifiable (1) psychologically by subjective rating or (2) technically by a physical noise descriptor, for example, the equivalent continuous A-weighted sound pressure level ($L_{Aeq.T}$). For a given person, the correlation coefficient between descriptor and related ratings usually does not exceed 0.5 due to the influence of other factors in determining annoyance. *See* equivalent continuous A-weighted sound pressure.

†**Background noise**—Total noise from all sources of interference in a system used for the production, detection, measurement, or recording of a signal, independent of the presence of the signal.

NOTES:

1. Ambient sound detected, measured, or recorded with the signal is part of the background noise.

2. Interference resulting from primary electric power supplies (commonly described as a *hum* when heard separately) is included in the definition of background noise.

†**Day average sound level**—Time-average sound level between 0700 and 2200 hours. Unit, decibel (dB); abbreviation, DL; symbol, L_d.

NOTE: Day average sound level in decibels is related to the corresponding day sound exposure level, L_{Ed}, according to

$$L_d = L_{Ed} - 10\ \lg(54{,}000/1)$$

where 54,000 is the number of seconds in a 15-hour day.

†**Day-night average sound level**—Twenty-four-hour average sound level for a given day, after addition of 10 decibels to levels from midnight to 0700 hours and from 2200 hours (10 p.m.) to midnight. Unit, decibel (dB); abbreviation, DNL; symbol, L_{dn}.

NOTES:

1. Day-night average sound level in decibels is related to the corresponding day-night sound exposure level, L_{Edn}, according to

$$L_{dn} = L_{Edn} - 10\ \lg(86{,}400/1)$$

[1]The asterisk (*) and the dagger (†) indicate the source of the definition, as explained at the end of the glossary.

where 86,400 is the number of seconds in a 24-hour day.

2. A frequency weighting is understood, unless another frequency weighting is specified explicitly.

†**Decibel**—Unit of level when the base of the logarithm is the tenth root of 10 and the quantities concerned are proportional to power. Unit symbol, dB.

NOTE: Examples of quantities that qualify are power (in any form), sound pressure squared, particle velocity squared, sound intensity, sound-energy density, and voltage squared. Thus, the decibel is a unit of sound-pressure-squared level; in common practice, however, called sound pressure level, unless an ambiguity results from so doing.

†**Effective perceived noise level**—Level of the time integral of the antilogarithm of one-tenth of tone-corrected perceived noise level over the duration of an aircraft flyover, the reference duration being 10 seconds. Unit, decibel (dB); abbreviation, EPNL; symbol, L_{EPN}.

NOTE: The integral is usually approximated by summation, over the top 10 decibels of an aircraft noise signal of the antilogarithms of one-tenth of tone-corrected perceived noise level at successive 0.5 second intervals.

†**Equivalent continuous sound level**—Ten times the logarithm to the base 10 of the ratio of time-mean-square instantaneous A-weighted sound pressure, during a stated time interval T, to the square of the standard reference sound pressure. Unit, decibel (dB); respective abbreviations, TAV and TEQ; respective symbols, L_{AT} and L_{AeqT}.

NOTES:

1. A frequency weighting other than the standard A-weighting may be employed if specified explicitly. A frequency weighting that is essentially constant between limits specified by a manufacturer is called "flat."

2. In symbols, time-average (time-interval equivalent continuous) A-weighted sound level in decibels is

$$L_{AT} = 10 \lg \left\{ \frac{(1/T)\int_0^T p_A^2(t)dt}{p_0^2} \right\}$$

$$= L_{AeqT}$$

where p_A^2 is the squared instantaneous A-weighted sound pressure signal, a function of elapsed time t; in gases reference sound pressure p_0 = 20 μPa; T is a stated time interval.

3. In principle, the sound pressure signal is not exponentially time weighted, either before or after squaring.

***Frequency weighting**—Modification of the spectrum of an acoustical signal by means of an analog or digital filter having one of the standardized response characteristics known as A, B, C, etc., defined in IEC 61672-1. The A-weighting filter is the one most commonly used. *See* weighting network.

***Hearing conservation program**—A system to identify noise-exposed workers and monitor their exposure and audiometric function.

***Hearing loss**—Increase in the threshold of audibility due to disease, injury, age, or exposure to intense noise. **Conductive hearing loss:** Hearing loss caused either by blockage of the external ear or by disease or damage in the middle ear, so that the signal amplitude reaching the inner ear is reduced. **Noise-induced hearing loss (NIHL):** Cumulative hearing loss associated with repeated exposure to noise. **Sensorineural hearing loss:** Hearing loss due to a lesion or disorder of the inner ear or of the auditory nervous system. **Nonoccupational hearing loss:** Hearing loss caused by exposure outside of the occupational environment.

***Hertz (Hz)**—A unit of frequency measurement representing cycles per second.

†**Muffler**—Duct designed to reduce the level of sound. The sound-reducing mechanisms may be either absorptive, reactive, or a combination of both.

***Newton (N)**—A unit of force. The force of one Newton accelerates a 1 kg mass at 1 m/s^2.

***Noise dose**—(1) According to the definition given by Occupational Safety and Health Administration (OSHA), noise dose is the ratio, expressed as a percentage of (a) the time integral, over a stated time or event, of the 0.6 power of the measured "S" (slow) exponential time-averaged, squared A-weighted sound pressure and (b) the product of the criterion duration (8 hours) and the 0.6 power of the squared sound pressure corresponding to the criterion sound pressure level (90 dB). (2) According to the definition given by OSHA, noise dose is the percentage of actual exposure relative to the amount of allowable exposure, and for which 100 percent and above represents exposures that are hazardous. The noise dose is calculated using:

$$D = \sum_{i=1}^{n} C_i / T_i \times 100\%$$

where C_i is the total time of exposure at a specified noise level, and T_i is the exposure time at which noise for this level becomes hazardous.

†**Noise**—(a) Undesired sound, by extension is any unwarranted disturbance in a useful frequency band, such as undesired electric waves in a transmission channel or device. (b) Erratic, intermittent, or statistically random oscillation.

NOTES:

1. If ambiguity exists as to the nature of the noise, a term such as "acoustic noise" or "electric noise" should be used.

2. Since definitions 3.25 (a) and (b) are not mutually exclusive, it is usually necessary to depend on context for the distinction.

***Noise-induced hearing loss**—*See* hearing loss.

†**Noise-induced permanent threshold shift**—Permanent hearing loss resulting from noise exposure. Abbreviation: NIPTS.

†**Noise-induced temporary threshold shift**—Temporary hearing loss resulting from noise exposure. Abbreviation: NITTS.

***Octave band**—A frequency band with upper and lower frequency limits in the ratio of 2. *See* one-third octave band.

***One-third octave band**—A frequency band with upper and lower frequency limits in the ratio of $2^{1/3}$.

***One-third octave filter**—A filter with upper and lower passband limits in the ratio of $2^{1/3}$ centered at one of the preferred frequencies given in ISO 266. Should meet the attenuation characteristics of IEC 61260 and ANSI S1.11–1986.

***Pascal (Pa)**—unit of pressure corresponding to a force of 1 Newton acting uniformly on an area of 1 square meter. 1 Pa = 1 N/m^2.

†**Perceived noise level**—Frequency-weighted sound pressure level obtained by a stated procedure that combines the sound pressure levels in the 24 one-third octave bands with midband frequencies from 50 to 10 kHz. Unit, decibel (dB); abbreviation, PNL; symbol, L_{PN}.

NOTE: Procedures for computing perceived noise level are stated in Federal Aviation Regulation Part 36, *Noise Standards: Aircraft Type and Airworthiness Certification*, Appendix B, and in International Civil Aviation Organization Annex 16, Volume 1, *Aircraft Noise*, 3rd ed., July 1993.

†**Perceived noisiness**—Prescribed function of sound pressure levels in the 24 one-third octave bands with nominal midband frequencies from 50 to 10 kHz used in the calculation of perceived noise level. Unit, noy; abbreviation, *n*.

NOTE: The prescribed function is given in Federal Aviation Regulation Part 36, *Noise Standards: Aircraft Type and Airworthiness Certification*, Appendix B, and in International Civil Aviation Organization Annex 16, Volume 1, *Aircraft Noise*, 3rd ed., July 1993.

***Permissible exposure level (PEL)**—Regulatory limit of sound exposure. The OSHA (Occupational Safety and Health Administration) PEL is a noise dose of 1.0 based on an 8-hour A-weighted sound exposure level at 90 dB with a 5-dB exchange rate. European PEL is generally an 8-hour A-weighted sound exposure level at 85 dB with a 3-dB exchange rate.

†**Phon**—Unit of loudness, judged or calculated.

***Radiation efficiency; radiation factor**—The ratio of the sound power radiated by a vibrating surface, with a given time-mean-square velocity, to the sound power, which would be emitted as a plane wave by the same vibrating surface with the same vibration velocity. The radiation factor is given by the following equation:

$$\sigma = \frac{P_S}{\rho c S_S \overline{\overline{\upsilon^2}}}$$

where P_s is the airborne sound power emitted by the vibrating surface, ρc is the characteristic impedance of air, S_s is the area of the vibrating surface, and $\overline{\overline{\upsilon^2}}$ is the squared rms value of the vibratory velocity averaged over the area S_s. Unit; none; symbol, σ. *See* sound power.

***Single-event sound pressure level**—Time-integrated sound pressure level of an isolated single sound event of specified duration *T* (or specified measurement time *T*) normalized to T_0 = 1 s. It is given by the formula:

$$L_{p,1s} = 10 \log_{10}\left[\frac{1}{T_0}\int_0^T \frac{p^2(t)}{p_0^2}\,dt\right]$$

$$= L_{peq,T} + 10 \log\left(\frac{T}{T_0}\right) \text{dB},$$

where $p(t)$ is the instantaneous sound pressure, p_0 is the reference sound pressure, and $L_{peq.T}$ is the equivalent continuous sound pressure level. Unit, decibel (dB); symbol: $L_{p,1s}$.

†**Sound exposure**—Time integral of squared instantaneous frequency-weighted sound pressure over a stated time interval or event. Unit, pascal-squared second; symbol, *E*.

NOTES:

1. If frequency weighting is not specified, A-frequency weighting is understood. If other than A-frequency weighting is used, such as C-frequency weighting, an appropriate subscript should be added to the symbol (e.g., E_C).

2. Duration of integration is implicitly included in the time integral and need not be reported explicitly. For the sound exposure measured over a specified time interval, such as 1 hour, a 15-hour day, or a 9-hour night, the duration should be indicated by the abbreviation or letter symbol, for example 1-hour sound exposure (1HSE or E_{1h}) for a particular hour; day sound exposure (DSE or E_d) from 0700 to 2200 hours; and night sound exposure (NSE or E_n) from 0000 to 0700 hours plus from 2200 to 2400 hours.

3. Day-night sound exposure (DNSE or E_{dn}) for a 24-hour day is the sum of the day sound exposure and 10 times the night sound exposure.

4. Unless otherwise stated, the normal unit for sound exposure is the pascal-squared second.

†**Sound exposure level**—Ten times the logarithm to the base 10 of the ratio of a given time integral of squared instantaneous A-weighted sound pressure, over a stated time interval or event, to the product of the squared reference sound pressure of 20 micropascals and reference duration of 1 second. The frequency weighting and reference sound exposure may be otherwise if stated explicitly. Unit, decibel (dB); abbreviation, ASEL; symbol, L_{AE}.

NOTE: In symbols, (A-weighted) sound exposure level is

$$L_{AE} = 10 \ \lg\left\{ \frac{\left[\int_0^T p_A^2(t)dt \right]}{p_0^2 t_0} \right\}$$

$$= 10 \ \lg(E / E_0)$$

$$= L_{AT} + 10 \ \lg(T / t_0),$$

where p_A^2 is the squared instantaneous A-weighted sound pressure, a function of time t; for gases p_0 = 20 μPa; t_0 = 1 s; E is sound exposure; $E_0 = p_0^2 t_0$ = (20 μPa)^{2}s is reference sound exposure.

†**Sound intensity**—Average rate of sound energy transmitted in a specified direction at a point through a unit area normal to this direction at the point considered. Unit, watt-per square meter (W/m^2); symbol, I.

NOTES:

1. Sound intensity in the specified direction is given by the expression

$$I = (1 / T)\int_0^T pvdt,$$

where
T = time, which should be long compared with the reciprocal of the lowest frequency of interest;
p = instantaneous sound pressure;
v = component of instantaneous particle velocity in the specified direction; and
t = time.

2. In the case of a free plane or spherical wave having time-mean-square pressure p^2, velocity of propagation c, in a medium of density p, the intensity in the direction of propagation is given by

$$I = p^2/pc.$$

†**Sound-level meter**—Device used to measure sound pressure level with a standardized frequency weighting and indicated exponential time weighting for measurements of sound level, or without time weighting for measurements of time-average sound pressure level or sound exposure level.

†**Sound power**—Sound energy radiated by a source per unit of time. Unit, watt (W), symbols, P or W.

†**Sound power level**—Ten times the logarithm to the base 10 of the ratio of a given sound power in a stated frequency band, to the reference power of 1 picowatt (1 pW). Unit, decibel (dB), abbreviation PWL; symbols, L_P or L_W.

†**Sound pressure**—Root-mean-square instantaneous sound pressure at a point during a given time interval. Unit, pascal (Pa).

NOTE: In the case of periodic sound pressures, the interval is an integral number of periods or an interval that is long compared to a period. In the case of nonperiodic sound pressures, the interval should be long enough to make the measured sound pressures essentially independent of small changes in the duration of the interval.

†**Sound pressure level**—(a) Ten times the logarithm to the base 10 of the ratio of the time-mean-square pressure of a sound, in a stated frequency band, to the square of the reference sound pressure band, to the square of the reference

sound pressure in gases of 20 μPa. Unit, decibel (dB); abbreviation, SPL; symbol, L_p. (b) For sound in media other than gases, unless otherwise specified, reference sound pressure is one micropascal (1 μPa).

NOTE: A sound pressure level with reference to a pressure of 1 μPa is numerically $10 \lg(20^2/1^2) = 26$ decibels greater than the sound pressure level for the same sound pressure but with reference to 20 μPa.

†**Speech interference level**—One-fourth of the sum of the band sound pressure levels for octave bands with nominal midband frequencies of 500, 1,000, 2,000, and 4,000 Hz. Unit, decibel; abbreviation, SIL; symbol, L_{SI}.

***Tire/road noise**—Unwanted sound generated by the interaction between a rolling tire and the surface on which it is rolling. Also known as *tire/pavement noise*.

†**Time-average sound level**—See description for equivalent continuous sound level.

***Turbulence**—A fluid mechanical phenomenon that causes fluctuation in the local sound speed relevant to sound generation in turbo machines (pumps, compressors, fans, and turbines), pumping and air-conditioning systems, or propagation from jets and through the atmosphere.

***Weighting network**—Electronic filter in a sound-level meter that approximates, under defined conditions, the frequency response of the human ear. The A-weighting network is most commonly used. *See* frequency weighting.

† Reprinted from ANSI S1.1-1994 (R 2004) American National Standard Acoustical Terminology, © 1994, with the permission of the Acoustical Society of America, 35 Pinelawn Road, Suite 114E, Melville, NY 11747, USA.

* *Handbook of Noise and Vibration Control.* ISBN 978-0-471-39599-7, edited by Malcolm J. Crocker. Copyright John Wiley, 2007. Reprinted with permission of John Wiley & Sons, Inc.

Please note: No rights are granted to use content that appears in the work with credit to another source.

Appendix J

Biographical Sketches of Committee Members

George C. Maling, Jr. (*chair*) is Managing Director Emeritus of the Institute of Noise Control Engineering of the USA (INCE/USA), past president of the INCE Foundation, managing editor of *Noise/News International*, and vice president for communications of the International Institute of Noise Control Engineering. In 1958 he became a consultant to the International Business Machines Corporation (IBM), and he joined the company in 1965. In 1992 he retired as senior engineer, having worked on numerous projects related to noise control engineering, including research, standards, and product design. During his IBM years he worked on several national and international standards and served a term as chair of the American National Standards Committee S1, which at the time included noise measurement standards. Dr. Maling is the author of more than 80 technical papers and several articles in handbooks—most recently a chapter on noise for the *Springer Handbook of Acoustics* (2007). He has also edited numerous conference proceedings for the INTER-NOISE and NOISE-CON series of conferences. A fellow of the Institute of Electrical and Electronics Engineers (IEEE), the American Association for the Advancement of Science (AAAS), the Acoustical Society of America (ASA), and the Audio Engineering Society, he received the Silver Medal in Noise from ASA in 1992 and the Rayleigh Medal from the Institute of Acoustics (United Kingdom) in 1999. He served as president of INCE/USA in 1975 and received the Distinguished Noise Control Engineer Award from that organization in 2001. He received the INCE/USA Distinguished Service Medal in 2009. Dr. Maling was elected to the National Academy of Engineering in 1998. He received his Ph.D. in physics (1963), an electrical engineering degree (1958), an M.S.E.E. (1954), and a B.S. (1954), all from Massachusetts Institute of Technology. He also received an A.B. in physics (1954) from Bowdoin College.

Robert J. Bernhard received his B.S.M.E. from Iowa State University in 1973, his M.S.M.E. from the University of Maryland, College Park, in 1976, and his Ph.D. in engineering mechanics from Iowa State University in 1982. He then joined the faculty of the School of Mechanical Engineering of Purdue University. From 1994 to 2004, he was director of the Ray W. Herrick Laboratories at Purdue, and from 1998 to 2007, he was director of the Institute for Safe, Quiet, and Durable Highways. From 2004 to 2007, he was associate vice president for research at Purdue. In August 2007 he left Purdue for Notre Dame, where he became vice president for research and professor of aerospace and mechanical engineering. Dr. Bernhard's areas of expertise include tire noise, traffic noise, numerical noise control design methods, noise source identification, active noise and vibration control, and machinery noise control. He is a past president of the Institute of Noise Control Engineering of the USA (INCE/USA) and has been the secretary general of the International Institute of Noise Control Engineering since 2000. He is a fellow of the American Society of Mechanical Engineers and the Acoustical Society of America and was named a Distinguished Noise Control Engineer by the Institute of Noise Control Engineering in 2003.

Robert D. Bruce, principal engineer at Collaboration in Science and Technology Inc. (CSTI Acoustics), is a registered professional engineer and is board certified by the Institute of Noise Control Engineering of the USA (INCE/USA). He has served as chair of local chapters of the Acoustical Society of America (ASA) in Boston and Houston, the ASA Technical Committee on Noise, and the INCE Technical Advisory Group on Industrial Machines and Processes. In 1986 he was president of INCE/USA. Mr. Bruce has given lectures on industrial noise control at the National Academy of Sciences Acoustical Society Lecture Series in 1971 and at universities throughout the country. His career has focused on the prediction and control of noise in the workplace, and he has written 12 chapters in books, mostly on industrial noise measurement, prediction, and control. He has also authored or coauthored more than 35 publications on industrial noise control. A fellow of ASA and a long-time member of the American Industrial Hygiene Association, Mr. Bruce received electrical engineering and S.M. degrees from the Massachusetts Institute of Technology in 1966 and a B.S.E.E. from Lamar State College of Technology in Beaumont, Texas, in 1963.

Beth A. Cooper, an internal hearing-conservation consultant to the National Aeronautics and Space Administration Office of the Chief Health and Medical Officer, provides support for the agency's occupational health and engineering communities and manages the development, promotion, and public distribution of multimedia training resources for hearing conservationists and noise control professionals. From 1999 to 2007, as manager of the Glenn Research Center Acoustical Testing Laboratory (ATL), Ms. Cooper provided noise control design, testing, and training support for science experiment payloads for the International Space Station. She managed the conceptual design, construction, accreditation, and operations of ATL, the only laboratory accredited by the National Voluntary Laboratory Accreditation Program of the U.S. Department of Commerce for sound pressure level determinations in accordance with ISO 11201. Ms. Cooper has served as director of communication of the National Hearing Conservation Association and is a member of the ANSI S12 Accredited Standards Committee on Noise and Working Group #11 on Hearing Protector Attenuation. She has been a member of the the Institute of Noise Control Engineering of the USA (INCE/USA) Board of Directors, vice president for board certification, and general chair of NOISE-CON 2003. She frequently speaks at workshops and seminars on hearing conservation, with a special focus on multimedia presentation techniques and tools for hearing conservation training. She has a B.S. in mechanical engineering from the University of Hartford and an M.S. in acoustics from the Pennsylvania State University.

Patricia Davies received her B.Sc. in mathematics from the University of Bristol in 1977 and her M.Sc. and Ph.D. in sound and vibration from the University of Southampton in 1981 and 1985, respectively. She remained at the Institute of Sound and Vibration Research until December 1986, doing postdoctoral research on statistical modeling of shock propagation through structures. She is currently a professor of mechanical engineering at Purdue University and director of the Ray W. Herrick Laboratories, where she conducts research on sound perception, signal processing, and nonlinear system identification. She has coauthored more than 120 journal and conference papers and supervised the research of 25 M.S. thesis and Ph.D. students. She also cofounded a perception-based engineering research center that conducts collaborative research by engineering and psychology professors at Purdue. One goal of this research is to integrate the ways people perceive and are affected by noise from machinery into the design of engineering systems; for example, by coupling engineering stimulus prediction models with sound perception and human decision-making models, connections can be established between the characteristics of the engineered system and its impact on people. Dr. Davies is a member of the Institute of Noise Control Engineering (and 2007–2009, president), the Acoustical Society of America, and the American Society for Engineering Education.

Carl E. Hanson is cofounder of Harris Miller Miller & Hanson Inc., one of the leading noise and vibration consulting firms in the United States. Prior to that, he worked at Bolt Beranek and Newman Inc. as group leader for surface transportation consulting services. Dr. Hanson specializes in noise and vibration control engineering projects, particularly related to rail transportation. He is active in a wide range of rail transportation projects, including noise control designs of vehicles and facilities, compliance tests, environmental assessment, community measurement programs, and expert testimony. Dr. Hanson is a consultant for architects, engineers, and planners on projects for railroads, rapid transit, state agencies, and the federal government and has conducted research and consulting projects in Europe. He was the lead author of two guidance manuals used throughout the United States, *Transit Noise and Vibration Impact Assessment* (Federal Transit Administration, 1995, 2006) and *High Speed Ground Transportation Noise and Vibration Impact Assessment* (Federal Railroad Administration, 2005). He is a licensed professional engineer in four states, an active participant on committees of the Transportation Research Board and American Railway and Maintenance-of-Way Association, and a member of the international committee for the International Workshop on Railway Noise. He earned a Ph.D. in acoustics (1970) and an M.S. in mechanical engineering (1967), both from the Massachusetts Institute of Technology, and a B.S. in aero engineering (1965) from the University of Minnesota.

Robert D. Hellweg Jr., an independent consultant and senior consultant with Epsilon Associates, Inc., was senior member of the technical staff—acoustic engineer—with Hewlett-Packard (HP) Company (formerly Compaq and formerly Digital Equipment Corporation [DEC]) from 1981 until his retirement from HP in 2007. During his years with HP/Compaq/DEC, he reduced noise emitted by computer products, led the company's work on acoustic standards, and coordinated acoustical activities. He also served (and continues to serve) on several national and international standards committees. He was chair of American National Standards Committee S12 from 2002 to 2009 and is currently vice chair. From 1972 to 1981 he was an environmental protection engineer for the Illinois Environmental Protection Agency; as head of noise technical operations and standards, he developed statewide noise regulations and determined practical noise reduction techniques for meeting regulatory limits. He is past president (2002), past secretary (1997–2000), and a member of the board of directors (2000–2004) of the Institute of Noise Control Engineering of the USA (INCE/USA) and a member of the board of directors of the INCE Foundation. He has also led the Information Technical Industry Council and Ecma-International technical committees on product noise. Mr. Hellweg is a fellow of the Acoustical Society of America, an INCE board-certified noise control engineer, and a licensed professional engineer. He received a B.S. (1966) and an M.S. (1971) in aeronautical and astronautical engineering from the University of Illinois.

Gerald C. Lauchle earned a B.S. (1968) and an M.S. (1970) in aerospace engineering and a Ph.D. in engineering acoustics (1974), all from the Pennsylvania State University. He was subsequently appointed to a faculty position at his alma mater, where he taught, conducted research, and served the university for 38 years. He retired in 2006 as Professor of Acoustics Emeritus, but he continues to consult in hydrodynamics and acoustics, with a strong emphasis on the physics and control of flow-induced noise. Dr. Lauchle has supervised 23 master's theses and 17 Ph.D. dissertations. The author or coauthor of 80 refereed journal articles, parts of six books, 38 nonrefereed journal articles, 142 reports, 56 workshops, 89 professional meeting presentations, and more than 100 other presentations, he also holds two patents and has one pending. He is a fellow of the Acoustical Society of America and a board-certified member of the Institute of Noise Control Engineering of the USA (INCE/USA). In 2002 he received the INCE/USA Martin Hirschorn IAC Award for coauthoring the best paper on a new or improved cost-effective noise control process. He has chaired several INCE/USA committees, served on the board of directors (1997–2000, 2006–2007), and was technical chair of the Sources and Propagation Committee (1998–2002), vice president for technical activities (2003), executive vice president (2004, 2006–2007), and president (2005). Dr. Lauchle has been an associate editor of *Noise Control Engineering Journal* and the *Journal of the Acoustical Society of America*.

Richard H. Lyon has been working in acoustics, vibrations, and dynamics since the early 1950s. He graduated from Evansville College (now the University of Evansville) in 1952 and earned his Ph.D. in physics from Massachusetts Institute of Technology (MIT) in 1955. In 1956 he joined the faculty of the Electrical Engineering Department at the University of Minnesota, and in 1959 he was promoted to associate professor. In 1960 he joined Bolt Beranek and Newman (BBN) Inc., where he worked on problems of sound structure interaction and excitation of structures by turbulence for industry, the National Aeronautics and Space Administration, and the U.S. Department of Defense. In 1967 he became BBN corporate vice president. In 1970 Dr. Lyon was appointed professor of mechanical engineering at MIT, where he led research on noise propagation and machinery noise and taught courses in basic and applied acoustics. He retired from MIT in 1995 and began working full time at RH Lyon Corp (RHLC). In 2005 the RHLC product design, machinery diagnostics, and structural acoustics activities were joined with Acentech Inc., where Dr. Lyon, as chief scientist, continues his work on transducer design and the design and diagnostics of products, primarily with regard to sound and vibration.

Ian A. Waitz is Jerome C. Hunsaker Professor and head of the Department of Aeronautics and Astronautics at the Massachusetts Institute of Technology (MIT) and director of the Partnership for Air Transportation Noise and Emissions Reduction (PARTNER), a Center of Excellence sponsored by the Federal Aviation Administration (FAA), the National Aeronautics and Space Administration (NASA), and Transport Canada. His principal areas of interest are modeling and evaluation of climate and impacts of aviation on local air quality and noise and assessing technological, operational, and policy options for mitigating these impacts. Professor Waitz has written approximately 75 technical publications, including a report to Congress on aviation and the environment. He holds three patents and has been a consultant for many organizations. From 2002 to 2005, he was deputy head of the Department of Aeronautics and Astronautics at MIT. He has also been an associate editor of the *AIAA Journal of Propulsion and Power*. In 2003 Professor Waitz received a NASA Turning Goals into Reality Award for noise reduction, and in 2007 he was awarded the FAA 2007 Excellence in Aviation Research Award. He is a fellow of the American Institute of Aeronautics and Astronautics and a member of the American Society of Mechanical Engineers and the American Society for Engineering Education. He teaches graduate and undergraduate courses in thermodynamics and energy conversion, propulsion, and experimental projects. He was honored with the 2002 MIT Class of 1960 Innovation in Education Award and an appointment as an MIT MacVicar Faculty Fellow in 2003. Professor Waitz received a B.S. in aerospace engineering from the Pennsylvania State University in 1986, an M.S. in aerospace engineering from George Washington University in 1988, and a Ph.D. in aeronautics from the California Institute of Technology in 1991.

Appendix K

Expert Panels

PANEL ON COST-BENEFIT ANALYSIS

Ian A. Waitz, Chair
Professor, Aeronautics and Astronautics
Massachusetts Institute of Technology
Cambridge, MA

Robert J. Bernhard
Vice President for Research
University of Notre Dame
Notre Dame, IN

Katherine Harback
Center for Advanced Aviation Systems Development
MITRE Corporation
McLean, VA

PANEL ON COMPETITIVENESS OF U.S. PRODUCTS

Robert D. Hellweg, Jr., Chair
Consultant
Wellesley, MA

Loren A. DeVries
Staff Engineer
John Deere Technology Center
Moline, IL

Carol J. Drutowski
The Toro Company
Minneapolis, MN

Michael J. Lucas
Principal Engineer
Ingersoll-Rand
Rotary Compressor Division
Davidson, NC

PANEL ON NOISE R&D INFRASTRUCTURE

Richard H. Lyon, Chair
President
RH Lyon Corporation
Belmont, MA

Krish K. Ahuja
Regents Professor, School of Aerospace Engineering
Georgia Institute of Technology
Director and General Manager
Georgia Tech Ireland
Westmeath, Ireland

Paul R. Donavan
Illingworth & Rodkin, Inc.
Petaluma, CA

Gregory C. Tocci
Senior Principal Consultant
Cavanaugh Tocci Associates, Inc.
Sudbury, MA

Richard F. Topping
President
RF Topping Consultants LLC
Westborough, MA

PANEL ON HAZARDOUS NOISE

Robert D. Bruce, Chair
CSTI Acoustics
Houston, TX

John Casali
John Grado Professor
Grado Department of Industrial and Systems Engineering
Virginia Polytechnic Institute and State University
Blacksburg, VA

Beth A. Cooper
Manager, Auditory Demonstration Laboratory
National Aeronautics and Space Administration, Glenn Research Center
Cleveland, OH

Richard L. McKinley
Principal Engineer
Air Force Research Laboratory
Wright Patterson AFB, OH

Kenneth M. Eldred
Consultant
East Boothbay, Maine

Lawrence S. Finegold
Research Psychologist
Finegold & So, Consultants
Centerville, OH

Carl E. Hanson
Senior Vice President
Harris Miller Miller & Hanson Inc.
Burlington, MA

Ben H. Sharp
General Manager, Research & Consulting
Wyle Laboratories
Arlington, VA

PANEL ON METRICS FOR COMMUNITY NOISE

Patricia Davies, Chair
Director, Ray W. Herrick Laboratories and
Professor of Mechanical Engineering
Purdue University
West Lafayette, IN

Appendix L

Workshop Agendas

WORKSHOP ON TECHNOLOGY FOR A QUIETER AMERICA

Organized and Hosted by the National Academy of Engineering

Washington, DC
August 13–15, 2005

Welcoming Remarks
Proctor Reid, Program Office, National Academy of Engineering

Introduction to Plenary Session
George Maling, Chair, Steering Committee on Technology for a Quieter America

Overview of the Aviation Portfolio Management Tool
Katherine Harback, MITRE Corporation

Impact of Product Noise on Manufacturing Competitiveness
Russell Hutchinson, Association of Equipment Manufacturers

NIOSH Perspectives on Preventing Occupational Hearing Loss
Mark Stephenson and William Murphy, National Institute for Occupational Safety and Health

Occupational Noise Exposure—A Regulatory Perspective
John Seiler, Mine Safety and Health Administration

Community Noise around Airports
Arlene Mulder, Mayor, Village of Arlington Heights, Illinois

Community Noise near Highways
Karl Dreher, California Department of Transportation

Noise in Urban Areas
Arline Bronzaft, Mayor's Office, New York City

Noise in Naturally Quiet Areas
Robert Rossman, National Park Service

Education of and Demand for Noise Control Specialists
Robert Bernhard, Purdue University

Aviation and the Environment: Navigating the Future
Carl Burleson, Federal Aviation Administration

Programs for Reduction of Aircraft Noise: Source Reduction and Operational Techniques
Richard Wlezien, National Aeronautics and Space Administration, Vehicle Systems Program

Department of Transportation Programs for the Reduction of Surface Transportation Noise (Rail and Highway)
Arnold Konheim, U.S. Department of Transportation

Engineering Progress and Challenges in Quiet Highway Development
Mark Swanlund, Federal Highway Administration

Community Noise from Industrial Plants
Eric Wood, Acentech

DISCUSSION SESSIONS (panel discussions)

Issues with Respect to Manufacturing Competitiveness—Both Export and Import Issues
Bennett Brooks, Brooks Acoustics Corporation
Robert Hellweg, Jr., Hewlett Packard

Occupational Noise Exposure, Hearing Protection Devices, Impact On: Productivity, Communications, Safety, Quality of Life
Lee Hagar, Sonomax Hearing Healthcare, Inc.

Metrics for a Cost-Benefit Analysis of Noise Reduction (Brainstorming a Methodology of This Type of Analysis)
Katherine Harback, MITRE Corporation

Potentially Hazardous Noise for Users of Consumer Products—Personal Music Devices, Children's Toys, Recreational Vehicles
William Martin, Oregon Health Sciences University

The Nature and Extent of Complaints about Noise (Suburban, Urban, Rural) and Public Demand for Quiet Environments and Products
Les Blomberg, Noise Pollution Clearinghouse

Technical Issues with Respect to Metrics/Descriptors for Community Noise (Annoyance, Activity Interference, Noticeability)
Nicholas Miller, Harris Miller Miller & Hanson Inc.

Noise Insulation (from Aircraft and Highway, etc.) in Homes to Control Noise Exposure and Improve Quality of Life
Ben Sharp, Wyle Acoustics Group

Noise Control Engineering Education and Workforce Development
David Wormley, Pennsylvania State University

Annoyance from Noise as a Quality-of-Life Issue, and Its Relationship to Other Sources of Annoyance
Larry Finegold, Finegold & So, Consultants

Future Directions in the Design of Noise Barriers, Quiet Vehicles, and Quiet Pavements
Gregg Fleming, John A. Volpe National Transportation Systems Center

Noise Standards and Design Issues for Rooms (Schools, Hospitals, Offices, etc.)
Richard Peppin, Scantek, Inc.

Technology for the Design of Products with Lower Noise and Better Sound Quality
Gordon Ebbitt, Carcoustics
Richard Topping, TIAX LLC

COST-BENEFIT ANALYSIS OF TRANSPORTATION NOISE CONTROL TECHNOLOGY

Organized by the National Academy of Engineering and Hosted by the Volpe Center

Cambridge, Massachusetts
February 22–23, 2007

Opening Remarks
Ian Waitz, Massachusetts Institute of Technology, Subcommittee Chair
Gregg Fleming, Volpe Center, Workshop Host

An Environmental Economist's Perspective on Cost-Benefits Analysis
Sabrina Lovell, formerly with the Environmental Protection Agency

The O'Hare Residential Sound Insulation Program and Acceptance in Communities, and Sound Insulation Ordinance
Arlene Mulder, Mayor, Village of Arlington Heights, Illinois

The Federal Highway Administration's Noise Program and Rules for Noise Barrier Construction
Mark Ferroni, Federal Highway Administration

Federal Highway Administration and State Activities in the Design of Quiet Pavements: Construction, Maintenance, and Life Cycle Issues
Mark Swanlund, Federal Highway Administration

Reductions in Noise Emissions from Porous Highways: Current State of the Technology in the USA and Europe
Paul Donavan, Illingsworth Rodkin

Discussion of European Activities Related to Cost-Benefit Analysis and Highway Noise
George Maling, Institute of Noise Control Engineering of the U.S.A.
Ulf Sandberg, Swedish National Road and Transport Research Institute

Cost-Benefit Analysis and Transportation Noise
Jon Nelson, The Pennsylvania State University

Discussion of Construction Process, Costs, Maintenance, Performance Characteristics, and Noise Levels of Rubber-Modified Asphalt Highways
Michael Blumenthal, Rubber Manufacturers Association

Future Technology for Design of Quiet Tires and European Specifications for Tire/Road Noise
Ulf Sandberg, Swedish National Road and Transport Research Institute

Federal Highway Administration Traffic Noise Model (TNM): Cost-Benefit Analyses and Tire/Pavement Noise Effects
Judy Rochat, Volpe Center

IMPACT OF NOISE ON COMPETITIVENESS OF U.S. PRODUCTS

Organized and Hosted by the National Academy of Engineering

Washington, DC
June 20–21, 2007

Opening Remarks
Robert Hellweg,Jr., Workshop Chair
George Maling, Study Committee Chair

Foreign Requirements on Industrial Machinery and Consumer Products Used Indoors. Safety Issues as Well as Lower Levels for IT Equipment and Consumer Products. Comments on Eco-Labels. Foreign Requirements on Industrial Machinery Used Outdoors.
David Rowe, Ingersoll Rand Portables
Matt Nobile, IBM
Mac Mezache, Copeland Corporation
George Maling, Study Committee Chair

American Participation in International Standards Activities. International Standards for Determination of Noise Emission of Industrial Machinery and Consumer Products
Paul Schomer, Schomer & Associates
Rich Harmening, Trane
Carol Drutowski, Toro
George Maling, Study Committee Chair

Role of NIST/NVLAP in Accreditation of Laboratories for Noise Emission Verification; Foreign Testing Laboratories and Their Relation to NIST/NVLAP
Betty Ann Sandoval, National Institute of Standards and Technology
Ileana Martinez, National Institute of Standards and Technology

Commerce Department U.S. Trade Representative's Office Resources to Aid Companies Faced with Foreign Noise Requirements
Robert Straetz, U.S. Department of Commerce
Sarah Bovim, Whirlpool Corporation

Noise Labeling Programs: Voluntary and Mandated. Marketing Quiet Products—Technical Accuracy and Consumer Accessibility. What Are We Learning from European Labeling Programs?
Victor Vukorpa, Whirlpool Corporation
Matt Nobile, IBM

Impact of Noise Requirements on American Exporters of Office Equipment, IT Machinery, and Consumer Products. Changing Customer Expectations Regarding Sound Levels and Sound Quality.
Marco Beltman, Intel
James Walters, Air Conditioning/Refrigeration Institute
Ken Feith, U. S. Environmental Protection Agency
Nathan Mouw, Whirlpool Corporation

Impact of Noise Requirements on American Exporters of Indoor/Outdoor Power and Industrial Equipment
Loren DeVries, John Deere
Richard Wood, Carrier
Rich Harmening, Trane
Tom Disch, Briggs and Stratton
Dan Kato, Cummins Power

WORKSHOP ON NOISE R&D INFRASTRUCTURE

Organized and Hosted by the National Academy of Engineering

Washington, DC
June 11–12, 2008

Opening Remarks
Richard Lyon, Subcommittee Chair

Overview of the Technology for a Quieter America Project
George Maling, Study Committee Chair

Summary of the Aircraft Noise Day of the CAETS Workshop on Transportation Noise Sources in Europe, June 2–4, 2008, Southampton, United Kingdom
Krish Ahuja, Georgia Institute of Technology

Features to Emulate in a Future Quiet Aircraft—Recommendations
Krish Ahuja, Georgia Institute of Technology

Overview of the Above CAETS Workshop, Days 2 and 3
George Maling, Study Committee Chair

Aircraft Noise Control—Challenges and Opportunities
Joe Posey, National Aeronautics and Space Administration, Langley Research Center

Aircraft Noise Prediction—Conventional to Revolutionary
Casey Burley, National Aeronautics and Space Administration, Langley Research Center

Can a Gearbox Ever Be Silent?
Rajendra Singh, The Ohio State University

Current Federal Highway Administration Noise Research Activities
Adam Alexander, Federal Highway Administration

Rail Transportation Noise Control Technology
Carl Hanson, Harris Miller Miller & Hanson, Inc.

Technology for a Quieter America—Building Acoustics
Gregory Tocci, Cavanaugh Tocci Associates, Inc.

Building Acoustics: Sound-Absorptive Materials, Damping Materials, and Noise Transmission
Stuart Bolton, Purdue University

Electro-acoustic Systems
Chuck McGregor, Eastern Acoustic Works

New Technologies for a Quieter America: Stationary Machinery and Equipment
Michael Lucas, Ingersoll Rand Corporation

Signal Processing and Data Analysis Issues in Sound Quality Design and Assessment
Patricia Davies, Purdue University

In-Ear Digital Active Noise Reduction
William Saunders, Adaptive Technologies, Inc.

Inverse Holographic Methods Impacting New Technology for a Quieter America
Earl Williams, Naval Research Laboratory

Computational Aeroacoustics
Christopher Tam, Florida State University

Computational Tools for Design and Noise Control
Phil Shorter, EST Group

Numerical Methods for Noise Control: An Educational Perspective
Stephen Hambric and Anthony Atchley, The Pennsylvania State University

Noise Control Research and Development: A Program for Producing Demonstrations of Practical Value and Adding Designs for Quieter and Better-Sounding Products
Richard Lyon, RHLyon Corporation

Noise and Vibration Control Research at the Ohio State University
Rajendra Singh, The Ohio State University

Current Topics in Noise Control Research at Purdue University
Patricia Davies, Purdue University

Noise Control Research and Development at the Pennsylvania State University
Stephen Hambric and Anthony Atchley, the Pennsylvania State University

European and Asian Research on Noise Control
Paul Donavan, Illingworth Rodkin, Inc.

European Noise Research
George Maling, Study Committee Chair

National Science Foundation Research (an Internet search)
George Maling, Study Committee Chair

ENGINEERING RESPONSES TO HAZARDOUS NOISE EXPOSURES

Organized and Hosted by the National Academy of Engineering

Washington, DC
August 14–15, 2008

Opening Remarks
George Maling, Study Committee Chair
Robert Bruce, Workshop Chair

The Occupational Noise Problem in the USA—Its Costs and the Number of Noise-Exposed Workers
Robert Bruce, CSTI Acoustics

The Scientific Basis for the 85-dB Criterion and 3-dB Exchange Rate versus the Different Exposure Limits and Exchange Rates Used in the USA and Elsewhere
Mark Stephenson, National Institute for Occupational Safety and Health

Impulsive Noise in Industry and in the Community: Considerations for Measuring Impulsive Noise
Bill Murphy, National Institute for Occupational Safety and Health

Available Data to Develop Damage-Risk Criteria of Impulsive Noise and Validation Data
Armand Dancer, French-German Research Institute

Review of Engineering Controls for Occupational Noise Including Equipment for Which There are Acceptable Controls and Benefits of Reduced Noise Exposure
Dennis Driscoll, Associates in Acoustics, Inc.

Panel Discussion of Employees' Concerns, Accident Investigations, Workplace Benefits, or Reduced Noise Exposures
Scott Schneider, Laborer's Health and Safety Fund of North America
John Casali, Virginia Polytechnic Institute and State University
Mike Bobeczko, Sukut, Inc.

Buy Quiet Programs' Engineering Specifications for Noise Emissions and Ensuring an Immission Specification Is Met
Beth Cooper, National Aeronautics and Space Administration, Glenn Research Center
Bob Anderson, Anderson Consulting Associates

Nonoccupational Hazardous Noise. Recreational Equipment, Personal Music Devices, Toys, Buses, etc. Focus on Children
Brian Fligor, Harvard Medical School

Engineering Advances in Hearing Protection
Richard McKinley, Air Force Research Laboratory

IMPROVED METRICS FOR COMMUNITY NOISE

Organized and Hosted by the National Academy of Engineering

Washington, DC
September 25–26, 2008

Opening
Patricia Davies, Subcommittee Chair
George Maling, Study Committee Chair

Utility and Credibility of Dosage-Response Relationships for Transportation Noise Regulation
Sanford Fidell, Fidell Associates

Sleep Disturbance Metrics (Mostly from Aircraft Operations)
Nicholas Miller, Harris Miller Miller & Hanson Inc.

A-Weighting and a Possible Replacement Metric
Paul Schomer, Schomer and Associates

Community Response to Low-Frequency Aircraft Noise
Sanford Fidell, Fidell Associates

Comments on American Classic Papers and European Approaches to Metrics
George Maling, Study Committee Chair

Noise Model Issues
Paul Schomer, Schomer and Associates

Sound Quality Metrics and Their Potential Use in Assessment of Environmental Noise
Patricia Davies, Purdue University

Metrics for Impulsive Noise Sources
Kenneth Plotkin, Wyle Laboratories

Noise Metrics in Low-Ambient-Noise Communities and Other Environments (Rural Settings, Recreational Areas, etc.)
Nick Miller, Harris Miller Miller & Hanson Inc.

Environmental Noise Measurement and Metrics
James Thompson, Brüel and Kjær, Inc.

Effect of Noise on Learning: Quantification of Effects Outside DNL 65
Mary Ellen Eagan, Harris Miller Miller & Hanson Inc.

U.S. EDUCATION IN NOISE CONTROL ENGINEERING[1]

Organized by Noise Control Foundation Workshop held during NOISE-CON 2007

Reno, Nevada
October 23, 2007

Opening Remarks
George Maling, Chair

Noise Control Engineering Education
Robert Bernhard, University of Notre Dame

[1]This workshop was not sponsored by the National Academy of Engineering; it was held in conjunction with NOISE-CON 07, the 2007 National Conference and Exposition on Noise Control Engineering. NOISE-CON 07 was organized by the Institute of Noise Control Engineering of the USA, Inc.

Acoustics and Noise Control Engineering at Brigham Young University
Scott Sommerfeldt, Brigham Young University

The Challenge of a Noise Control Education at a Research University
Kenneth Cunefare, Georgia Institute of Technology

Practice and Science Track Courses in Noise and Vibration Control
Raj Singh, Ohio State University

How Can We Fulfill the Demand for Industry- and Academia-Desired Engineers with Expertise in Acoustics, Vibration, and Noise Control?
Patricia Davies, Purdue University

Is Noise Control Engineering Education a Sustainable Resource?
Anthony Atchley, The Pennsylvania State University

Noise Control Engineering Education for Specialists and Generalists
Dave Holger, Iowa State University

Engineering Skills Required to Design Low-Noise Products
Michael Lucas, Ingersoll Rand

Industry Needs for Noise Control Engineers
Dan Kato, Cummins Engine

Education and the Sustainability of Noise Control Engineering Education
Paul Donavan, Consultant

Education for Noise and Vibration Control Engineering and Architectural Acoustics
Eric Wood, Acentech, Inc.

Noise Control Education to Support Aerospace Noise Control Needs
Evan Davis, Boeing Aircraft

Noise Control Courses for the Working Stiff
Courtney Burroughs, Consultant

HOW DO WE STIMULATE COLLECTIVE ACTION TO MOTIVATE THE PUBLIC TO DEMAND QUIET?[2]

Organized by the Noise Control Foundation

Dearborn, Michigan
July 29, 2008

Forum Opening
George Maling, Chair

What Tools Do People Need to Help Them Achieve Quieter Communities?
David Bell, Noise Regulation Report

What the Public Should Know
Beth Cooper, NASA Glenn Research Center

Federal Government's Role in Public Education on Noise
Catrice Jefferson, U.S. Environmental Protection Agency

Let's Educate Children on the Adverse Effects of Noise and Society Will Follow
Arline Bronzaft, Council on the Environment of New York City

Creating Demand through the Engagement of Noise Control Engineers
Mandy Kachur, INCE/USA Vice President for Public Relations

Creating Consumer Demand through the Use of Simple, Uniform Product Noise Declarations
Matthew Nobile, IBM Hudson Valley Acoustics Laboratory

---

[2]This workshop was not sponsored by the National Academy of Engineering; it was held in conjunction with NOISE-CON 08, the 2008 National Conference and Exposition on Noise Control Engineering. NOISE-CON 08 was organized by the Institute of Noise Control Engineering of the USA, Inc.